P9-CQL-764

edible
and
poisonous
mushrooms
of canada

COVER: The *Morchella esculenta* (left) and *Amanita muscaria* (right) are detail from drawings by Henry Jackson. They are part of a collection of the National Gallery of Canada and were donated to the Gallery by Dr. Naomi Jackson Groves.

AMANITA caesarea (Scop.) Pers.
Fine collection made near
Quebec City by Dr. R. Pomerleau.
July 1951
Henry A. C. Jackson
1951

Amanita caesarea. **Painting by Henry A. C. Jackson.**

edible and poisonous mushrooms of canada

J. WALTON GROVES
Addendum by S. A. Redhead
Biosystematics Research Institute
Ottawa, Ontario

Research Branch
Agriculture Canada
Publication 1112
1979

Available in Canada through

Authorized Bookstore Agents
and other bookstores

or by mail from

Canadian Government Publishing Centre
Supply and Services Canada
Ottawa, Canada K1A 0S9

Catalogue No. A43-1112/1979
ISBN 0-660-10136-X

Canada: $21.95
Other countries: $26.35

Price subject to change without notice

Ottawa, 1962
Reprinted 1972
Revised 1975, 1979
Reprinted 1981
Reprinted 1986

CONTENTS

Page

ACKNOWLEDGMENTS

Many people have been generous with their co-operation and assistance in the preparation of this book. I would especially like to thank Dr. A. H. Smith of the University of Michigan who has read much of the manuscript, supplied some of the photographs, and made many helpful suggestions. Mrs. Sheila C. Thomson, who was formerly employed in this section, gave invaluable assistance during the early stages of preparation by aiding with identifications and preparation of descriptions. Parts of the manuscript were read by colleagues Dr. Mildred Nobles and Dr. Ruth Macrae. My wife, Dr. Naomi Groves, read the entire manuscript and offered many valuable suggestions. Mr. Henry A. C. Jackson kindly gave permission to use his painting of *Amanita caesarea* as a frontispiece.

Grateful acknowledgment is made to the following for permission to use their photographs.

Dr. A. H. Smith — Figures 52, 56, 64, 79, 145, 155, 160, 198, 255, 266, 274, 302, 340, 351, 361, 388, 395.

Mr. K. A. Harrison, Kentville, N.S. — Figures 124, 156, 180, 182, 196, 262, 291, 332, 335, 353.

Dr. and Mrs. B. T. Denis, Quebec City — Figures 105, 107, 110, 112, 342, 343, 344, 347, 348, 349.

Dr. Maria Pantidou, Ottawa — Figures 317, 319, 324.

National Museum of Canada — Figures 90, 91, 92, 126, 127, 128, 303, 304, 327, 328, 329, 384, 386, 387, 400.

Bio-Graphic Unit, Canada Department of Agriculture — Figures 174, 175, 396, 397.

INTRODUCTION

Mushrooms appeal to different people in different ways. The biologist is attracted by the variety of species, their place in the economy of nature and their interrelations with other plants and animals; the artist or photographer delights in their infinite variety of form and color; the medical research worker may look to them hopefully as a possible source of new drugs; but to most people the quality that first arouses interest in them is their use as food. In Roman times edible mushrooms were renowned as a delicacy and today in some parts of the world they constitute an important part of the food supply of the people.

One of the questions often asked a mycologist is "How do you tell an edible mushroom from a poisonous one?" It would seem that the questioner expects some simple test or rule of thumb by which an instantaneous diagnosis can be made. It is curious that this attitude toward mushrooms should exist because it is not manifested toward other plants. People rarely ask how to tell an edible berry from a poisonous one nor do they expect to be given a simple test to distinguish between edible and poisonous leaves.

Although several reasons for this attitude might be suggested, perhaps one may be connected with the comparatively late development of precise knowledge of the structure and life history of mushrooms and other fungi. For a long time even botanists did not look upon the fungi as plants in themselves but regarded them as a sort of excrescence on decaying vegetable matter. Their apparently sudden appearance and disappearance often late in the season without visible seeds or means of reproduction, their frequent association with decaying organic matter, their vivid colors, fantastic shapes, and in some instances their poisonous properties, caused them to be regarded as objects of mystery and sometimes even to be associated with the supernatural.

One common superstition concerned the fairy rings, those dark green circles in the grass where mushrooms appear. We now know that these are caused by the circular growth outward of the fungus in the soil, but they were once believed to mark the spot where the fairies held their midnight revels.

Another well-known example of magical power attributed to a fungus occurs in *Alice in Wonderland* where a bite of one side of a certain mushroom would make you grow taller and a bite of the other side would make you grow shorter, so that by a little judicious nibbling it was possible to adjust oneself to any desired dimension.

Mycologists take a more realistic and less fanciful view of the mushrooms but to most people these are still a very unfamiliar and somewhat mysterious group of organisms, and perhaps it is because of this background of mystery that some magical test is expected to distinguish good mushrooms from bad.

Actually there is only one test to find out whether a mushroom or any other plant is poisonous and that is to eat it. If it makes you sick or kills you it is poisonous, and it is mainly through such human experience that we have built up our knowledge of which plants are edible and which poisonous.

Throughout man's process of becoming civilized he has probably at some time or another tried eating everything that looked edible. From the records of experience of other people in trying fungi we know that certain species are poisonous and that certain other species are edible and desirable, just in the same way that we know that certain berries are poisonous and others are edible. The edible qualities of a good many species of mushrooms are still unknown and reports about some species are conflicting.

Conflicting reports may arise from a variety of reasons. Sometimes illness arising from some other cause is wrongfully attributed to mushrooms. It is a common practice to fry mushrooms in butter but the use of too much butter may cause illness and the mushrooms be blamed, or some other dish eaten at the same meal as the mushrooms may be the real cause of the illness. Sometimes misidentification of mushrooms may occur and the wrong species be blamed for causing illness. Apparently allergy may also be involved and some species may sicken certain individuals but not others. It is possible, too, that certain geographic races of mushrooms may produce poisons and other races may not. Possibly a species may be edible when it is young and fresh and may become poisonous when it is overmature and has started to decay. On the other hand, the method of preparation or cooking may destroy a poisonous substance that is present in the uncooked specimen.

Any one of these reasons may result in reports at one time that a mushroom is edible and at another time that it is poisonous, but until there is clear-cut and convincing evidence that a species is edible, it should be regarded with suspicion.

Thus any rules about eating mushrooms resolve themselves into one bit of common sense — eat only the species you know and avoid all the others. Just as one can easily learn to identify a wild raspberry or blueberry and avoid unknown berries, he can learn to identify several common species of mushrooms such as the meadow mushroom *Agaricus campestris*, the chantarelle *Cantharellus cibarius*, the shaggy mane *Coprinus comatus*, the 'delicious lactarius' *Lactarius deliciosus*, the parasol mushroom *Lepiota procera*, the morel *Morchella esculenta*, and the giant puffball *Calvatia gigantea*, to mention only a few.

Unrecognized species should not be eaten, and because of the very great danger from the deadly species of *Amanita*, the characters of this genus should be learned and its species avoided. However, by collecting and studying different species the number that can be identified and eaten will gradually increase. Some people will be satisfied to know a very few species; others will want to try different ones. Obviously one should proceed cautiously when trying any species for the first time because of the possibility of allergic reactions, but by making certain that amanitas are avoided and by using only species that have been identified and are known to be edible, one may enjoy many tasty mushroom dishes in safety.

Although much of the interest in mushrooms arises from their use as food, other aspects also attract the collector. The wide variety they exhibit in form and color, the great number of species that occur, and their interesting relationships with other plants and types of habitat are all features that invite further study.

One of the things that adds to the interest of mushroom collecting is the fact that one may visit the same locality time after time and continue to find different species. Some of the common ones will be found repeatedly of course, but something different is likely to be found at any time under varying weather conditions. Some species seem to produce fruiting bodies only rarely, perhaps only once in several years, so that there is always the possibility of coming upon a rare and unusual species, even on familiar ground.

Mushrooms are also interesting from the standpoint of their place in the economy of nature. One of their chief functions is to aid in the breakdown of dead organic material and to return the essential elements to the soil. When this function is appreciated, their frequent association with decay is understood and any feeling of repulsion toward them disappears. Some of the species are found only with certain trees where they form associations with the tree roots that are termed 'mycorrhiza.' Some trees cannot thrive without the presence of their fungus associate. Attention has also been directed in recent years to the mushrooms as a possible source of antibiotic substances that might prove useful in medicine. Investigations are being carried out to see if the hallucinogenic mushrooms of Mexico might prove to be a source of a non-habit-forming tranquilizing drug that would be valuable in neuropsychiatric research. Investigation of the mushrooms from these and similar angles is only beginning.

It should be realized that in our mushroom flora we have many more species that must be omitted from a book of this nature than can be included. Consequently, caution must be exercised in making identifications. If the characters of a particular specimen under examination do not agree in all respects with the description, there is a good chance that the mushroom may be a species not in the book, hence for safety's sake it should not be eaten.

PARTS OF A MUSHROOM

Most people have a general idea of what a mushroom is, but the term has never been precisely defined and has different meanings to different people. Perhaps the most generally accepted usage is to apply the term mushroom to a fungus fruiting body with a more or less evident stalk, bearing an expanded cap at the apex, with a series of thin, radiating, gill-like or blade-like structures on the lower surface of the cap. Some would consider that only one or two species such as the meadow mushroom and cultivated mushroom are true mushrooms, whereas others would call almost any large fleshy fungus a mushroom. From a scientific standpoint it is probably best to use the term mushroom to apply to the whole group of gill-bearing fungi and it is used in that sense in this book.

In classifying fungi or other plants, botanists try to group together those forms that are thought to be closely related, and in general it is believed that the fungi bearing gills, or more properly lamellae, are more closely related to each other than to those not bearing lamellae. The species bearing lamellae are grouped by mycologists in a family, the Agaricaceae, and therefore the use of the term mushroom for these fungi gives it a popular meaning roughly equivalent to the family Agaricaceae.

'Toadstool' is another popular term that is frequently used and it too means different things to different people. To some it means any fungus except the field mushroom or cultivated mushroom, to others it is any inedible or poisonous fungus. Scientifically the term has no meaning at all since species related closely in a botanical sense may be either edible or poisonous. We can avoid confusion by dropping the term 'toadstool' altogether and speaking only of edible and poisonous mushrooms.

The structure that we call a mushroom is in reality only the fruiting body of the fungus. The vegetative part of the plant consists of a system of branching threads and cord-like strands that ramify through the soil, manure, or other material on which the fungus may be growing. This vegetative part is called

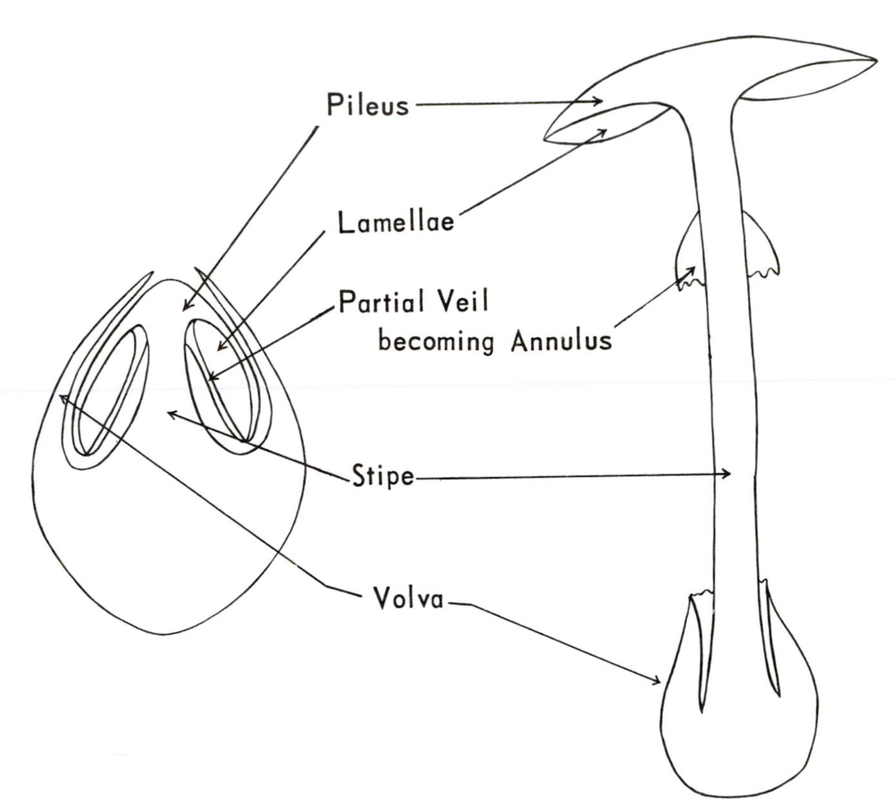

Figure 1. Diagram of a section of an amanita illustrating the principal parts of a mushroom. Young plant on left, mature plant on right.

the mycelium; it is used by commercial mushroom growers to plant their beds and is commonly called mushroom spawn. After a period of growth and accumulation of food reserves and under favorable conditions of temperature and moisture, the mycelium will produce the fruiting structure that we call the mushroom.

The principal parts of a mushroom are illustrated by a diagram representing a section through an amanita (Figure 1). The fruiting body consists of a stem-like part called the stipe, which supports an expanded, umbrella-shaped cap or pileus. On the under side of the pileus are the gills or lamellae. In some mushrooms, particularly those occurring on wood, the stipe may be lacking and the pileus is then said to be sessile (Figure 10, p. 5). The pileus is usually

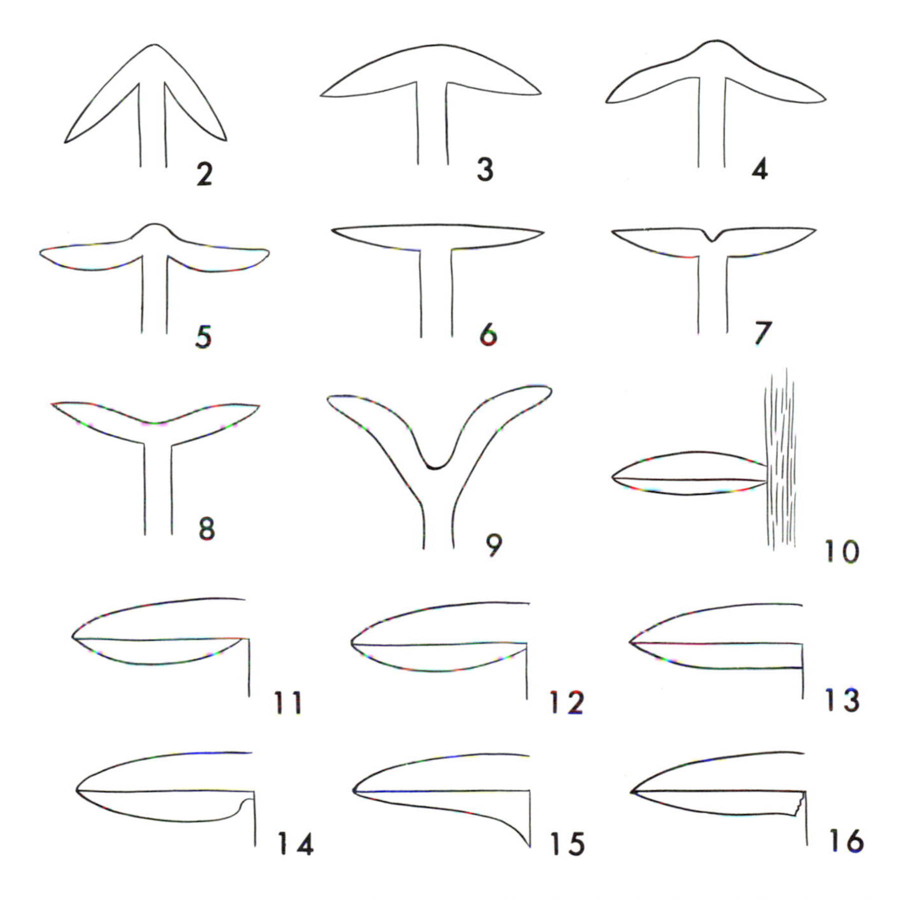

Figures 2-9. Diagrams illustrating various terms used to describe the shape of the pileus: 2, conical; 3, convex; 4, campanulate; 5, umbonate; 6, plane; 7, umbilicate; 8, depressed; 9, infundibuliform.

Figure 10. Diagram illustrating a sessile fruiting body; the stipe is lacking. In this instance the fruiting body is represented as growing on a tree trunk.

Figures 11-16. Diagrams illustrating various terms used to describe the attachment of the lamellae: 11, free; 12, adnexed; 13, adnate; 14, sinuate; 15, decurrent; 16, seceding.

circular but may vary from being somewhat irregular to fan-shaped or ear-shaped, or sometimes shelf-like. The lamellae radiate from the stipe to the margin of the pileus and may vary from their typical form as one or more series of knife-blade-like structures to scarcely more than slight folds on the under surface of the pileus.

In some mushrooms the young plant is at first completely enclosed in a sheath of tissue called the universal veil or volva. As growth proceeds, the volva is torn open and the young mushroom emerges, leaving remnants of the volva as a sheath surrounding the base of the stipe. This universal veil or volva is not present in all mushrooms but is an important character to look for in recognizing the dangerous genus *Amanita.* Two types of volva are found in *Amanita.* In one type, such as is found in *A. virosa,* the volva tears across the top and remains as a loose, cup-like sheath around the base of the stipe. In the other type, as in *A. muscaria,* the sheath is not loose but more or less grows together with the rest of the tissue of the fruit body. As the fruit body grows, the volva tears around the margin of the pileus rather than across the top. Part is left adhering to the surface of the pileus where it becomes torn into patches as the pileus expands, and part remains attached to the base of the stipe where it may form a series of rings or patches on the stipe or form a boot-like cup closely adhering to the base. This second type of volva is more difficult to recognize in the field and careful examination should be made to determine whether or not it is present.

In some kinds of mushrooms the lamellae in the young stage are enclosed by a layer of tissue that extends from the margin of the pileus to the stipe. This tissue is known as the partial veil and it usually tears around the margin of the pileus as the latter expands, and remains attached to the stipe where it forms a ring or annulus. It may sometimes tear at the stipe and remain attached to the margin of the pileus, which would then be described as appendiculate. Some genera such as *Amanita* have both a universal veil and a partial veil, others such as *Agaricus* may have only the partial veil, and in yet others such as *Clitocybe* both the universal veil and partial veil may be absent.

The presence of a partial veil is frequently used as a character to distinguish genera and its presence is usually indicated by the occurrence of the annulus in the mature fruit body. Care must be taken in determining this character because in some species the annulus is very delicate and may soon disappear. It is advisable to examine young specimens to determine whether or not an annulus is present.

The pileus, lamellae, stipe, volva, and annulus constitute the principal parts of a mushroom that can be seen with the naked eye. Their variations in form, color, texture, surface covering and so on are all important in the recognition of species. Other characters can be observed only with the microscope, and the research taxonomist is coming to place more and more reliance on these microscopic characters both as a means of distinguishing between similar species and also of providing characters that indicate relationships between species or groups of species.

For the purpose of this book relatively little emphasis is being placed on microscopic characters but some mention of them must be made in order to understand the function of the mushroom fruiting body. A mushroom reproduces by means of spores and the fruiting bodies are organs developed to promote the dissemination of the spores.

Spores of very varied forms are produced by fungi in general but in the mushrooms they usually consist of a single, minute cell, rarely more than 1/50 millimeter or 0.0008 inch in length and usually much smaller. They are too small to be seen singly by the naked eye but in mass appear as a white or colored powder. Their size, shape, and surface markings, if any, are important in identifying species but these features can be seen only with the microscope.

The measurements of spores are usually expressed in μ (microns). One μ (micron) equals one one-thousandth of a millimeter. Thus when we say a spore is 10μ in length we mean 10/1000 or 1/100 of a millimeter in length and since a millimeter is about 1/25 of an inch, it would take 2500 such spores to equal one inch.

Minute objects such as these are measured by placing a glass disk marked with a scale in the eyepiece of the microscope. The scale can be carefully calibrated with a special slide that is ruled very accurately in tenths and hundredths of a millimeter. It is then easy to calculate what each division on the eyepiece scale measures and the spores can be measured directly in ordinary slide mounts.

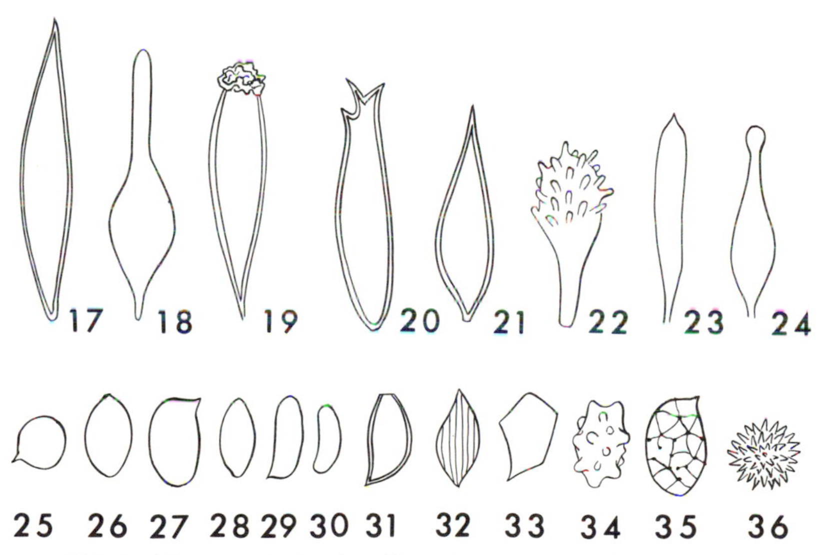

Figures 17-24. Semidiagrammatic drawings illustrating some types of cystidia: 17, fusiform-lanceolate; 18, flask-shaped; 19, capitate-encrusted; 20, horned; 21, ventricose-fusiform; 22, clavate with projections at the apex; 23, cylindric, obtuse with mucronate tip; 24, ninepin-shaped (lecythiform).

Figures 25-36. Semidiagrammatic drawings illustrating various types of spores: 25, globose; 26, ovoid; 27, ellipsoid; 28, ellipsoid-fusiform; 29, cylindric; 30, allantoid; 31, ellipsoid with truncate apex; 32, longitudinally striate; 33, angular; 34, tuberculate; 35, reticulate; 36, echinulate.

When a spore germinates it puts forth a slender thread called a hypha. This thread grows rapidly, develops cross walls or septa and becomes many-celled, then branches repeatedly, producing a mat of hyphae termed the mycelium. This is the vegetative part of the fungus from which the fruiting body or mushroom arises. The mushroom is composed of hyphae consisting of interwoven, branched, septate threads, although in some groups such as *Lactarius* and *Russula* there are also globular cells termed sphaerocysts in the tissue.

The structure of the lamellae is of special interest because it is here that the spores are produced. If a section of a lamella is examined it is seen to be roughly wedge-shaped or triangular. The spores are produced on the outer surface in a definite layer called the hymenium whereas the central part is composed of more or less interwoven hyphae and is called the trama (Figure 42, p. 9). The hymenium consists of basidia, paraphyses, and sometimes cystidia. The basidia are the cells on which the spores are produced; each one has four little stalks at its apex and a spore develops on the tip of each stalk. Between the basidia there are cells somewhat similar in shape but lacking the stalks and never producing spores. These are the paraphyses. Their function is apparently to hold the basidia far enough apart that their slightly sticky spores will not become entangled and prevented from shooting forth. In some mushrooms there are also specialized cells of varying shape and size that project from the hymenium and are called cystidia. They may be rounded, pointed, thick- or thin-walled, sometimes encrusted, and variously shaped, but they are usually constant for each species. Cystidia may also occur on the surface of the pileus or on the stipe. Their exact function is still uncertain but when present they can be of great assistance in the identification of the species.

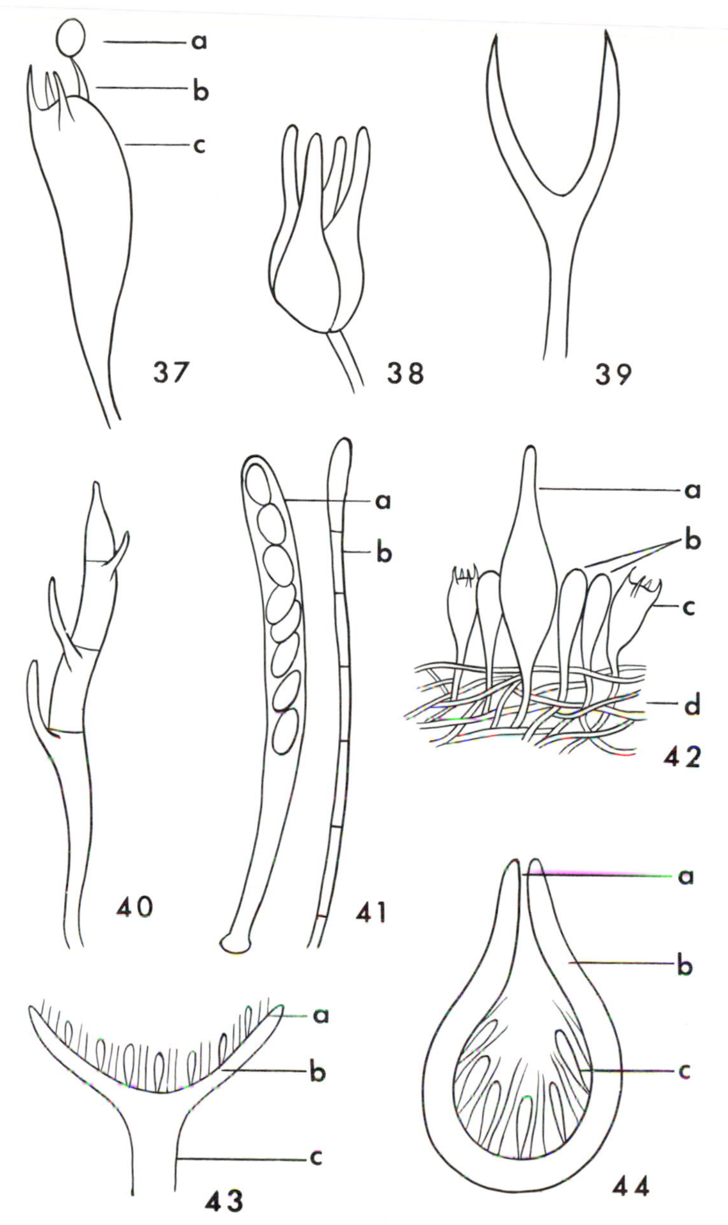

Figures 37-40. Semidiagrammatic drawings illustrating various types of basidia: 37, basidium of Agaricaceae (*a* spore, *b* sterigma, *c* basidium); 38, basidium of Tremellaceae with longitudinal septa; 39, deeply forked basidium of Dacrymycetaceae; 40, basidium of Auriculariaceae with transverse septa.

Figure 41. Semidiagrammatic drawing of: *a* ascus containing ascospores, *b* paraphysis.

Figure 42. Diagram illustrating a small section of the lamella of a mushroom: *a* cystidium, *b* paraphyses, *c* basidium, *d* trama.

Figure 43. Diagram of a transverse section of an apothecium: *a* hymenium composed of asci and paraphyses, *b* hypothecium, *c* stipe.

Figure 44. Diagram of a transverse section of a perithecium: *a* ostiole, *b* perithecial wall, *c* asci and paraphyses.

COLLECTING MUSHROOMS

The equipment required for collecting mushrooms is relatively simple. A basket, preferably a wide and fairly deep one, a sharp knife, and a supply of paper or plastic bags and waxed paper are about all the essentials. A good hand lens with a magnification of 10 to 14 diameters is desirable, and also a pencil and notebook if one wishes to take notes in the field. Details of location, habitat, whether or not associated with particular trees and so on are best noted down on the spot, and it is well to note such characters as color, taste, odor, etc., in the field. If the specimens are collected carefully and examined promptly, detailed notes can be made after returning from the collecting trip, but it is important that accurate data about the specimens be obtained as soon as possible.

The method of collecting will vary slightly depending on the purpose for which the specimens are collected. Those intended for study and identification should be gathered very carefully. It is important to collect the whole fruiting body including the base of the stipe, and in order to obtain this it may be necessary to dig the mushroom. The fruit bodies may be placed separately in paper bags, preferably first wrapped in waxed paper. This is best done by laying the specimen on a sheet of waxed paper, rolling the paper into a cylinder, and twisting the ends. The cylinders can then be stacked in the basket or placed in a large paper bag without danger of the specimens being damaged or the collections mixed.

If the specimens are to be photographed great care must be taken to avoid damaging them and it is advisable to keep them standing upright in the collecting basket.

Collections of different species should be kept separate. Specimens that look the same and are growing in clusters or very closely associated may be assumed to be the same species, but this is not a safe assumption for specimens growing singly and scattered over an area of several square yards. This is especially true in genera such as *Russula*, *Cortinarius*, and *Inocybe* where specimens that look very similar may have quite different microscopic characters. It is bad practice to put different mushrooms in the same basket unwrapped. One fruit body may shed spores on to the lamellae of another fruit body of a different species and so create much confusion.

Whenever possible, a number of fruit bodies of the same species, including very young stages, should be collected. Overmature specimens or those badly infested with insect larvae are best discarded. Such specimens should certainly be discarded if the collection is intended to be used for food.

Perhaps the most satisfactory way of working is to collect in the morning and examine the specimens in the afternoon. However, if this is not possible, most of the larger fleshy forms can be kept satisfactorily overnight, especially if stored in a cool place, but delicate forms such as species of *Coprinus* will not keep overnight and must be examined at once.

When the collector is certain that he knows an edible species and can recognize it in the field he may then proceed to collect it for food. For this purpose it is advisable to cut off the stem well above ground level to avoid getting dirt in the specimens, but one should always be certain that the specimen is not an *Amanita* and that there is no volva buried in the ground. Young, sound specimens should be chosen and overmature ones discarded. Cut the fruit bodies in two and if the flesh shows tunnel-like pinholes indicating infestation with insect larvae these specimens should be discarded also.

When mushrooms are being collected for food keep the species separate. Species differ in texture and some may require longer cooking than others. Those of similar texture can be cooked together satisfactorily but if one makes a practice of keeping each species separate when collecting, he is more likely to examine each fruit body more carefully and is less likely to collect a poisonous one by mistake.

FOOD VALUE OF MUSHROOMS

Mushrooms have long been esteemed for their pleasant flavor but the question has often been raised as to whether or not they possess any nutritive value as well, and contradictory statements as to their food value have appeared from time to time. Several chemical analyses of the composition of mushrooms have been published and these provide good evidence that mushrooms are indeed a valuable source of food.

Figures vary to some extent with different species and different analyses but in general the water content is about 89 per cent, protein 3 per cent, fat 0.4 per cent, carbohydrates 6 per cent, and minerals about 1 per cent. Although mushrooms do not compare with meat as a source of protein and, some of the protein appears to occur in an indigestible form, they are good sources of such minerals as iron and copper.

Mushrooms have been found to be high in vitamin C, ascorbic acid, niacin and pantothenic acid. These vitamins are well retained during cooking and in canned or dried mushrooms.

In contrast, the caloric value of mushrooms is low, roughly 30 calories per 100 grams or about one-quarter pound, but the calorie value will, of course, be increased if they are cooked in excess fat.

In general it seems safe to say that in addition to their value as flavor, mushrooms compare favorably with most vegetables as to nutritive value and vitamin content.

MUSHROOM POISONING

The fear of mushroom poisoning is so great among many people that it arouses a feeling of dread of all mushrooms. This attitude is justified to a considerable extent because of the high percentage of fatalities among cases of poisoning by amanitas, and unless a person can recognize the genus *Amanita* he is well advised to avoid unknown mushrooms. No case of mushroom poisoning should ever be regarded lightly and medical assistance should be sought at once.

The following summary of mushroom poisoning is mainly taken from the account by Pilat (1954) who considered that there are at least seven different types of mushroom poisoning.

Unquestionably the most dangerous type of poisoning is caused by mushrooms in the *Amanita phalloides* group. *A. phalloides* itself is a greenish olive species with radiating blackish fibrils on the pileus. It apparently does not occur in Canada, but our white *A. virosa* is equally deadly. Ramsbottom (1953) estimated that 90 per cent of the recorded deaths from fungus poisoning have been caused by species of this group. The folly of the superstition that a mushroom that peels is safe is well illustrated here because these *Amanita* species peel readily. The mushrooms of the *A. phalloides* group are so deadly that even small amounts may prove fatal. The danger is increased by the fact that there is apparently no unpleasant taste and no symptoms are manifested until 8 to 12 hours, or sometimes even longer, after the mushrooms are eaten. By this time the poison has been absorbed into the blood stream and the usual procedures such as pumping out the stomach are of no avail.

The general symptoms of this type of poisoning are severe abdominal pains, vomiting, cold sweats, diarrhea and excessive thirst. After persisting for some time the symptoms usually subside for a while and then recur more intensely; the liver is affected as well as the nervous system. There may be delirium, deep coma, and finally death. The patient suffers great pain.

Early investigations on the nature of the poison showed that there were at least two poisonous substances in *A. phalloides*. One of these was destroyed by heat but the other was not, and this latter was responsible for most of the poisoning cases. It was called amanita toxin and later study has shown that this is a complex of three substances, α-amanitine, β-amanitine, and phalloidine. All of these are very poisonous.

Although in cases of poisoning by this group of mushrooms the percentage of fatalities is very high, three methods of treatment have been used with some apparent success.

A serum has been produced at the Institut Pasteur in Paris by immunizing horses and it is said to give good results if injected hypodermically or intravenously and used early. However because of the rare and sporadic occurrence of this type of poisoning, supplies of fresh serum are not readily available.

The second method is to give injections of glucose in normal saline. This treatment is based on the fact that in *Amanita* poisoning there is a pronounced lowering of blood sugar with consequent damage to the liver and kidneys and the injections may help to restore the amount of sugar and modify the effects of the poison. Normal saline alone has also been used. It is better to give these hypodermically than by mouth because vomiting is usually associated with this type of poisoning.

The third method sounds fantastic but some success has been claimed for it. It is based on the idea that the gastric juices of a rabbit will neutralize the poison of the *Amanita*. It is recommended that the stomachs of three rabbits and the brains of seven be chopped up finely and made into a paste or pellets and eaten raw by the patient. Sugar or jam can be added to make it more palatable, and the sugar itself may be beneficial. This method has received some publicity in the press but it is difficult to know whether or not it is really effective.

The second type of poisoning is caused by species such as *Amanita muscaria* and *A. pantherina* and is believed to be due to a substance called mycoatropine. The symptoms usually appear soon after eating the mushrooms, within one-half to four hours. The most characteristic symptoms are nervous excitement, hallucinations and behavior suggesting alcoholic intoxication. This may be followed by coma and sometimes death, although the percentage of recovery from this type of poisoning is much greater than with the *A. phalloides* type. *A. pantherina* is considered to be more dangerous than *A. muscaria.*

Treatment consists in the administration of emetics and purgatives to clean out the digestive tract and then in treating the delirium with chloral hydrate or potassium bromide and providing a heart stimulant.

The third type of poisoning is due to the substance called muscarine and is caused by some *Inocybe* species and *Clitocybe dealbata. Amanita muscaria* also contains muscarine but its main effects are now believed to be due to mycoatropine. The symptoms of muscarine poisoning are profuse sweating, vomiting, diarrhea, pains in the stomach, distortion of vision and slowing down of the heart. Death rarely occurs in this type of poisoning and if it does it is due to the effect on the heart. Atropine is an antidote for muscarine poisoning.

A fourth type of poisoning may be caused by *Entoloma lividum* and a few other species. It is a violent gastrointestinal disturbance usually occurring within one to two hours after the mushrooms have been eaten. The symptoms include vomiting, diarrhea, acute pain and profuse perspiration. The symptoms may persist for a long time and make the patient very weak. Not much appears to be known about the poison involved.

The fifth type of poisoning is caused by some of the acrid species of *Russula* and *Lactarius.* It has the effect of a very violent purgative and causes vomiting and stomach pains. It is claimed that the poisonous substances can be removed from these mushrooms by boiling them in several changes of water, but these species are not recommended as food.

The sixth type is the poisoning caused by some Discomycetes and believed to be due to helvellic acid. The most important fungus in this group is *Gyromitra esculenta* and reports about this fungus are very contradictory. There seems to be no doubt that many people eat this species frequently, apparently without any harm. On the other hand there are well-authenticated cases of poisoning and even of deaths caused by it. It would appear that the danger is greatest with overmature or slightly decomposed specimens, but danger certainly exists and on no account can this fungus be recommended as food.

Finally there is a peculiar type of poisoning that has been said to be associated with *Coprinus* species eaten at the same time that alcohol has been consumed. It is apparently not dangerous, but in about twenty minutes to two hours the face may become very red and then violaceous and the color may spread to the neck and body. The tip of the nose and the ear lobes remain pale. There is a sensation of heat and the pulse beat becomes very rapid. The symptoms disappear in a short time and apparently there are no ill effects.

There has been some controversy about this type of poisoning. Child (1952) described experiments in which Coprini were fed to a person with and without alcohol and no effects were observed, but when *Panaeolus campanulatus* was eaten by itself the type of symptoms described above appeared. He suggested that the reports of poisoning caused by a combination of *Coprinus* and alcohol might be due rather to accidental inclusion of *Panaeolus* with the *Coprinus*. However his experiments were rather limited in scope and European mycologists still insist that these symptoms have appeared in well-authenticated cases where there was no possibility of a misidentification of *Panaeolus*. The question may be regarded as still open but it would probably be advisable to avoid eating *Coprinus* and consuming alcohol at the same time.

In any case of mushroom poisoning, medical assistance should be summoned immediately. As a first-aid measure the stomach and intestines should be emptied by inducing vomiting or administering purgatives or an enema. Parts of mushrooms vomited up, or the remains of the dish eaten should be preserved so that the species responsible for the poisoning can be identified. If any fresh mushrooms of the original gathering remain they would be still more useful for this purpose.

IDENTIFICATION

At first, mushrooms may all look rather similar but as we observe them more closely and become more interested in them, differences and similarities are perceived and more and more species are recognized. Some people will be content with learning to identify a few common species, others will want to know more and may even want to make a special study of some groups or genera.

Correct identification, then, becomes a matter of observing carefully the characters possessed by the fungus, comparing them with descriptions and illustrations or with other specimens, and assessing the value of differences and similarities observed. There is no rigid concept or set of rules by which one can say whether or not an observed difference represents a real difference between species or simply variation among individuals of the same species. Size of the pileus and length of the stipe, for example, are characters that are usually fairly constant within limits for any particular species, but some individuals may be found in which these or other characters may far exceed the ordinary limits. Colors may fade, heavy rains may wash scales off the pileus, a delicate annulus may disappear very early, and so on. On the other hand, spore color is a very constant character as are many of the microscopic characters.

No single book contains descriptions of all the known species of mushrooms and no one person can, with certainty, identify every mushroom he finds. Much remains to be learned about the species comprising our mushroom flora, and for the amateur the identification of mushrooms offers a challenge and an interest that he can pursue as far as he wishes.

Probably the most usual way for a beginner to start learning to identify mushrooms is by association with some more experienced collector who can point out the common species and the characters by which they are recognized. This is a good way for the beginner to start and it often results in arousing his interest and curiosity to the point where he will want to be able to identify other species for himself.

At this point the necessity for consulting books on mushrooms will become apparent and these books usually contain keys, descriptions and illustrations to aid in identification. A key is a guide to identification that is constructed by presenting a choice between two characters or groups of characters. The student decides which of the characters is possessed by the fungus under study, and by eliminating the species not possessing the characters, he eventually narrows the choice to one species. Difficulties may arise because the characters are not clear-cut, or the specimen is inadequate to show all the characters, or the species under investigation is not found in the key being used. It is sometimes impossible to decide with certainty which choice should be made and both will have to be followed up until a definite elimination can be made.

Some keys are designed to show relationships so that related forms key out close to each other but, since relationships are indicated by similarities, and eliminations are usually made by noting differences, keys of this type are often difficult to use in making identifications. The keys in this book are intended primarily to aid in identification and are not designed to indicate relationships.

When a specimen keys out it should then be compared with the detailed description of the species and, if possible, with good illustrations. It is not recommended that identification of mushrooms be attempted by simply comparing the specimens with illustrations. Undoubtedly many correct identifications of mushrooms have been made in this way but there are so many species of mushrooms that are superficially similar in appearance that this method is likely to lead to serious errors.

In making an identification of a species, the first thing that must be determined is the color of the spores. This can best be seen in a spore deposit or spore print (Figure 45). To obtain a spore print cut off the stipe close to the pileus, lay the pileus on a piece of white paper with the lamellae downward and leave it for several hours. Better results will be obtained if it is covered with a glass or dish of some sort to protect it from air currents. Some people have used black paper in order to better show up white spore deposits, but pale cream, pale pink, or lilac spores may appear to be white if deposited on black paper, and since white spores can always be seen on white paper if viewed at an angle, only white paper should be used. Satisfactory spore prints of some of the firmer species can be obtained in the field by cutting off the stipe, laying the pileus on a piece of white paper with the lamellae down, wrapping it carefully in waxed paper, and laying it flat on the bottom of the basket. A good spore deposit may be obtained by the time one returns from the collecting trip.

If the spore deposit is white to yellow, the amyloid reaction of the spores should be determined. This reaction is determined by placing some of the spores on a glass slide and adding a drop of a solution (Melzer's Reagent) made up of 1.5 grams potassium iodide, 0.5 grams iodine, 20 grams distilled water and 20 grams chloral hydrate. If the spores turn gray-blue to blackish blue they are amyloid and if there is no reaction they are nonamyloid. The reaction can be observed with the microscope if one is available, or, if not, it can be seen by simply holding the slide over white paper. The reaction should be noted within a few minutes of making the test and it is more reliable if tested on spores that have first been dried. A few species, particularly in *Lepiota*, will give a falsely amyloid or pseudoamyloid reaction and the color will be reddish brown.

This reaction is of considerable importance in taxonomic work. For example, the genus *Leucopaxillus* contains some species that were formerly in *Clitocybe* and some that were in *Tricholoma*. These are believed to form a natural group and can be recognized by the amyloid reaction of the spores. In

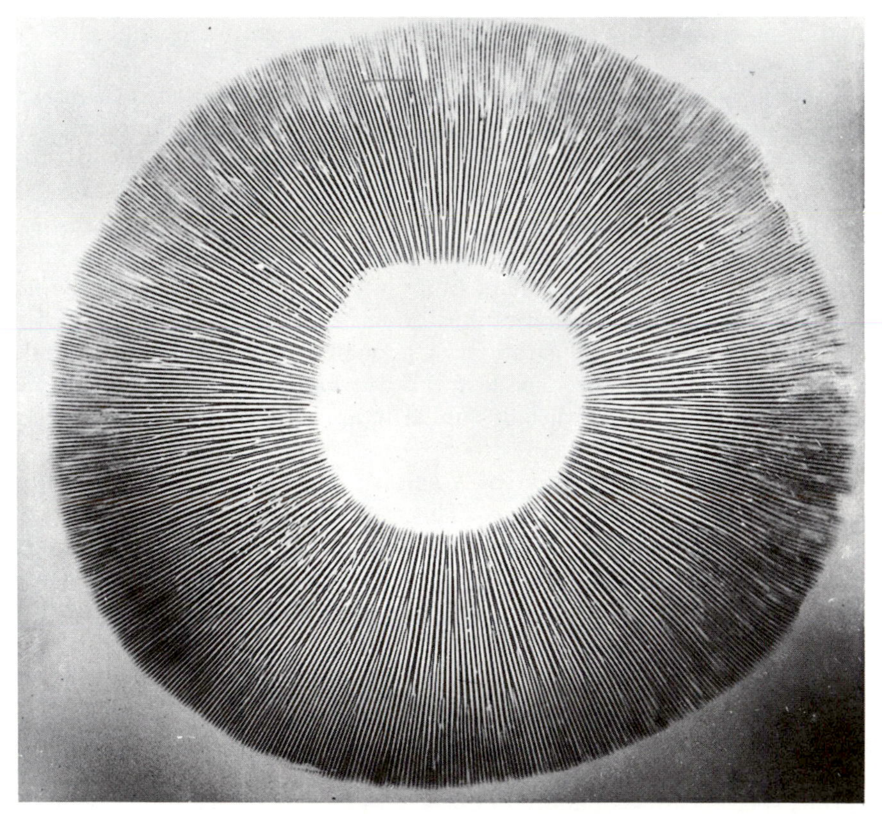

Figure 45. Spore print of a mushroom.

this instance the reaction is used to delimit a genus. Sometimes it is not considered to indicate a difference of generic rank but is useful in distinguishing species. For example *Amanita flavoconia* and *A. frostiana* have been confused at times but they can readily be separated by this reaction since the spores of *A. flavoconia* are amyloid and those of *A. frostiana* are not.

Notes should then be taken on the color and size of the fruit bodies, whether or not the surface is smooth, hairy, scaly, etc., and whether or not it is dry, viscid, or hygrophanous. This latter term describes a condition in which the flesh absorbs moisture and becomes darker, taking on a watery appearance, and then fades considerably on drying. It is often a useful field character when one has become familiar with the characteristic appearance. The odor and taste should be noted, although amanitas should not be tasted. If a juice or latex is present it should be tasted, and the color and any changes in color noted. It is especially important in *Cortinarius* to record the color of the young lamellae before the spores have matured. Features of the stipe that should be noted are the size, color, and consistency, whether or not there is an annulus or volva present, whether or not the surface is smooth, fibrillose, scaly, etc., or if it is viscid or dry.

In short, the collection should be studied carefully and all the characters noted while the material is still fresh. Notes that are made in the form of a concise description are likely to be better than those made by following a form or chart. Probably most people will make their final identifications from fresh material but if good notes are made on fresh specimens that are then carefully dried, it is often possible and even in some instances preferable, to make identifications from dried specimens.

In preparing dried specimens it is important to dry them as quickly as possible without scorching or cooking them. A good free circulation of air is essential. Probably the best method is to spread the mushrooms on a screen or series of screens that can be suspended or placed in a frame over the source of heat.

In this book relatively little emphasis has been placed on the use of microscopic characters in the identification of species. However, for those who have a microscope, the size and shape of the spores have been given throughout, and in some cases also the cystidia are described.

After the specimens have been studied and their characters noted one can then attempt to identify them by using the key. The key should be used as a guide only, and when a species keys out it should be checked carefully with the description and illustrations. Mushrooms are living organisms and show variations between individuals and between collections that have grown under different environmental conditions. The descriptions in this book are based as far as possible on normal and typical specimens and they attempt to describe the range of variation ordinarily encountered. It should be kept in mind also that the descriptions of a great many species have, of necessity, been omitted and hence a specimen that keys out to a certain species but does not agree with the description of that species should not be so identified. The chances are good that it is a species not described in this book.

Suppose we have a specimen to be identified. The first choice in the key (p. 29) is based on the character of the lamellae and in this specimen they are thin, well developed and crowded, so we go to choice 2. We check the spore deposit and find it is colored, so we go to choice 25. Here there are five choices and we find our specimen has a pink spore deposit, so we go to choice 26. It has a central stipe, which takes us to choice 27, and the lamellae are free from the stipe, which carries us to 28. There is no trace of an annulus or volva, so we come to *Pluteus* and turn to the key to the species of this genus (p. 165). The pileus is brown and the lamellae do not have a black edge so we arrive at *Pluteus cervinus*, and on comparing our specimen with the description we find it matches and we can conclude that we have identified the species.

If the specimen had had a yellow pileus and stipe we would similarly have checked *Pluteus admirabilis* but a specimen with a yellow pileus and white stipe would have caused difficulty. Another reference book might show that there is a species of *Pluteus* with a yellow cap and white stipe, *P. leoninus*, that is not described in this book. But it would have been wrong to assume that we had found *P. admirabilis* because it keyed out here, when the description of *P. admirabilis* called for a yellow stipe and our specimen had a white one.

A key thus has its limitations but if used critically and with caution it should be of great assistance in identifying the specimens collected.

If other methods of identification fail it is sometimes possible to send specimens to specialists and have them identified. In the Canada Department of Agriculture we do not have the staff or the time available to undertake identifications on a large scale. However, for the beginner who may wish to check on the identification of some common species of which he is uncertain, or for the more advanced student who thinks he may have found some rare or unusual species of special interest, we try to provide all the assistance possible.

Fresh specimens may be sent in by mail and if they are not too long in transit and are not overmature, worm-eaten, or decayed, they often arrive in good condition. Specimens wrapped in tissue paper or newspaper which will absorb excess moisture arrive in better condition than those wrapped in waxed paper or packed in tin cans. Under these latter conditions moisture accumulates and bacterial decay sets in. Species of *Coprinus* do not travel well by mail and usually arrive as an inky mess.

Specimens of mushrooms should never be preserved in fluid or sent in for identification in fluid. They are almost impossible to identify when preserved in this way.

It is much preferable when sending material in for identification to take careful notes on the fresh condition, then dry the specimens and send them in the dried condition. If properly dried they will keep indefinitely and with good notes they can be identified just about as readily as in the fresh condition. Furthermore, there is the advantage to us that if they prove to be an interesting species the dried specimens can be placed in the herbarium and become part of the permanent scientific record.

NOMENCLATURE

Undoubtedly one of the things that discourage the amateur from taking up the study of fungi is the difficulty of mastering the names. Relatively few species have common or vernacular names and attempts to create common names for them by translating the Latin names have not been very successful. In reality the Latin names are not so difficult as would first appear and after a little experience in associating them with actual specimens they become as familiar as do the Latin names of common flowers to enthusiastic gardeners.

Obviously we must apply names to fungi and other plants in order to refer to them and, since fungi do not recognize national boundaries, some system of naming must be followed that can be used by scientists of all countries. The system that has been adopted was devised by the Swedish botanist Carl Linnaeus in the eighteenth century and is called the binomial system of nomenclature. In this system individual plants that are considered to be of the same kind are grouped together to form a species, related species are placed in a genus, related genera in a family, related families in an order, related orders in a class, and related classes in a division. The whole hierarchy constitutes the plant kingdom.

The name of any individual plant consists of two words, the name of the genus and the name of the species, the latter usually a descriptive adjective. When we name a plant in this way we are at the same time classifying it in relation to other plants. Our knowledge of the relationships of plants to each other is continually increasing so that our ideas about the classification of plants change accordingly and this leads inevitably to changes of the names.

Ideally, of course, one species should have one correct name, but as long as our system of classification is imperfect and the relationships of plants not fully understood, botanists will differ in their ideas about the classification and relationships and so will differ in their choice of names. Furthermore, many mistakes can and do occur with resultant confusion concerning names. Some common errors of this type arise from different botanists describing the same plant under different names or describing different plants under the same name, or applying a name to a different plant from that intended by the original author. In order to settle disputes and clear up confusion of this sort, it is necessary for botanists to agree on a set of rules determining the choice of a name.

From time to time botanists hold an international congress and the authority of this body is generally recognized in the drawing up of a set of rules of nomenclature and in making revisions deemed necessary. The official title of the set of rules is now the *International Code of Botanical Nomenclature*.

The *Code* has nothing to do with classification. Each botanist is free to study a plant and come to a decision himself regarding its relationships with other plants, but when he has reached such a decision the *Code* determines the correct name to use for the plant. It is impossible to discuss the *Code* in any detail here, but some of the more important rules might be noted.

In the first place it is necessary to have a starting point for our nomenclature and although for the higher plants this is Linnaeus' *Species Plantarum*, 1753, for most of the fungi it has been decided to start with the *Systema Mycologicum* published in 1821 by the Swedish mycologist E. M. Fries.

In order to have any claim to recognition, a name must be validly and effectively published. It is not sufficient to put a name on a specimen in a herbarium or botanic garden, or to mention it at a public meeting or refer to it in a thesis. It must be made available to botanists everywhere. Names published before the official starting point are regarded as not validly published. A name must be accompanied by a description and since 1935 a name is not considered to be validly published unless it is accompanied by a Latin diagnosis.

Sometimes more than one name may be validly published for the same plant and in that case the earliest name is considered to be the correct one.

These are, perhaps, the most important rules but there are others which, if they are not complied with, render a name illegitimate. If a name is found to be illegitimate it cannot be used and the earliest legitimate name must be chosen for the plant. If none exists, the plant must be given a new name.

It is usual when writing the scientific name of a plant to give also the name of the author who proposed the name. If the species is later transferred to another genus, the name of the original author is placed in parentheses followed by the name of the author who made the transfer. This practice has sometimes been criticized as a form of self-advertisement on the part of the authors, but that is not the purpose. It is rather to supply a reference to the source of the name, thus enabling taxonomists to check on the validity and legitimacy of the names and determine exactly to what plants they apply, and in this way these author references are invaluable to the research taxonomist.

A primary aim of the *Code* is, of course, to bring about stability of names and one of the most important means of achieving this is the use of the type concept. When an author describes a new species he is expected to designate some particular specimen as the type of that species. If he does not do so, some specimen must subsequently be chosen as the type. The name is then permanently fixed to that specimen and when we apply the name to any other specimen we are, in effect, saying that it belongs to the same species as the type. If, as sometimes happens, it is found that a name has been applied to plants belonging to more than one species, the name must be retained for those that match the type and the others must be given another name. Sometimes an author may make an error in describing a species or misinterpret structures he has observed; the concept of the species is then determined by the type specimen and not by what the author said about it.

Similarly when a genus is established, one species is taken as the type of the genus and the application of the name is determined by that species. For example, if it is considered that unrelated species may have been placed in the same genus and it is necessary to divide the old genus into two or more genera, the original name must be retained for the type species and others that may be considered congeneric with it. A good example of this is seen in the mushrooms. In his *Systema Mycologicum*, Fries placed nearly all the mushrooms in the genus *Agaricus* but he divided the genus into a number of sections such as *Lepiota*, *Tricholoma*, *Pholiota*, *Psalliota*, etc. Later authors raised these sections to the rank of genera, but the old name *Agaricus* had to be retained for one of these sections, depending on what was considered to be the type species of *Agaricus*. Since the common meadow mushroom, *Agaricus campestris*, is taken as the type, the name *Agaricus* must be used for it and its close relatives. The name *Psalliota*, which was used by Fries for this section and by some later authors as a generic name, then becomes a synonym of *Agaricus* and is no longer a legitimate name.

In this book a number of species may be found under unfamiliar names. Some of the changes are the result of advances in our knowledge and of consequent improvements in the classification. For example, it is believed that it is a better expression of relationships to remove the species with granulose caps from *Lepiota* to *Cystoderma*, and the species with viscid caps from *Lepiota* to *Limacella*. Other changes are necessitated in order to comply with the rules rather than because of changes in our ideas of classification. Examples of these changes are the use of *Agaricus* rather than *Psalliota*, *Volvariella* rather than *Volvaria*, and *Lepiota molybdites* rather than *L. morgani* for the green-spored *Lepiota*. Such changes are regretted but only by careful adherence to the rules and the acceptance of changes necessitated thereby, will we eventually attain a stable nomenclature.

CLASSIFICATION

Fungi, the class of plants to which mushrooms belong, may be defined in a general way as plants lacking true leaves, stems, and roots, lacking chlorophyll, and reproducing by spores. They are usually divided into four main subclasses.

The first of these is called the Phycomycetes. The fungi referred to this group are characterized in general by the absence of cross walls or septa in the hyphae composing the mycelium and by the production of spores within a sac, usually a more or less swollen cell, termed a sporangium. The Phycomycetes include forms such as the common bread mold, the potato-blight fungus, the downy mildews, many aquatic fungi and many minute, one-celled forms. None of the Phycomycetes will be discussed in this book.

The second subclass is called the Ascomycetes. In this group the hyphae have cross walls and the spores are produced in a specialized cell called an ascus (Figure 41, p. 9). The production of spores in the ascus is regarded as a sexual process. In the young ascus two nuclei fuse and then typically divide three times, forming eight spores which are forcibly discharged when they are mature. The asci may be produced directly on the mycelium or developed within more or less specialized fruiting bodies. Those Ascomycetes in which the fruiting bodies bearing the asci are structures that are closed, or that open by a narrow pore or beak, are known as Pyrenomycetes (Figure 44, p. 9);

whereas those in which the asci are arranged in a layer exposed to the air are known as Discomycetes (Figure 43, p. 9). The Ascomycetes as a whole comprise many thousands of species including yeasts, molds, powdery mildews, ergots, numerous leaf-spotting and wood-staining fungi and many others. In this book only a very few of the larger fleshy Discomycetes, and one Pyrenomycete, are discussed.

The third subclass is called the Basidiomycetes. In this group the hyphae also have cross walls and the spores are produced on a specialized cell known as the basidium (Figure 37, p. 9). In the basidium two nuclei fuse and then typically divide twice, giving rise to four nuclei, but in contrast to the Ascomycetes where the spores mature within the ascus and are discharged when mature, in the Basidiomycetes the nuclei migrate to the tips of little stalks and the spores develop outside the basidium itself and are forcibly shot off the stalks when they are mature. Most of the fungi described in this book belong in the Basidiomycetes.

The fourth subclass is known as the Fungi Imperfecti. This group is not strictly comparable to the other three since it comprises those forms in which a perfect or sexual state is unknown or lacking and reproduction takes place by vegetative means, usually by some form of asexual spore. These asexual spores are often called conidia and many forms that used to be grouped with the Fungi Imperfecti are now known to be conidial states of Ascomycetes, Phycomycetes, or Basidiomycetes. However there are still a great many whose perfect or sexual state is unknown and it is possible that some of these forms reproduce so efficiently with conidia alone that they have lost the ability to develop the sexual state. Hence it is necessary to retain this subclass, although it does not represent a natural grouping. None of the Fungi Imperfecti are discussed in this book.

BASIDIOMYCETES

The mushrooms and most of the larger fungi that will be found by the amateur collector belong in the subclass Basidiomycetes and it is necessary to consider the classification of this group in a little more detail. As has been noted, the fundamental character of the group is that the spores are produced outside the mother cell rather than within it. The typical basidium is a more or less club-shaped single cell with four little stalks or sterigmata arising at the apex and the spores develop on the tips of these stalks (Figure 37, p. 9). When mature, the spores are forcibly discharged from the sterigmata by the pressure of surface tension of a drop of water that is excreted from the tip of the sterigma below the spore. This is the type of basidium found in most of the larger fleshy fungi such as the agarics, boletes, hydnums, clavarias, and polypores.

The basidia are arranged in a definite layer in the fruiting body in such a way that the spores can be discharged into the open air. This layer is called the hymenium and these fungi are further divided into families based on the shape of the hymenium. In the Agaricaceae, or mushrooms, the hymenium covers the surface of thin blades or gill-like structures termed lamellae which are more or less radially arranged on the under surface of the pileus. In the Hydnaceae it covers the surface of tooth-like structures. In the Boletaceae the hymenium lines the inner surface of tubes and the fruiting body is soft, fleshy, and mushroom-like whereas in the Polyporaceae it also lines the surface of tubes but the fruiting body is tough, corky to leathery, or woody. In the Clavariaceae it is smooth and covers the entire fruiting body; whereas in the Thelephoraceae it is also smooth but the fruiting body is more or less differentiated into an upper, sterile surface, and a lower, fertile surface.

The above brief account is an outline of the traditional basis for distinguishing these families of Basidiomycetes but as a result of recent studies of microscopic structure, chemical reactions, cultural characters and so on, modern taxonomists are pretty generally agreed that it is not a satisfactory classification from a scientific standpoint. For example, *Lenzites* has lamellae but is obviously much more closely related to the polypores than to the other mushrooms, and *Gomphidius* has lamellae but is apparently more closely related to the boletes than to the mushrooms. Many other examples could be given. However, for the purpose of this book the traditional classification is quite satisfactory.

In the Gasteromycetes or puffballs, the basidia are produced in a closed fruiting body and not arranged in a hymenium. The spores are not forcibly discharged into the air but are disseminated by wind, rain, and insects.

Other groups of Basidiomycetes may exhibit variations in the form and structure of the basidium itself. In the Auriculariaceae the basidium becomes transversely septate, the cross walls forming four cells from each of which a sterigma arises (Figure 40, p. 9). In the Tremellaceae the basidium becomes longitudinally septate and the walls are at right angles; consequently four cells are formed, from each of which a sterigma arises (Figure 38, p. 9). In the Dacrymycetaceae the basidium becomes deeply divided or forked with sterigmata at the apices of the forks (Figure 39, p. 9).

One large group of Basidiomycetes is not discussed in this book. This group consists of the rusts and smuts, which live as parasites on other plants and in which the basidium has cross walls and is produced directly on germination of a specialized, thick-walled resting spore rather than in a fruit body.

The above classification may be presented briefly as follows:

FUNGI

Phycomycetes — hyphae lacking cross walls; spores borne in sporangia
Ascomycetes — hyphae with cross walls; spores produced in asci
    Plectomycetes — asci produced directly on the mycelium
    Pyrenomycetes — asci produced in a closed fruiting body or perithecium
    Discomycetes — asci produced in an apothecium, a fruiting body with an exposed fruiting surface

Basidiomycetes — hyphae with cross walls; spores produced on a basidium

Hemibasidiomycetes — basidia produced directly on germination of a resting spore (the rusts and smuts)

Eubasidiomycetes — basidia usually produced in a fruiting body, not on germination of a resting spore

Tremellales — basidia septate or deeply forked; fruiting bodies usually more or less gelatinous

Auriculariaceae — basidia transversely septate

Tremellaceae — basidia longitudinally septate

Dacrymycetaceae — basidia deeply divided, forked

Hymenomycetales — basidia one-celled, arranged in a hymenium; fruiting bodies fleshy to membranous to leathery or woody

Agaricaceae — hymenium covering lamellae

Hydnaceae — hymenium covering teeth

Boletaceae — hymenium lining tubes; fruit body soft and fleshy

Polyporaceae — hymenium lining tubes; fruit body tough, corky, leathery, or woody

Clavariaceae — hymenium smooth, covering entire fruiting body

Thelephoraceae — hymenium smooth; fruiting body differentiated into sterile and fertile surfaces

Gasteromycetales — basidia one-celled, borne in a closed fruiting body, not arranged in a hymenium

Phallaceae — spore mass slimy, evil smelling, at maturity raised on a stalk-like receptacle

Lycoperdaceae — spore mass powdery, remaining enclosed by the peridium

Nidulariaceae — fructification a cup-shaped or vase-shaped structure containing several peridioles (hard, egg-like bodies) within which the spores are produced

Fungi Imperfecti — fungi lacking sporangiospores, ascospores, or basidiospores; no sexual organs present, reproduction by vegetative spores

---

Figures 46-55

46. *Cantharellus cibarius.*
47. *C. cinnabarinus.*
48. *C. clavatus.*
49. *C. floccosus.*
50. *C. tubaeformis.*
51. *C. umbonatus.*
52. *C. multiplex.*
53. *Lactarius camphoratus.*
54. *L. affinis.*
55. *L. affinis.*

46
47
48
49
50
51
52
53
54
55

Figure 56. *Lactarius cinereus.*

Figures 57-66

57. *Lactarius deliciosus.*
58. *L. deliciosus.*
59. *L. controversus.*
60. *L. griseus.*
61. *L. helvus.*
62. *L. hygrophoroides.*
63. *L. lignyotus.*
64. *L. trivialis*
65. *L. indigo.*
66. *L. indigo.*

57
58
59
60
61
62
63
64
65
66

Figure 67. *Lactarius deceptivus.*

Figure 68. *Russula delica.*

KEY TO THE GENERA OF MUSHROOMS

1. Lamellae fold-like, thick on edge, forked .... *Cantharellus*
1. Lamellae well developed, crowded to distant but not fold-like 2

2. Spore deposit white 3
2. Spore deposit colored 25

3. Lamellae free from stipe 4
3. Lamellae attached to stipe 6

4. Volva and annulus present .... *Amanita*
4. Volva present; annulus absent .... *Amanitopsis*
4. Volva absent; annulus present 5

5. Pileus viscid .... *Limacella*
5. Pileus not viscid .... *Lepiota*

6. Annulus present 7
6. Annulus not present 8

7. Cuticle of cap granulose to warty .... *Cystoderma*
7. Cuticle smooth or scaly but not granulose .... *Armillaria*

8. Fruit body soft and fleshy, not reviving when moistened 9
8. Fruit body tough, corky to leathery, more or less reviving when moistened 21

9. Stipe excentric, lateral, or absent .... *Pleurotus*
9. Stipe central 10

10. Lamellae of waxy consistency 11
10. Lamellae not waxy 12

11. Spores smooth .... *Hygrophorus*
11. Spores echinulate .... *Laccaria*

12. Trama of fruit body composed of both filamentous and globular cells; texture brittle; lamellae stiff and easily broken; spores amyloid 13
12. Trama of filamentous cells only; lacking above combination of characters 14

13. Milky juice present .... *Lactarius*
13. Milky juice absent .... *Russula*

14. Stipe cartilaginous in texture, different from the pileus 15
14. Stipe fleshy or fibrous, somewhat similar to the pileus in consistency or tougher 17

15. Stipe somewhat horny in consistency; lamellae decurrent .... *Xeromphalina*
15. Stipe not horny; lamellae adnate to adnexed 16

16. Margin of pileus incurved; pileus becoming expanded .... *Collybia*
16. Margin of pileus straight; pileus usually somewhat bell-shaped .... *Mycena*

17. Lamellae decurrent 18
17. Lamellae not decurrent 19

18. Spores amyloid .... *Leucopaxillus*
18. Spores not amyloid .... *Clitocybe*

19. Spores not amyloid .... *Tricholoma*
19. Spores amyloid 20

20. Pileus hygrophanus; lamellae with harpoonlike cystidia ..... *Melanoleuca*
20. Pileus rarely hygrophanous, usually dull colored, large and fleshy; lamellae without harpoonlike cystidia ..... *Leucopaxillus*

21. Lamellae split along edge ..... *Schizophyllum*
21. Lamellae not split along edge 22

22. Lamellae serrate-torn on edge ..... *Lentinus*
22. Lamellae not serrate-torn on edge 23

23. Lamellae crisped, thick ..... *Trogia*
23. Lamellae entire 24

24. Stipe central; pileus membranous to somewhat fleshy, reviving ..... *Marasmius*
24. Stipe excentric, lateral, or wanting; pileus tough, fleshy-leathery to corky ..... *Panus*

25. Spore deposit greenish ..... see *Lepiota molybdites*
25. Spore deposit lilac to grayish lilac ..... *Pleurotus* (see *Laccaria ochropurpurea*)
25. Spore deposit pinkish 26
25. Spore deposit yellowish to rusty or brown 33
25. Spore deposit purplish to purple-brown or blackish 45

26. Stipe lateral or lacking ..... *Phyllotopsis*
26. Stipe central 27

27. Lamellae free from the stipe 28
27. Lamellae attached to the stipe 29

28. Volva present; annulus lacking ..... *Volvariella*
28. Both annulus and volva lacking ..... *Pluteus*

29. Lamellae decurrent ..... *Clitopilus*
29. Lamellae adnate to adnexed 30

30. Lamellae sinuate 31
30. Lamellae not sinuate 32

31. Spores angular ..... *Entoloma*
31. Spores not angular, slightly rough ..... *Tricholoma*

32. Margin at first incurved; pileus convex ..... *Leptonia*
32. Margin at first straight; pileus usually more or less conical to campanulate ..... *Nolanea*

33. Stipe excentric, lateral or wanting ..... *Crepidotus*
33. Stipe central 34

34. Veil composed of cobweb-like filaments; spore deposit dark brown ..... *Cortinarius*
34. Veil membranous or lacking 35

35. Annulus present 36
35. Annulus lacking 37

36. Pileus with a mealy-granulose surface ..... *Phaeolepiota*
36. Surface of pileus smooth or scaly, not mealy-granulose ..... *Pholiota*

37. Lamellae separating readily from the pileus trama ..... *Paxillus*
37. Lamellae not separating readily from the pileus trama 38

38. Trama of pileus composed of filamentous and globose cells; spores amyloid 39
38. Trama of pileus filamentous; spores not amyloid 40

39. Milky juice present .... *Lactarius*
39. Milky juice lacking .... *Russula*

40. Stipe fleshy to fibrous 41
40. Stipe cartilaginous 43

41. Growing on wood; lamellae usually bright colored, yellow to rusty .... *Flammula*
41. Growing on the ground; lamellae dull brownish to pallid 42

42. Pileus viscid, smooth .... *Hebeloma*
42. Pileus more or less fibrillose to silky or scaly, or splitting .... *Inocybe*

43. Lamellae decurrent .... *Tubaria*
43. Lamellae not decurrent 44

44. Pileus convex to plane; margin at first incurved .... *Naucoria*
44. Pileus more or less conical; margin straight; cuticle composed of vesiculose cells .... *Conocybe*
44. Fruiting bodies fairly large, white, staining brown; spores pale yellow .... see *Collybia maculata*

45. Lamellae waxy, decurrent .... *Gomphidius*
45. Lamellae not decurrent 46

46. Lamellae deliquescing .... *Coprinus*
46. Lamellae not deliquescing 47

47. Annulus present 48
47. Annulus lacking 50

48. Lamellae free from the stipe .... *Agaricus*
48. Lamellae attached to the stipe 49

49. Lamellae mottled .... *Panaeolus*
49. Lamellae not mottled .... *Stropharia*

50. Cuticle of pileus filamentous .... *Naematoloma*
50. Cuticle of pileus composed of vesiculose cells 51

51. Pileus plicate-striate and paraphyses coprinoid .... *Pseudocoprinus*
51. Not with above combination of characters 52

52. Lamellae mottled .... *Panaeolus*
52. Lamellae not mottled .... *Psathyrella*

CANTHARELLUS

Cantharellus is an important genus for the mycophagist because many of the species are large and conspicuous, and fairly easily recognized without much danger of being confused with poisonous species. The chanterelle, *C. cibarius*, is one of the species frequently used and highly recommended for food and should be on the list of every amateur collector.

The genus is characterized by the thick, fold-like lamellae which are decurrent, usually distant and more or less forked. The stipe is continuous with the pileus and there is no veil. In some species the lamellae are poorly developed and little more than wrinkles. The genus then approaches *Craterellus* of the Thelephoraceae in which the hymenium is smooth. *Cantharellus clavatus*, in which the lamellae are mere wrinkles, has been placed in *Craterellus* by some authors. On the other hand, in species in which the lamellae are better developed and less fold-like, the genus approaches *Clitocybe*. *Cantharellus umbonatus* with well-formed lamellae is likely to be sought in *Clitocybe*, and *Clitocybe aurantiaca* has been called *Cantharellus aurantiacus* by many authors.

For those who are interested in the problems of relationships, it may be noted that modern taxonomists tend to the view that *Cantharellus*, as here constituted, does not comprise a natural group of species. For example, *C. multiplex*, with its peculiar warted spores, has been made the type of a new genus, *Polyozellus;* and *C. umbonatus*, with amyloid spores, has been made the type of a new genus, *Cantharellula*. Although it is evident from modern taxonomic studies that these and other species are not closely related to *C. cibarius*, which is the type species of *Cantharellus*, it seems preferable to retain *Cantharellus* in the more traditional sense for the purpose of this book. If *Cantharellus* were to be used in the strict sense for only those species closely related to *C. cibarius*, it might be considered to be more closely related to the Clavariaceae than to the Agaricaceae.

Key

1. Fruiting bodies entirely red .... *C. cinnabarinus*
1. Fruiting bodies not entirely red 2
2. Fruiting bodies more or less yellow to brown 3
2. Fruiting bodies not yellow 5
3. Fruiting bodies large, vase-shaped, yellowish, with reddish or reddish orange scales .... *C. floccosus*
3. Fruiting bodies not as above 4
4. Fruiting bodies bright chrome-yellow to egg-yellow, firm, fleshy .... *C. cibarius*
4. Fruiting bodies brownish yellow to ochraceous brown, thin, pliant; lamellae drying grayish .... *C. tubaeformis*
5. Fruiting bodies gray, usually with a small umbo; lamellae well developed, white; flesh reddening when wounded .... *C. umbonatus*
5. Fruiting bodies purplish flesh color to blackish 6
6. Growing in dense cespitose masses; spores nearly globose, warty .... *C. multiplex*
6. Usually separate to gregarious or sometimes slightly cespitose; spores narrow-ellipsoid, wrinkled .... *C. clavatus*

CANTHARELLUS CIBARIUS Fr. Edible

Figure 46, page 25; Figure 411, page 295

Chanterelle

PILEUS 1–4 in. broad, fleshy, firm, convex or sometimes top-shaped, becoming expanded and then depressed in the center, often irregularly wavy or lobed, chrome-yellow to egg-yellow, fading in age, slightly fibrillose to glabrous, not striate, dry. FLESH firm, whitish to yellowish, taste mild to somewhat peppery, odor fruity or sometimes lacking. LAMELLAE decurrent, distant, forked, thick, blunt on the edge, narrow, yellow. STIPE 2–3 in. long, ½–1 in. thick, narrower toward the base, solid, glabrous, concolorous with the pileus or paler. SPORES elliptical, smooth, tinged yellowish in mass, 8–11 × 4–6 μ.

Scattered or in groups, or sometimes in small clusters, on the ground in open woods, either coniferous or deciduous. July–Sept.

Not many mushrooms are sufficiently well known to possess a common name but this species, which is highly prized as food, especially in Europe, has many names in different languages of which the best known is the chanterelle. The European plants are said to have a fruity odor resembling apricots and, although this appears to be sometimes lacking in North American chanterelles, they are none the less desirable for the table. Because of its firm texture this species may require longer cooking than some of the more tender ones.

It is an important mushroom, for it occurs fairly commonly, is widely distributed, and is sufficiently distinctive in appearance that it is not likely to be confused with any other species. It is one that the beginner may easily learn to recognize and collect with confidence. Care should be taken to distinguish between this species and the poisonous *Clitocybe illudens*, which is somewhat similar in color but has thin, close to crowded lamellae, and usually grows in large clusters. *Clitocybe aurantiaca* is another species of doubtful reputation that might be confused with it, but it is more orange in color and also has thin, close lamellae.

Cantharellus subalbidus Smith & Morse is a western species that is similar in stature and appearance to *C. cibarius* but it is whitish in color and the spores are white rather than yellowish.

CANTHARELLUS CINNABARINUS Schw. Edible

Figure 47, page 25

PILEUS ½–1½ in. broad, rarely larger, fleshy, firm, convex, obtuse becoming expanded-depressed, often irregular, cinnabar-red, fading when old or on drying, glabrous, margin often wavy or lobed. FLESH thin, whitish, reddish at the surface, odor and taste mild. LAMELLAE long-decurrent, distant, forked, thick, blunt on the edge, narrow, varying from red to yellowish or pinkish. STIPE ¾–1½ in. long, ⅛–¼ in. thick, equal or tapering downward, sometimes compressed at the apex, tough, fleshy, solid or sometimes stuffed, smooth, concolorous with the pileus or paler. SPORES white or faintly pink in mass, oblong-elliptical, smooth, (7) 8–10 (11) × 3.5–5 μ.

Scattered or in groups in open deciduous woods or along roadsides. July–Oct.

This is the only *Cantharellus* that is entirely red, and, although it fades considerably on exposure to wind and sun, it is not likely to be confused with any other species. It is said to be common in some localities but has only rarely been collected in the Ottawa district.

CANTHARELLUS CLAVATUS Fr. — Edible

Figure 48, page 25

PILEUS 1–4 in. broad, occasionally larger, fleshy, firm, top-shaped, obconic, or becoming depressed and cup-shaped with a flaring margin, usually irregular and lobed, purplish flesh color to greenish yellow, surface cottony to slightly scaly. FLESH firm, rather tough, thick, whitish, odor and taste mild. LAMELLAE long-decurrent, rather distant, narrow and ridge-form, forking, flesh color to pale purplish umber. STIPE ½–2 in. long, ¼–¾ in. thick, gradually expanding into the pileus, usually tapering downward, solid, purplish flesh color to pallid, white-floccose below. SPORES pale ochraceous in mass, narrow-elliptical to cylindrical, wrinkled, 10–14 × 4–6 μ.

In groups or clusters on the ground in coniferous woods. July-Sept.

The shape and color of this species are highly distinctive although when growing luxuriantly it might be confused with *C. multiplex*. The size and shape of the spores provide a certain and easy way to distinguish these two.

The lamellae are sometimes very narrow and poorly developed and for this reason some authors have placed this fungus in *Craterellus* of the Thelephoraceae, but usually they are sufficiently well developed that it would seem to be better placed in *Cantharellus*.

C. pseudoclavatus Smith is a western species that is very similar to *C. clavatus* but has smooth spores. *C. brevipes* Peck is also close to *C. clavatus* but has larger spores, 13–16 × 5–6 μ.

CANTHARELLUS FLOCCOSUS Schw. — Edible for most people

Figure 49, page 25

PILEUS 2–4 in. broad, sometimes broader, deeply funnel-shaped, vase-shaped or trumpet-shaped, 3–6 in. high, firm, yellowish at first, becoming reddish or orange, floccose-scaly, the scales more or less reddish to reddish orange, margin sometimes wavy. FLESH white or whitish, odor and taste mild. LAMELLAE long-decurrent, close to subdistant, narrow and ridge-form, blunt on the edge, forked, ochraceous or reddish yellow. STIPE short, expanding into the pileus, ½–1 in. thick, glabrous, pale ochraceous, whitish at base, at first solid, becoming hollow, the base sometimes abruptly narrowed and often deep in the ground. SPORES ochraceous in mass, elliptical, smooth, 11–15 (21) × (6) 6.5–7.5 (8) μ.

In groups on the ground in coniferous woods. July–Oct.

This mushroom with its large, brightly colored, vase-shaped fruiting bodies is one of the most striking fungi to be found in the woods. It is not likely to be confused with any other species, although there is a western species, *C. kauffmannii* Smith, that is somewhat similar in stature and also has scales on the pileus. The scales on *C. kauffmannii*, however, are brownish and never yellow or orange.

This fungus should not be eaten in quantity unless small amounts are first tested. Although it is usually considered edible and very good, there is some evidence that it may cause illness in certain individuals.

CANTHARELLUS MULTIPLEX Underw.

Figure 52, page 25

PILEUS ½–2½ in. broad, fleshy-pliant, somewhat fan-shaped to funnel-shaped, purplish to black, drying black, surface uneven, rough, margin irregular, lobed or contorted, more or less inrolled, paler to vinaceous brown. FLESH brittle, purplish, taste mild, odor aromatic. LAMELLAE long-decurrent, distant, narrow and ridge-form, often connected by cross veins and somewhat net-like, ashy gray. STIPE ½–1½ in. long, ⅛–½ in. thick, central or excentric, more or less fused irregularly toward the base, equal or tapering downward, solid, concolorous with pileus, black at the base, glabrous, often somewhat grooved above. SPORES white, irregularly warted, subglobose, 4–6 × 3.5–5 μ.

Densely clustered, arising from a compact, blackish base and growing in masses on the ground in coniferous woods. July–Oct.

Judging from the published reports, this rare and striking fungus has seldom been collected in North America. In the herbarium at Ottawa there are several collections including a part of the original or type collection. There are also two specimens from Japan. Those who are interested in looking for rare species should keep a special watch for this one.

In many respects, notably the warted spores, it differs from all other species of *Cantharellus*, and Murrill has erected a new genus, *Polyozellus*, for it. However, for our purpose it seems preferable to leave it in *Cantharellus*.

It might be confused with luxuriant forms of *C. clavatus* but the spores will distinguish it at once.

We have no information concerning its edibility.

CANTHARELLUS TUBAEFORMIS Fr. — Doubtful

Figure 50, page 25

PILEUS ¾–2 in. broad, at first convex and obtuse, becoming depressed, sometimes nearly infundibuliform, usually not perforated at the center but may become so in age, brownish yellow to yellowish ochraceous, minutely silky-hairy, margin irregular and wavy. FLESH thin, whitish ochraceous, odor and taste none. LAMELLAE decurrent, distant, narrow, ridge-form, blunt on the edge, forked, at first yellowish ochraceous, becoming grayish at maturity. STIPE 1–2½ in. long, ⅛–¼ in. thick, equal, glabrous, brownish yellow to ochraceous, whitish at the base, solid or stuffed, sometimes hollow in age. SPORES creamy white in mass, ovoid to subglobose or broadly elliptical, smooth, 7–10.5 × 6–8 (9) μ.

In groups on the ground in swampy places, usually among sphagnum. July–Sept.

Kauffman thought that the specific name *tubaeformis* was misleading because the stipe in this species is solid and not tube-like. However, the name is derived from *tuba*, a trumpet, and refers to the shape of the pileus rather than to the stipe. It is not so deeply trumpet-shaped as *C. floccosus* and is a thinner, more pliant, and less brightly colored plant.

Considerable uncertainty exists concerning the taxonomy and nomenclature of this and related species. There appear to be four similar but probably distinct fungi which have been called respectively, *C. tubaeformis* Fr., *C. infundibuliformis* Fr., *C. lutescens* Fr., and *Craterellus lutescens* Pers. ex Fr.

The *Craterellus* is a rather bright yellow species with a smooth or slightly wrinkled hymenium, and when it dries the pileus becomes gray to blackish and the hymenium pale yellow. The spores are 9–12 × 7–8.5 μ. *Cantharellus lutescens* is yellowish orange to brownish ochre. The lamellae are relatively well formed, orange-buff in young specimens, more grayish in older ones, and becoming gray when dried. The spores are (9) 10–12 (13) × 6–8.5 (10) μ.

C. tubaeformis and *C. infundibuliformis* are not so brightly colored and have slightly smaller, more subglobose spores. The spores of *C. infundibuliformis* are said to be tinged yellowish to salmon in mass whereas those of *C. tubaeformis* are whitish to cream. This seems to be the principal difference between the two species, although the stipe of *C. tubaeformis* is said to be solid at first, becoming hollow in age, whereas the stipe of *C. infundibuliformis* is hollow from the first. These fungi need further study to determine whether or not the differences are constant and sufficiently great to warrant regarding them as distinct species. If they should prove to be variations of the same fungus, *C. tubaeformis* would be the correct name of the species.

Apparently there is a fungus occurring in the western United States that has a yellow spore deposit and grows on wood. This may be the true *C. infundibuliformis* or may perhaps be an undescribed species. If this proves to be *C. infundibuliformis* probably all the material of the eastern form occurring in swamps is *C. tubaeformis*.

CANTHARELLUS UMBONATUS Fr. — Edible

Figure 51, page 25

PILEUS ½–1½ in. broad, pliant, at first convex or topshaped, becoming plane to depressed, usually with a small umbo, bluish gray, gray-brown, or blackish gray, smooth or slightly flocculose, margin even or wavy. FLESH thin, white, often changing to reddish when wounded or in age, taste and odor mild. LAMELLAE slightly decurrent, forked, close, narrow, blunt on edge but not ridge-form, white or stained reddish. STIPE 1–3 in. long, ⅛–¼ in. thick, equal or slightly tapering upward, whitish or pale gray, usually slightly silky, stuffed or solid. SPORES white, smooth, fusiform to fusiform-elliptical, narrow, (8) 10–12 (14) × 3–4 (5) μ.

In groups among moss, usually *Polytrichum*. July–Oct.

On account of the relatively thin, close lamellae, this species is rather difficult to place at first, and the beginner is inclined to look for it in *Clitocybe*. Once it is recognized as a *Cantharellus*, however, it is easy to identify and will not be confused with any other species in this genus. The grayish, umbonate pileus, the reddening of the flesh and lamellae, and the habitat among mosses are all distinctive characters. It is said to be edible but because of its small size is not likely to tempt many.

Singer has considered the amyloid spores and well-formed lamellae of this species to be sufficiently distinct to separate it from *Cantharellus* and he has made it the type of a new genus *Cantharellula*. Other species that Singer has included in *Cantharellula* are *Clitocybe ectypoides* Peck and *C. cyathiformis* (Bull. ex Fr.) Kummer.

LACTARIUS

The principal distinguishing character of *Lactarius* is the presence of a latex or milky juice. This latex can best be demonstrated by cutting or breaking the lamellae or flesh of young specimens. It is sometimes difficult to demonstrate in old specimens or under very dry conditions, but the apex of the stipe where it meets the lamellae is a good place to try. A few other mushrooms do have a latex but they do not resemble *Lactarius* in stature. *Lactarius* species have a characteristic, rather stiff stature and brittle texture that results from the tissue of the fruit body being composed of many large, round cells termed sphaerocysts, as well as the usual filamentous hyphae found in other mushrooms. The spores are invariably ornamented with more or less prominent warts and spines or with a raised network. This ornamentation is strongly amyloid, and the pattern of the ornamentation as observed under very high magnification is important in critical identification of species. The spores should be measured in side view since those seen in end view will appear to be globose.

These characteristic spores and the presence of sphaerocysts in the tissues are features that distinguish the genera *Lactarius* and *Russula* from all other mushrooms and these two genera are sometimes placed in a separate family, the Russulaceae. *Lactarius* is, of course, distinguished from *Russula* by the presence of the latex.

Lactarius is an important and interesting genus for the amateur collector. There are many species and they may be found over a long period throughout the summer and fall. Many of the species are large and attractive and a great many can be identified with reasonable certainty from macroscopic characters.

In collecting *Lactarii* it is important to note the color of the latex and any color changes that occur when the latex is exposed to the air, and whether or not these changes occur slowly or rapidly. The taste of the latex is also an important character, and this may be ascertained simply by touching the latex with the tip of the tongue. It may be mild or acrid and burning or sometimes astringent. Sometimes the burning sensation develops slowly. The color of the pileus and stipe, and whether or not these are viscid, should be noted.

Although *Lactarius* is usually classified among the white-spored genera, many of the species have colored spores and it is advisable to obtain a spore print in making identifications. The color of the spore deposit is considered to be a constant character and valuable in determination of species.

Some of the species such as *L. deliciosus* are well known to be of excellent quality for eating, but others are doubtful. It is probably better to avoid all those with an acrid taste even though some are said to be harmless after cooking. *L. rufus* has been reported to be poisonous and since there are a number of reddish forms that might be confused with it, these should be tried very cautiously, and all acrid, reddish fruit bodies should be discarded. Species in which the latex turns lilac should also be avoided. In *Lactarius*, as in other mushrooms, the species should be determined before any are used as food.

Key

1. Latex colored from the first 2
1. Latex white at first, unchanging or becoming colored on exposure to the air 4

2. Latex blue .... *L. indigo*
2. Latex not blue 3

3. Latex orange-red or carrot-colored .... *L. deliciosus*
3. Latex dark crimson-red .... *L. subpurpureus*

4. Latex white at first, changing color on exposure to the air, at least on the bruised flesh 5
4. Latex white, unchanging 13

5. Latex changing to lilac or violet 6
5. Latex not changing to lilac or violet 7

6. Pileus glabrous, brownish gray .... *L. uvidus*
6. Pileus tomentose, especially on margin, dull yellow .... *L. representaneus*

7. Latex changing to yellow 8
7. Latex not changing to yellow 10

8. Pileus glabrous, grayish to tawny reddish .... *L. chrysorheus*
8. Pileus tomentose, especially toward margin 9

9. Pileus yellow; stipe scrobiculate spotted .... *L. scrobiculatus*
9. Pileus white; stipe not scrobiculate .... *L. resimus*

10. Latex drying greenish or gray-green on the bruised flesh 11
10. Latex not drying greenish 12

11. Spore deposit yellowish .... *L. trivialis*
11. Spore deposit white .... *L. mucidus*

12. Latex slowly becoming pinkish on the bruised flesh; pileus dark brown, velvety .... *L. lignyotus*
12. Latex causing gray to nearly black stains on the lamellae; pileus olive-umber .... *L. necator*

13. Pileus viscid 14
13. Pileus not viscid 15

14. Pileus glabrous, yellowish .... *L. affinis*
14. Pileus tomentose, pinkish .... *L. torminosus*

15. Fruiting body with strong, aromatic odor, especially noticeable on drying 16
15. No aromatic odor when fresh or on drying 17

16. Pileus dark brownish red; latex white .... *L. camphoratus*
16. Pileus tawny gray to pale tan; latex watery or whey-like .... *L. helvus*

17. Taste mild 18
17. Taste acrid 19

18. Pileus glabrous .... *L. subdulcis*
18. Pileus pruinose-velvety; gills distant .... *L. hygrophoroides*

19. Taste mild at first, sometimes slowly becoming acrid or slightly bitter *L. subdulcis*
19. Taste definitely acrid 20

20. Pileus glabrous 21
20. Pileus not glabrous 22

21. Pileus ashy gray, darker in center .... *L. cinereus*
21. Pileus reddish .... *L. rufus*

22. Pileus gray, usually less than 1½ inches broad .... *L. griseus*
22. Pileus not gray, larger 23

23. Pileus reddish, minutely silky at first, soon glabrous .... *L. rufus*
23. Pileus white or whitish 24

24. Pileus with a cottony roll on the margin .... *L. deceptivus*
24. Pileus without a cottony roll on the margin 25

25. Lamellae crowded, becoming pinkish; taste slowly acrid .... *L. controversus*
25. Lamellae subdistant, becoming creamy yellowish; taste very acrid .... *L. vellereus*

LACTARIUS AFFINIS Peck — Not recommended

Figures 54, 55, page 25

PILEUS 2–6 in. broad, fleshy, firm, at first convex-umbilicate, then becoming expanded and depressed in the center, yellowish or ochraceous yellow, sometimes slightly flesh-tinted, glabrous, viscid, not zoned, margin at first inrolled, becoming arched. FLESH white, firm, fairly thick. LATEX white, unchanging, acrid. LAMELLAE adnate to decurrent, close to subdistant, rather broad, forked near the base, whitish to creamy yellowish. STIPE 1–3½ in. long, ½–¾ in. thick, equal, glabrous, viscid, concolorous with pileus or slightly paler, sometimes spotted, stuffed, becoming hollow. SPORES broadly ellipsoid to subglobose, whitish, 7.5–10 × 6.5–8.5 μ, ornamented with warts joined by bands or heavy lines to form a fairly complete reticulum.

On the ground in mixed woods, usually solitary, sometimes in groups. July–Oct.

The pileus, lamellae and stipe are all more or less the same color and this, together with the broad subdistant lamellae and acrid latex, characterizes the species. *L. insulsus* (Fr.) Fr. is more orange in color and the pileus is distinctly zoned.

LACTARIUS CAMPHORATUS (Bull. ex Fr.) Fr. — Edible

Figure 53, page 25

PILEUS ½–1½ in. broad, convex, often umbonate, becoming expanded and at length depressed, fulvous to dark brownish red, dry, glabrous, not zoned, sometimes slightly wrinkled and uneven, margin at first inrolled, becoming arched. FLESH thin, firm, fragile, tinged the color of the pileus or paler, odor fragrant and aromatic, especially on drying. LATEX white, mild, in dry weather often scant. LAMELLAE adnate to slightly decurrent, close, rather narrow, whitish to flesh colored, becoming reddish brown. STIPE ½–2 in. long, ⅛–⅜ in. thick, equal, glabrous to pruinose, spongy-stuffed, colored like the pileus or paler. SPORES subglobose, white, mostly 6.5–8.5 × 5.5–7.5 μ, ornamented with fairly coarse warts, separate or more or less confluent forming short ridges, or some joined by lines, sometimes partly reticulate.

It grows on the ground or on very rotten wood in mixed woods. July–Sept.

L. camphoratus is a fairly common species and is reported by Kauffman to be edible. The characteristic odor of this species is not of camphor as the name might suggest. It is similar to the odor of *L. helvus*. It is sometimes very faint in fresh specimens and becomes more pronounced on drying. *L. rufus* is similar in color but is larger, has acrid latex, and lacks the odor. *L. camphoratus* might also be confused with *L. subdulcis* but the latter is usually paler colored and also lacks the odor.

LACTARIUS CHRYSORHEUS Fr. — Suspected

Figure 81, page 47

PILEUS 2–3 in. broad, fleshy, at first convex, usually umbonate but varying to umbilicate, then becoming plane to depressed, grayish flesh colored to tawny reddish or fulvous, glabrous, viscid when moist, somewhat or not at all zoned, margin inrolled at first, then spreading. FLESH fairly thick, white, staining yellow from the latex, odor strong and pungent. LATEX at first white, changing to sulphur yellow, slowly acrid, sometimes bitter at first. LAMELLAE adnate to slightly decurrent, close, rather narrow, whitish to yellowish, becoming reddish brown in age or when bruised, some forked near the stipe. STIPE 1–3 in. long, ¼–½ in. thick, equal, glabrous to slightly hairy at base, concolorous with pileus or paler, stuffed, becoming hollow. SPORES broadly ellipsoid to subglobose, 6–9 × 5–7 μ, ornamented with fairly high spines and warts which may be separate, or form short ridges, or be joined by bands or lines to form a partial reticulum.

It grows on the ground, usually in coniferous woods. July–Oct.

This reddish species with bitter to acrid, white latex that very quickly changes to bright yellow is fairly common. There has been some doubt as to whether it should be referred to *L. chrysorheus* or *L. theiogalus* Fr. According to recent illustrations by Wakefield and Dennis (1950) and Neuhoff (1956) *L. theiogalus* is a smaller and more reddish brown fungus.

LACTARIUS CINEREUS Peck — Not recommended

Figure 56, page 26

PILEUS ¾–2½ in. broad, at first convex, umbilicate, becoming expanded and depressed to infundibuliform, ashy gray, darker in the center, glabrous, viscid, not zoned or occasionally slightly so, margin at first inrolled, then spreading. FLESH thin, white. LATEX white, unchanging, acrid. LAMELLAE adnate, close, narrow, white, some forking near the stipe, not becoming spotted from bruising. STIPE 1–2 in. long, ¼–⅝ in. thick, equal or tapering upward, glabrous, tomentose at base, concolorous with the pileus, spongy, becoming hollow. SPORES white, broadly ellipsoid to subglobose, 6–8 × 5–6 μ, ornamented with separate or more or less confluent warts, and a few lines but scarcely reticulate.

In groups on the ground in mixed woods. July–Sept.

There are several grayish species with acrid latex that are not easy to separate. In *L. cinereus* the lamellae do not become spotted or stained from bruising, and the spore deposit is white. *L. trivialis* (Fr. ex Fr.) Fr. has a yellowish spore deposit, and the lamellae become stained grayish green to brownish. The lamellae also become stained in *L. varius* Peck, *L. mucidus* Burl., and *L. parvus* Peck. The latter is a small species with spores about the size of those in *L. cinereus*, but the pileus is soon dry. *L. varius* and *L. mucidus* have large spores, about 8–10 μ long, but differ from each other in the structure of the cuticle of the pileus. In *L. mucidus* the cuticle is composed of elongated, very gelatinous hyphae and is very viscid, whereas in *L. varius* the cuticle is composed of interwoven, subgelatinous hyphae and is soon dry. These species are not recommended as food.

LACTARIUS CONTROVERSUS (Pers. ex Fr.) Fr. Not recommended

Figure 59, page 27

PILEUS 3–8 in. broad, umbilicate, becoming depressed and then infundibuliform, whitish or flesh colored, stained with brownish or flesh-colored spots, indistinctly zoned toward the margin, viscid, slightly tomentose, margin at first inrolled, then elevated. FLESH firm, white or slightly flesh colored. LATEX white, unchanging, slowly acrid. LAMELLAE slightly decurrent, crowded, narrow, whitish to pink flesh colored. STIPE 1–1½ in. long, ¼–1 in. thick, equal or tapering slightly toward the base, slightly floccose-pubescent, white or slightly stained, solid, sometimes excentric. SPORES nearly white or slightly flesh tinted, broadly ellipsoid to subglobose, 5–7.5 × 4–5.5 μ, ornamented with a few heavy bands forming a partial reticulum, and some separate warts.

In groups in moist woods, associated with aspens. Aug.–Oct.

This is a very large species with flesh colored lamellae and brownish to pinkish stains on the pileus. The spots on the pileus are sometimes not very conspicuous, but the pink lamellae are a distinctive feature. The edible qualities of this fungus are not known but it is not recommended because of the acrid latex.

LACTARIUS DECEPTIVUS Peck Doubtful

Figure 67, page 28

PILEUS 2–6 in. broad, firm, at first convex-umbilicate and becoming expanded-depressed to subinfundibuliform, white or with rusty stains, dry, not zoned, glabrous except the margin which is covered by a cottony roll of tomentum and more or less inrolled, finally more or less elevated. FLESH white, rather thick, firm. LATEX white, unchanging, acrid. LAMELLAE adnate to slightly decurrent, close to subdistant, rather broad, some forked, white or creamy yellow. STIPE 1–3 in. long, ⅜–1½ in. thick, stout, equal, pubescent to tomentose, white, solid. SPORES white, broadly ellipsoid to subglobose, 9–12 (14) × 7–9 (10) μ, ornamented with low to medium, separate, scattered warts.

It grows singly or in groups on the ground in woods, usually at the edges of bogs and on boggy ground. July–Sept.

The most striking feature of this mushroom is the cottony roll on the margin. This will distinguish it from *L. vellereus*, which it closely resembles. Mature specimens in which the cottony roll on the margin has largely disappeared are easily confused with *L. vellereus*. The larger spores of *L. deceptivus* will distinguish them. Specimens in which the latex is scanty or not evident might be confused with *Russula delica*.

It is said that the acrid taste disappears on cooking and that it is edible, but there is danger of confusing mature specimens with *L. vellereus*, which has been reported poisonous.

LACTARIUS DELICIOSUS (L. ex Fr.) Gray — Edible

Figures 57, 58, page 27; Figure 412, page 295

Delicious Lactarius

PILEUS 2–5 in. broad, fleshy, firm, at first convex-umbilicate, then expanded and depressed in the center, reddish orange, often with brighter, concentric zones, fading to grayish or gray-green, glabrous, viscid when moist, margin at first inrolled, then arched and spreading. FLESH whitish, stained orange when broken and then becoming greenish. LATEX orange, reddish orange, or carrot colored, mild. LAMELLAE adnate-decurrent, close, rather narrow, bright orange, becoming greenish when bruised. STIPE 1½–4 in. long, ½–¾ in. thick, equal or narrowed at the base, pruinose to glabrous, colored like the pileus or paler, often with orange spots, stuffed, becoming hollow. SPORES faintly yellowish, subglobose, 8–10.5 × 7–8.5 μ, ornamented with lines and ridges forming a more or less complete reticulum, a few separate warts.

It grows scattered or in groups on the ground under conifers in moist woods or boggy places. July–Oct.

The 'delicious lactarius' is one of the more important edible mushrooms. It is easily recognized by the orange latex and the color. The greenish stains that develop on the broken flesh are somewhat unattractive in appearance but do not affect the edible qualities. It is of good flavor and can often be found in abundance.

In the past this species has been confused with a very similar one recently recognized by Dr. A. H. Smith, *L. thyinos*, which can be distinguished by its viscid stipe, and more strongly decurrent and more distant lamellae. The two species can be recognized in the field with a little experience but since both are edible, critical determination is of no importance to those collecting them for food. The western species, *L. sanguifluus* Fr., might also be mistaken for *L. deliciosus* but can be distinguished by its dark blood-red to purplish red latex. *L. sanguifluus* is also edible.

LACTARIUS GRISEUS Peck — Not edible

Figure 60, page 27

PILEUS ½–1½ in. broad, rather flaccid, at first convex, becoming deeply depressed to infundibuliform, smoky gray, usually darker at center, not zoned, dry, tomentose, the hairs forming small, erect points, margin incurved at first, then arched. FLESH white, thin. LATEX white, unchanging, slowly acrid. LAMELLAE adnate to decurrent, close to subdistant, rather broad, white, then cream to yellowish. STIPE ½–2 in. long, 1/16–3/16 in. thick, equal, glabrous, whitish or grayish, paler than the pileus, stuffed then hollow. SPORES ellipsoid to subglobose, white, 6–8.5 × 5–6.5μ, ornamented with a nearly complete reticulum of heavy bands and occasional separate warts.

Usually in groups on very rotten wood or on the ground. July–Sept.

Because of its small size and acrid taste it is of no value as food, but it is a rather common species and will often be found by the collector. The gray color, small size, and tomentose pileus distinguish it.

LACTARIUS HELVUS (Fr.) Fr.

Figure 61, page 27

PILEUS 1–4 in. broad, rather fragile, convex at first, becoming plane to depressed, sometimes slightly umbonate, tawny gray, fading to pale tan, dry, minutely floccose-fibrillose, not zoned, margin at first inrolled, then spreading. FLESH watery whitish, odor strong, fragrant and aromatic, especially noticeable and persistent on drying. LATEX watery to whitish, unchanging, mild or very slightly acrid. LAMELLAE slightly decurrent, close, narrow, whitish, then yellowish flesh color. STIPE 1½–3 in. long, or sometimes longer, ¼–½ in. thick, equal, pruinose above to finely hairy at base, concolorous with the pileus, stuffed, then hollow. SPORES broadly ellipsoid to subglobose, 6–8.5 × 5–6 μ, ornamented with a broken reticulum of bands and ridges or fine lines, a few separate warts.

On the ground or in sphagnum, usually in swampy woods. Aug.–Sept.

This is a rather common species. It is remarkable for the strong odor, which is very pronounced in dried specimens and persists for a long time. Its edible qualities are not known.

LACTARIUS HYGROPHOROIDES B. & C. — Edible

Figure 62, page 27

PILEUS 1–3 in. broad, rather firm and brittle, at first convex to plane, slightly depressed in the center, then becoming deeply depressed to somewhat funnel-shaped, tawny reddish to yellowish, surface dry, pruinose-velvety, smooth to more or less rugose, margin inrolled, then spreading. FLESH rather thin, whitish, odor and taste mild. LATEX white, unchanging, mild, copious. LAMELLAE adnate to slightly decurrent, distant, broad, whitish to cream colored. STIPE ¾–2 in. long, ¼–½ in. thick, equal, rather short, minutely velvety to glabrous, more or less concolorous with the pileus, solid. SPORES white, broadly ellipsoid, 8–10 × 6–7.5 μ, ornamented with low warts and ridges, some separate but mostly joined by fine lines.

Scattered or in groups, usually in deciduous woods. July–Aug.

The bright reddish-brown color and distant lamellae are the principal distinguishing characters of this species. *L. volemus* is similar in color but the lamellae are close.

Figures 69-78

69. *Lactarius necator.*
70. *L. necator.*
71. *L. representaneus.*
72. *L. representaneus.*
73. *L. resimus.*
74. *L. scrobiculatus.*
75. *L. rufus.*
76. *L. rufus.*
77. *L. subpurpureus.*
78. *L. subpurpureus.*

69
70
71
72
73
74
75
76
77
78

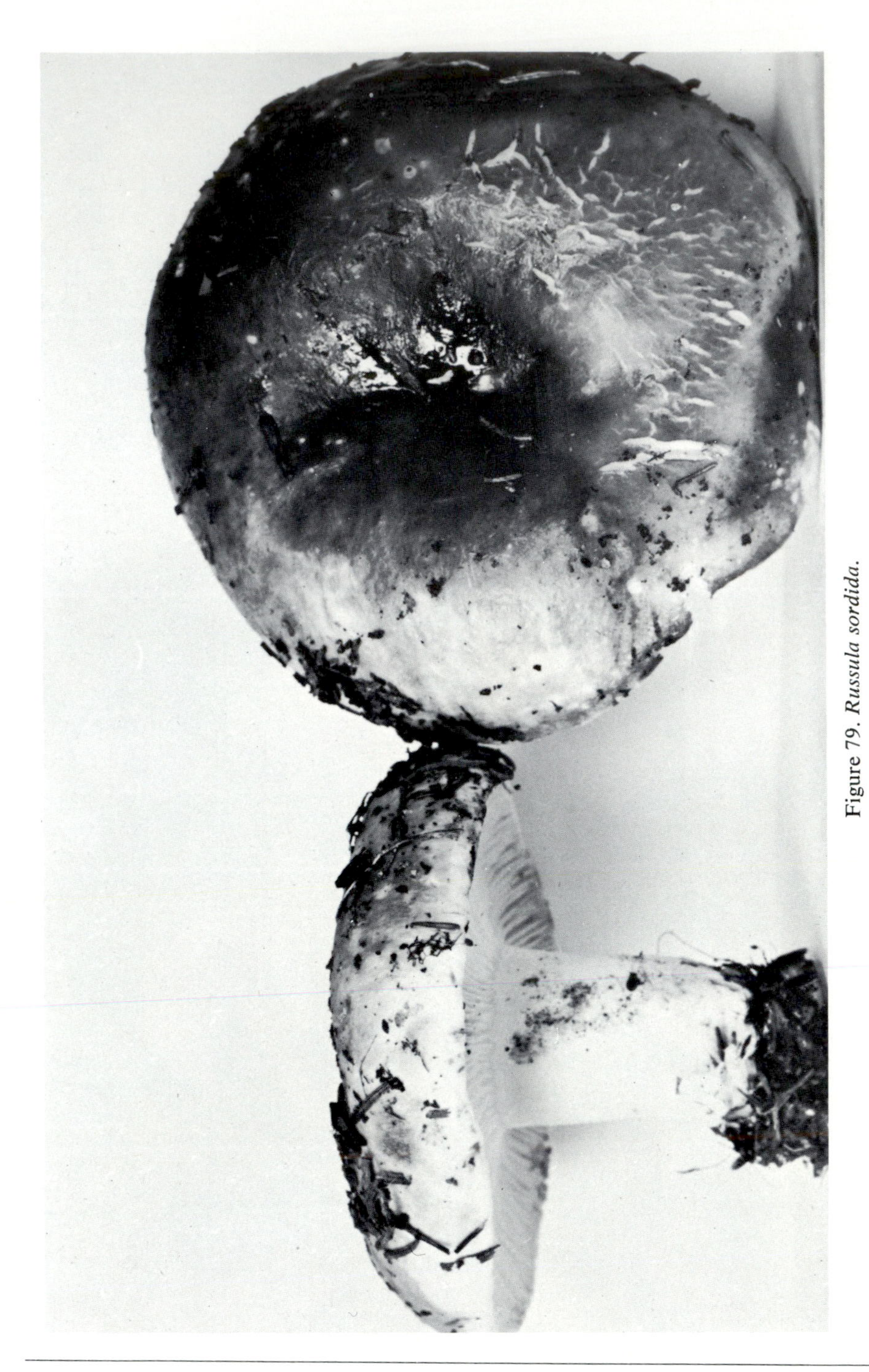

Figure 79. *Russula sordida.*

Figures 80-89

80. *Lactarius subdulcis.*
81. *L. chrysorheus.*
82. *L. torminosus.*
83. *L. mucidus*
84. *L. uvidus.*
85. *L. vellereus.*
86. *Russula alutacea.*
87. *R. alutacea.*
88. *R. aurantiolutea.*
89. *R. aurantiolutea.*

80
81
82
83
84
85
86
87
88
89

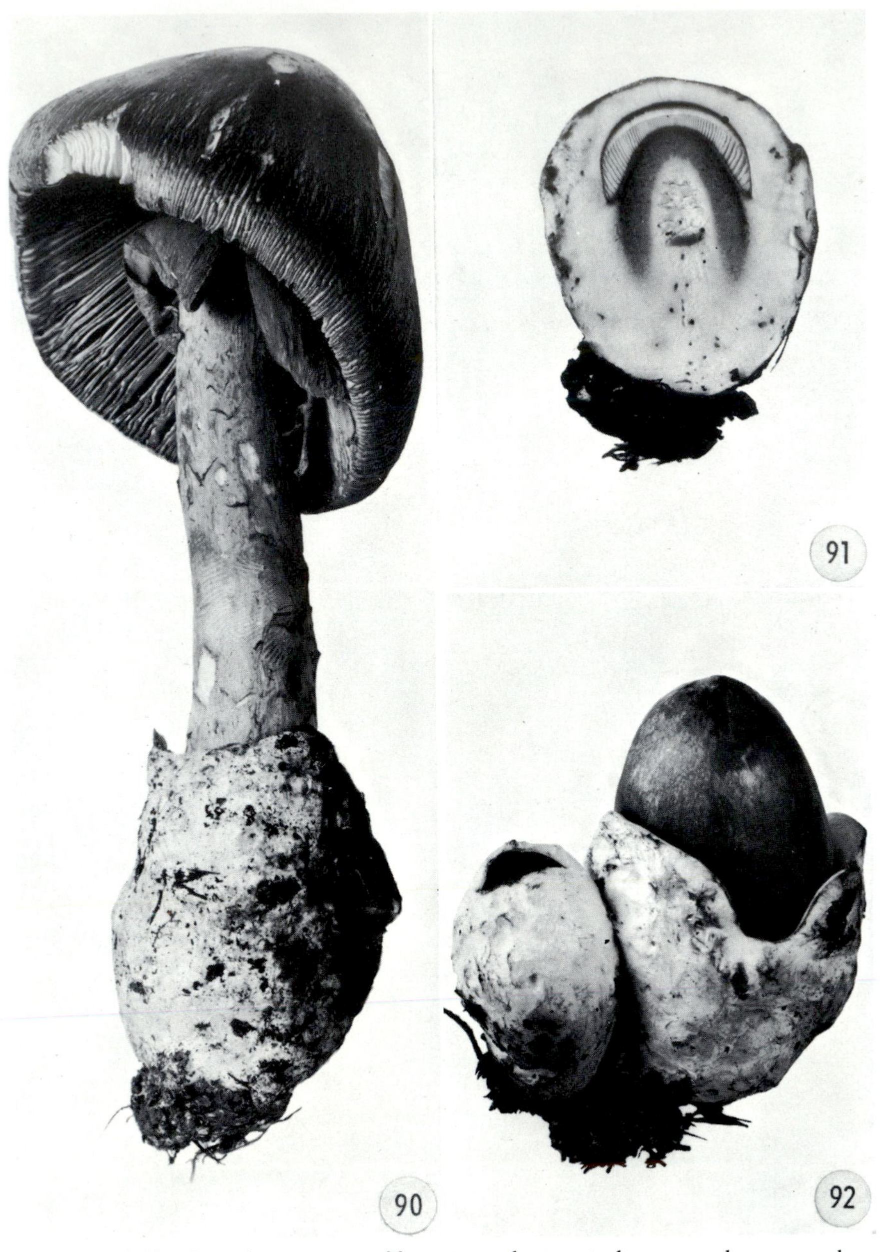

Figures 90-92. *Amanita caesarea.* 90, mature plant, note loose membranous volva; 91, section of young plant before volva has ruptured showing outline of young fruiting body within the volva; 92, young plants showing ruptured volva with young fruiting body emerging.

Figures 93-102

93. *Russula densifolia.*
94. *R. densifolia.*
95. *R. emetica.*
96. *R. emetica.*
97. *R. fallax.*
98. *R. flava.*
99. *R. foetens.*
100. *R. fragilis.*
101. *R. lutea.*
102. *R. nigricans.*

93
94
95
96
97
98
99
100
101
102

Figures 103-104. *Amanita muscaria.* 103, mature plant, note volva adhering to base of stipe in irregular rings and patches; 104, series of specimens illustrating the tearing of the volva to leave patches on the pileus and base of the stipe, and the tearing of the partial veil to form the annulus.

Figures 105-114

105. *Russula abietina.*
106. *R. aeruginea.*
107. *R. chamaeleontina.*
108. *R. decolorans.*
109. *R. integra.*
110. *R. mariae.*
111. *R. xerampelina.*
112. *Pleurotus serotinus.*
113. *Tricholoma pessundatum.*
114. *Marasmius siccus.*

105
106
107
108
109
110
111
112
113
114

Figure 115. *Amanita virosa:* one young fruiting body emerging from the volva and two mature plants. Note pure white color, membranous volva forming a sheath around the base of the stipe, and annulus hanging loosely around the stipe.

LACTARIUS INDIGO (Schw.) Fr.

Edible

Figures 65, 66, page 27

PILEUS 2–5 in. broad, at first convex, slightly umbilicate, becoming plane, depressed in the center, finally infundibuliform, indigo blue, fading and becoming paler with a silvery-gray appearance, zoned with concentric darker blue rings, glabrous, slightly viscid, margin at first inrolled, becoming arched and elevated. FLESH blue, becoming greenish in age. LATEX dark blue, mild. LAMELLAE adnate-decurrent, close, moderately broad, blue, becoming greenish in age. STIPE 1–2 in. long, 3/8–3/4 in. thick, equal, glabrous, concolorous with pileus or paler, sometimes bluish, spotted, stuffed, becoming hollow. SPORES yellowish, broadly ellipsoid to subglobose, 7.5–9.5 × 6–7.5 μ, ornamented with a nearly complete reticulum of light to heavy bands, and occasional separate warts.

Grows on the ground in woods. July–Sept.

This species is not common but when found cannot be mistaken for anything else. No other species has dark-blue latex.

LACTARIUS LIGNYOTUS (Fr. ex Fr.) Fr.

Suspected

Figure 63, page 27

PILEUS 1–3 in. broad, convex to plane, sometimes centrally depressed, usually umbonate, dark chocolate brown or sooty brown, evenly colored, dry, azonate, pruinose-velvety, the margin even to wavy and sometimes plicate. FLESH white, slowly becoming pinkish when wounded. LATEX white, slowly turning reddish, mild to slightly acrid. LAMELLAE adnate to decurrent, subdistant, broad, white then creamy to yellowish buff, the edges sometimes brown. STIPE 1 1/4–3 1/2 in. long, 1/8–3/8 in. thick, equal, plicate at the apex, pruinose-velvety, concolorous with pileus, stuffed. SPORES yellowish in mass, subglobose, 8–10 × 7.5–9 μ, ornamented with high, heavy bands and ridges forming a nearly complete reticulum.

It grows on the ground, usually in conifer woods, especially in bogs. July–Oct.

Reports concerning the edibility of this species have varied. Some people have said it is very good, others that it is poisonous. It should be tried cautiously if at all, and is probably best avoided.

The dark brown, velvety pileus of this fungus is very striking in appearance. The reddish stains develop slowly and are not conspicuous. It might be confused with *L. fuliginosus* Fr., which is paler, has a shorter stipe not plicate at the apex, and slightly smaller spores. *L. gerardii* Peck looks similar from the upper surface but can readily be recognized by its very distant lamellae.

LACTARIUS MUCIDUS Burl. — Not recommended

Figure 83, page 46

PILEUS 1–3 in. broad, at first convex, umbilicate, becoming expanded and depressed to infundibuliform, grayish, putty-colored at margin to brownish in center, viscid, glabrous, not zoned, margin inrolled at first, then spreading. FLESH white, not firm, thin. LATEX white, drying greenish gray to bluish gray, acrid. LAMELLAE adnate, close to subdistant, rather narrow, some forked, white, staining greenish gray, sometimes with a tinge of bluish when wounded. STIPE ¾–2 in. long, ¼–⅜ in. thick, equal or tapering upward, slightly viscid, glabrous or somewhat rivulose-wrinkled, colored like the pileus or paler, stuffed becoming hollow. SPORES white, ellipsoid to subglobose, $7.5–10 \times 6–8\ \mu$, ornamented with a coarse reticulum of lines and bands and a few separate warts.

On the ground, usually in conifer woods. July–Oct.

This is close to *L. trivialis* but is somewhat darker in color and the spores are smaller. The color change of the latex is slow and must be checked carefully or the fungus might be confused with *L. cinereus*. The spores of the latter are a little smaller. *L. varius* Pk. is scarcely viscid. It can be distinguished microscopically by the structure of the cuticle of the pileus. In *L. mucidus* the cuticle is composed of elongated, gelatinized hyphae, whereas in *L. varius* it is composed of much interwoven, very slightly gelatinized hyphae.

The edible qualities are not known but it is not recommended because of the acrid latex.

LACTARIUS NECATOR (Pers. ex Fr.) Lundell — Not recommended

Figures 69, 70, page 45

PILEUS 2–6 in. broad, firm, at first convex, umbilicate, becoming expanded and depressed in the center, olive brown to umber, darker on disk, not zoned, glabrous or with agglutinated fibrils, viscid when moist, margin at first yellow-villose. FLESH firm, thick, whitish. LATEX white, unchanging, acrid. LAMELLAE slightly decurrent, narrow, crowded, pale yellow, becoming black when bruised, gray on drying. STIPE 1–2¼ in. long, ½–1 in. thick, equal, glabrous, viscid when moist, colored like the pileus or paler, often with darker spots, stuffed or hollow. SPORES yellowish to cream colored, ellipsoid to subglobose, $7–9 \times 5.5–7\ \mu$, ornamented with a nearly complete reticulum of fairly heavy lines and occasional separate warts.

It grows on the ground, singly or in groups, in mixed woods. July–Sept.

This is an extremely unattractive mushroom, although it is said to be eaten in Europe. The acrid latex and repellent colors will probably deter most people from eating it. It was formerly known as *L. turpis* (Weinm.) Fr. but this name is illegitimate under the present rules of nomenclature.

It might be confused with *L. atroviridis* Peck which has more green in the color of the pileus and has a rough, scabrous surface.

LACTARIUS REPRESENTANEUS Britz. Not recommended

Figures 71, 72, page 45

PILEUS 3–6 in. broad, convex becoming plane, dull yellow, viscid, tomentose, not zoned, margin at first inrolled and strongly tomentose. FLESH firm, whitish, becoming lilac. LATEX very abundant, at first white, then watery, finally becoming lilac, slightly acrid. LAMELLAE adnate, slightly decurrent, close, moderately broad, dull yellowish, staining lilac when bruised. STIPE 2–2½ in. long. ¾–1¼ in. thick, equal, glabrous, pruinose at the apex, tomentose at the base, yellowish with brighter spots, hollow. SPORES white, broadly elliptic to subglobose, 9–11 × 7.5–9 μ, ornamented with a broken reticulum of fairly heavy bands and a few separate warts or short ridges.

On the ground in moist woods, singly or in groups. Aug.–Sept.

This species is not common but it is a very striking fungus with the yellow, hairy, viscid pileus and the abundant latex which becomes lilac-colored. The spotted stipe is also a striking character. *L. speciosus* Burl. is similar but has a zoned pileus and appears to be a more southern plant. The edible qualities of *L. representaneus* are not known but it is not recommended because of the acrid taste and also because other species in which the latex turns lilac such as *L. uvidus* are reported to be poisonous.

LACTARIUS RESIMUS Fr. Not edible

Figure 73, page 45

PILEUS 2–5 in. broad, deeply depressed to infundibuliform, white becoming tinged with yellow, not zonate or faintly so near the margin, viscid when moist, glabrous in the center, margin inrolled, then spreading, whitish tomentose. FLESH firm, white, strong odor when fresh. LATEX white at first, changing to sulphur-yellow, acrid. LAMELLAE whitish to pale cream, close, decurrent, some forked near the stipe. STIPE 1–1½ in. long, ⅜–¾ in. thick, whitish with yellowish stains or spots, equal, glabrous or pruinose above, stuffed becoming hollow. SPORES broadly ellipsoid to subglobose, white, 5.5–8 × 4.5–6 μ, ornamented with a broken reticulum of rather heavy bands and lines and occasionally a few separate warts.

On the ground associated with aspens. July–Sept.

This is a rather large white species with a hairy margin and acrid latex that turns yellow very quickly. No information regarding its edibility is available but it is not recommended because of the acrid latex.

LACTARIUS RUFUS (Scop. ex Fr.) Fr. Poisonous

Figures 75, 76, page 45

PILEUS 1½–4 in. broad, convex becoming depressed to infundibuliform, umbonate, bay-red to rufous, not fading, dry, not zoned, at first minutely flocculose-silky, soon glabrous, margin at first inrolled. FLESH thin, rather soft, whitish, tinged pink, odor none. LATEX white, unchanging, very acrid. LAMELLAE slightly decurrent, close, narrow, ochraceous, becoming rufous, sometimes forked. STIPE 2–3½ in. long, ¼–½ in. thick, equal, dry, glabrous to pruinose or hairy at the base, colored like the pileus or paler, stuffed, then hollow. SPORES broadly ellipsoid to subglobose, white 7–9 × 5–7 μ, ornamented with fairly heavy bands forming a nearly complete reticulum, some separate warts and short ridges.

On the ground in conifer woods, especially in spruce bogs. July–Sept.

Reported to be poisonous. It can be confused with *L. subdulcis* but is usually larger and more strongly acrid.

LACTARIUS SCROBICULATUS (Scop. ex Fr.) Fr. — Poisonous

Figure 74, page 45

PILEUS 2–6 in. broad, at first convex, becoming depressed to infundibuliform, pale yellow to ochraceous yellow, sometimes reddish yellow, varying from azonate to conspicuously zonate, viscid, more or less tomentose especially toward the margin, becoming glabrous or nearly so in old specimens, margin at first inrolled, then spreading. FLESH firm, white becoming yellow when wounded. LATEX white, changing quickly to sulphur-yellow, acrid. LAMELLAE adnate to slightly decurrent, crowded, rather narrow, sometimes forking near the stipe, whitish or yellowish. STIPE 1–3 in. long, ½–1 in. thick, equal, glabrous, colored like the pileus or paler, with brighter colored, depressed spots, hollow. SPORES white, broadly ellipsoid to subglobose, 7–9 × 6–7.5 μ, ornamented with a few heavy bands forming a wide, broken reticulum with separate warts or short ridges in the spaces.

In groups on the ground in moist woods, usually in coniferous woods. Aug.–Sept.

The most distinctive characters of this species are the hairy margin and the prominent depressed spots on the stipe. It is not common but is a striking species when found.

LACTARIUS SUBDULCIS (Bull. ex Fr.) Gray — Edible

Figure 80, page 47

PILEUS ¾–2 in. broad, at first convex, becoming depressed to infundibuliform, often papillate, brownish red to pale tan or reddish fulvous, not fading, dry, glabrous, not zoned, margin at first inrolled, then spreading. FLESH firm, whitish or tinged fulvous, odor none. LATEX white, unchanging, mild or slowly becoming acrid to bitterish. LAMELLAE adnate to decurrent, sometimes forked, close, rather narrow, whitish to yellowish flesh color or stained fulvous. STIPE 1–2¾ in. long, ⅛–¼ in. thick, equal or slightly tapering upward, glabrous or pubescent toward the base, colored like the pileus or paler, stuffed, becoming hollow. SPORES broadly ellipsoid to subglobose, white, 7–10 × 6–8 μ, ornamented with fine to medium, separate warts, verrucose.

In groups on the ground in woods, swamps, or wet places. June–Oct.

There are a number of small reddish species that are easily confused and the exact identity of *L. subdulcis* is not clear. Among the collections so labeled in the herbarium at Ottawa there are two distinct species, one with spores about 7–10 × 6–8 μ that are ornamented with warts and spines not forming a network, and the other with spores 5.5–8 × 5–6 μ and ornamented with a network of bands and ridges. Apparently the former is the true *L. subdulcis*.

It is reported to be edible but there is considerable danger of confusing it with the poisonous *L. rufus*. The latter is usually larger but small specimens might easily be taken for *L. subdulcis*. *L. rufus* is strongly acrid.

LACTARIUS SUBPURPUREUS Peck — Edible

Figures 77, 78, page 45

PILEUS 1–3 in. broad, fleshy, convex-umbilicate, then expanded, depressed to nearly infundibuliform, colors mixed and variable, dark red to hydrangea-pink, zoned with pink and spotted with green, with a grayish luster, fading, glabrous, slightly viscid when moist, margin at first inrolled, then arched, pruinose. FLESH whitish to pinkish, staining red, especially near the lamellae. LATEX dark crimson, mild. LAMELLAE slightly decurrent, subdistant, moderately broad, dark red, fading, becoming greenish in age. STIPE 1–2½ in. long, ¼–½ in. thick, equal, glabrous, hairy at the base, colored like the pileus, spotted with dark red, stuffed, then hollow. SPORES broadly ellipsoid to subglobose, yellowish, 8–11 × 7–9 μ, ornamented with heavy bands forming a nearly complete reticulum, a few separate warts and ridges.

It grows in moist woods, apparently associated with hemlock. July–Sept.

This is not a common species but it is a very striking one because of the dark red latex and the variegated colors of the pileus. The latex of *L. deliciosus* is more orange-red.

LACTARIUS TORMINOSUS (Schaeff. ex Fr.) Gray — Poisonous

Figure 82, page 47

PILEUS 1½–4 in. broad, fleshy, firm, at first convex, depressed in the center, becoming expanded to nearly infundibuliform, pale yellowish buff to rosy flesh color, often with more deeply colored zones, viscid, glabrous at the center, the margin inrolled and persistently covered with long whitish tomentum. FLESH firm, white to pale flesh colored. LATEX white, unchanging, very acrid. LAMELLAE decurrent, close, narrow, whitish to yellowish or at length tinged faintly pinkish, some forked near the stipe. STIPE 1–2½ in. long, ½–¾ in. thick, equal or slightly tapering downward, glabrous or pruinose, paler than the pileus, sometimes faintly yellowish spotted, stuffed, becoming hollow. SPORES white, broadly ellipsoid to subglobose, 7–10 × 5.5–8 μ, ornamented with heavy bands forming a fairly complete reticulum, a few separate warts or short ridges.

It grows on the ground in woods. July–Sept.

This species may be confused with *L. cilicioides* Fr. which is also reported to be poisonous. The latter is often more white but may be as deeply colored as *L. torminosus*. They can be distinguished with certainty by the spores which are smaller in *L. cilicioides*.

LACTARIUS TRIVIALIS (Fr. ex Fr.) Fr. — Suspected

Figure 64, page 26

PILEUS 2–6 in. broad, fleshy, at first convex, becoming plane or somewhat depressed, the margin decurved, then becoming arched, color variable, livid gray to smoky gray, usually tinted lilac or purplish, fading to pallid, sometimes pinkish brown on the disk, viscid, not zoned. FLESH pallid, rigid-fragile, thick. LATEX white, acrid, slowly staining the flesh and lamellae dingy grayish green. LAMELLAE adnate to short decurrent, close, narrow to moderately broad, some forked, creamy yellowish, staining grayish green or brownish when bruised or in age. STIPE 1½–4 in. long, ½–¾ in. thick, equal, surface even or somewhat wavy, concolorous with the pileus or paler, especially at the apex, hollow. SPORES yellowish, broadly ellipsoid, 9–12 × 8–10 μ, ornamented with fairly high warts joined by lines and ridges to form a partial reticulum.

In groups or scattered on the ground in coniferous or deciduous woods. Aug.–Oct.

This species can be distinguished from the other species of the *L. cinereus* group by the yellowish spore deposit.

LACTARIUS UVIDUS (Fr. ex Fr.) Fr. — Poisonous

Figure 84, page 47

PILEUS 1–3 in. broad, firm, convex becoming plane, depressed at the center, sometimes with a slight umbo, brownish gray, tinged with lilac, viscid, glabrous, usually not zonate but sometimes faintly spotted or zoned, margin inrolled at first, finally spreading. FLESH white, becoming lilac when wounded. LATEX white, changing quickly to lilac or violet, acrid and bitterish. LAMELLAE adnate to slightly decurrent, close, rather narrow, whitish to yellowish, quickly becoming lilac when bruised. STIPE 1½–3 in. long, ¼–½ in. thick, equal or nearly so, glabrous or tomentose at the base, viscid, whitish to yellowish, stuffed, becoming hollow. SPORES broadly ellipsoid to subglobose, white, rather variable in size, 7–12 × 6–8.5 μ, ornamented with rather high spines and warts, separate or more or less joined by lines and ridges forming a partial reticulum.

On the ground usually in low wet places, often among moss. July–Sept.

The species most likely to be confused with this is *L. maculatus* Peck which is usually larger, and has a zonate pileus and spotted stipe.

LACTARIUS VELLEREUS (Fr.) Fr. — Suspected

Figure 85, page 47

PILEUS 2–5 in. broad, convex, umbilicate, becoming expanded and deeply depressed to subinfundibuliform, white or whitish, tinged yellowish in places, dry, velvety to the touch, minutely tomentose under a lens, not zoned, margin at first inrolled, becoming elevated. FLESH white, firm, thick. LATEX white, unchanging or becoming creamy yellowish, finally staining the lamellae and flesh brownish, very acrid. LAMELLAE adnate to slightly decurrent, subdistant to distant, fairly broad, some forked, whitish becoming cream colored to yellowish, staining brownish when bruised. STIPE ½–2 in. long, ½–1¼ in. thick, equal or tapering downward, somewhat pruinose-velvety, white or whitish, solid. SPORES ellipsoid, white, 7.5–9.5 × 5–6.5 μ, ornamented with very fine, low, separate warts, nearly smooth.

On the ground in mixed woods, usually in groups and sometimes abundant. July–Sept.

It is most likely to be confused with mature specimens of *L. deceptivus* in which the cottony roll on the margin has more or less disappeared. It can be distinguished from *L. deceptivus* with certainty by the spores, which in the latter are larger and have more prominent markings on the walls.

A similar large white species, *L. piperatus* (L. ex Fr.) S. F. Gray, has very narrow, densely crowded, forked lamellae, and the pileus is not tomentose. It is very acrid. This species is reported as common in some parts of the United States but there are no Canadian specimens in the herbarium at Ottawa. *L. pergamenus* (Sw.) Fr. which is regarded by some as simply a variety of *L. piperatus* has very narrow, close lamellae.

L. subvellereus Peck is a less common species with a more southern distribution. The lamellae are closer than in *L. vellereus* and remain narrow.

RUSSULA

Russula is similar to *Lactarius* in having sphaerocysts present in the tissue of the fruiting body and in the broadly ellipsoid to subglobose, rough-walled, amyloid spores, but differs in the absence of a latex. The rather stiff stature and brittle texture that characterize these two genera are difficult to describe in words but are soon easily recognized in the field.

Russula is one of the largest and most important genera of the mushrooms. It is also one of the most difficult genera in which to make accurate identifications of species although it has been studied intensively by many mycologists. One reason for this is that there appear to be a great many species that are very similar and differ only in small characters so that many misidentifications have occurred and different authors will be found applying the same name to different fungi. Another reason is that *Russula* species frequently occur only singly and sporadically and it is difficult to study the range of variation of a species; consequently there is considerable difference of opinion as to the species limits in the genus.

Although *Russula* is usually classified among the white-spored genera, the color of the spore deposit varies in different species from pure white to cream, pale yellow, pale ochre or bright ochre. The exact color of the spore deposit is very important in the identification of species of *Russula* and a good deposit should be obtained from every collection. The spores, as in *Lactarius*, are typically broadly ellipsoid to subglobose, strongly amyloid, and ornamented with warts, spines, or a network of ridges, and the pattern of ornamentation is important in critical determination of the species. The attachment of the lamellae varies from adnexed to adnate or slightly decurrent and one section of the genus is characterized by having the lamellae alternately long and short.

The pileus is often brightly colored in shades of red, yellow, purple, green, or bluish, although there are some species with dull colors of white or brown. They may be dry or viscid, glabrous or pruinose to tomentose, and the margin may be more or less tuberculate-striate. The taste varies in different species from mild to acrid, bitter, nauseous, etc., and this should be noted in fresh specimens since it is an important character to aid in distinguishing species.

Some authors claim that all russulas are edible and that the acrid taste disappears on cooking. However, we do not recommend eating the acrid species and they should be tried very cautiously if at all. One species, *R. vesicatoria* Burl., described from Florida and also known on the west coast, is reported to cause blistering of the lips and tongue when tasted. Also species like *R. foetens*, which have a very unpleasant odor and taste, should certainly be avoided.

Only a few of the commoner species are described here. The collector will find many others that he will be unable to identify with this book.

Key

1. Lamellae alternating long and short 2
1. Lamellae equal 5

2. Flesh becoming black in age or when wounded 3
2. Flesh unchanging, fruit body white .... *R. delica*

3. Flesh first reddening, then blackening 4
3. Flesh blackening without any intermediate reddening .... *R. sordida*

4. Lamellae broad, subdistant .... *R. nigricans*
4. Lamellae narrow, very crowded .... *R. densifolia*

5. Spore deposit white or cream 6
5. Spore deposit yellow 13

6. Taste mild 7
6. Taste acrid 9

7. Pileus green or greenish to brownish on disk .... *R. aeruginea*
7. Pileus red or brownish red 8

8. Pileus dark crimson to maroon, sometimes paler to yellowish, dry; stipe usually red .... *R. mariae*
8. Pileus dull brownish red, drying greenish toward margin; stipe white staining yellowish to brownish at base .... *R. vesca*

9. Lamellae forked throughout; pileus pinkish to purplish becoming olivaceous or greenish .... *R. variata*
9. Lamellae equal, or forked rarely or only near the base 10

10. Pileus yellowish to brownish yellow; odor fetid .... *R. foetens*
10. Pileus more or less red, no odor 11

11. Pileus uniformly red or fading to whitish 12
11. Pileus red on margin, olivaceous to purplish in center, small .... *R. fallax*

12. Pileus mostly 1-2½ in. broad; flesh white under the pellicle .... *R. fragilis*
12. Pileus mostly 2-5 in. broad; flesh red under the pellicle .... *R. emetica*

13. Taste acrid 14
13. Taste mild 17

14. Pileus yellow to orange .... *R. aurantiolutea*
14. Pileus red 15

15. Spores deep ochraceous; margin of pileus striate; stipe often tinged red .... *R. tenuiceps*
15. Spores pale yellow, margin even or very slightly striate 16

16. Pileus large, bright, shining, red to purplish red or orange-red, fragile; taste slightly acrid; stipe often tinged reddish .... *R. paludosa*
16. Pileus medium size, rosy red, rather firm; taste very acrid; stipe never red .... *R. veternosa*

17. Pileus yellow 18
17. Pileus not yellow 19

18. Lamellae becoming gray on drying .... *R. flava*
18. Lamellae unchanging .... *R. lutea*

19. Lamellae becoming gray on drying 20
19. Lamellae unchanging 21

20. Pileus orange-red to red, no odor ..... *R. decolorans*
20. Pileus purplish red to olivaceous; fishy odor on drying ..... *R. xerampelina*

21. Spores ochraceous yellow 22
21. Spores pale yellow 24

22. Pileus mostly more than 2 in. broad ..... *R. alutacea*
22. Pileus mostly less than 2 in. broad 23

23. Pileus purplish to greenish, disk darker ..... *R. abietina*
23. Pileus reddish to purplish, disk paler ..... *R. chamaeleontina*

24. Pileus mostly more than 2 in. broad ..... *R. integra*
24. Pileus mostly less than 2 in. broad ..... *R. puellaris*

RUSSULA ABIETINA Peck — Edible

Figure 105, page 51

PILEUS 1–2½ in. broad, fleshy, thin, fragile, convex, becoming plane or slightly depressed, color variable, dull purple, greenish purple, or olive-green, the center always darker, sometimes nearly black, the margin paler to grayish, viscid, glabrous, with a separable pellicle, margin tuberculate-striate. FLESH white, rather fragile, mild. LAMELLAE whitish becoming pale yellow, narrowed toward the stipe, somewhat rounded behind and nearly free, subdistant, equal. STIPE 1–2½ in. long, ¼–½ in. thick, equal or slightly tapering upward, glabrous, white, stuffed or hollow. SPORES bright yellowish ochraceous, subglobose, about (7) 8–10 (11) × (6) 7–9 (9.5) μ, ornamented with warts that are mostly separate, or some confluent forming short ridges, a few joined by fine lines.

Gregarious under balsam fir. Aug.–Oct.

R. abietina is a small species with ochraceous spores, mild taste, and variable colors, usually more or less mixed purplish and greenish. *R. chamaeleontina* Fr. is similar but usually more reddish, tending to fade to yellowish in the center and has slightly smaller spores. Other somewhat similar species are *R. gracilis* Burl., which has yellow spores but is acrid, *R. fallax sensu* Kauffm., which has white spores and is also acrid, and *R. puellaris* Fr., which is mild but has pale yellow spores.

RUSSULA AERUGINEA Lindbl. — Edible

Figure 106, page 51

PILEUS 1½–3 in. broad, at first moderately firm, becoming fragile, convex becoming expanded and slightly depressed in the center, dull green, dark green, or smoky green, darker in the center and sometimes tinged brownish, paler on the margin, viscid when wet, slightly pruinose to pruinose-velvety when dry, the pellicle separable only on the margin, margin even or becoming slightly tuberculate-striate in age. FLESH thick on the disk, becoming thin at the margin, white or greenish, ashy under the pellicle, mild. LAMELLAE narrowly adnate to nearly free, close to subdistant, narrow, equal or with a few short ones, white, becoming cream colored. STIPE 1½–2 in. long, ¼–½ in. thick, nearly equal or slightly tapering downward, glabrous, white, firm to spongy-stuffed. SPORES creamy white, subglobose, 7–9 × 5.5–7 μ, ornamented with rather low, mostly separate warts and a few fine lines.

Gregarious or solitary on the ground in coniferous or mixed woods. July–Sept.

The green color, mild taste and creamy white spore deposit are the important field characters of this species. It is somewhat viscid when wet but is soon dry and more or less pruinose to minutely velvety. *R. virescens* Fr. is a green species with white spores and mild taste and the cuticle tending to become cracked on the margin. *R. olivascens* Fr. has yellow spores.

RUSSULA ALUTACEA (Pers. ex Schw.) Fr. — Edible

Figures 86, 87, page 47

PILEUS 3–6 in. broad, firm, convex, becoming depressed, dull red or dark reddish purple, sometimes fading to greenish, glabrous, viscid when wet, pellicle somewhat separable, margin even at first, becoming tuberculate-striate. FLESH firm, white, mild. LAMELLAE rounded behind, adnexed, nearly free, subdistant, fairly broad, ochraceous to tan colored, equal. STIPE 2–4 in. long, ½–1¼ in. thick, equal or nearly so, glabrous, white or tinged reddish, solid. SPORES ochraceous yellow, broadly ellipsoid to subglobose (7) 8–10 (11) × (6) 7–9 μ, ornamented with rather prominent separate warts and spines.

Usually solitary on the ground in frondose or mixed woods. Aug.-Sept.

This is one of the species about which authors do not agree concerning its identity. The name is used here for a medium to large, dull-reddish or purplish species usually with more or less red on the stipe, a mild taste, and ochraceous spore deposit.

RUSSULA AURANTIOLUTEA Kauffm. — Doubtful

Figures 88, 89, page 47

PILEUS 2–4 in. broad, thin, fragile, at first convex, becoming plane to slightly depressed, yellowish in the center to more orange on the margin, glabrous, viscid, pellicle separable to the disk, margin even, becoming slightly tuberculate-striate. FLESH white, yellowish under the pellicle, acrid. LAMELLAE narrowly adnate, close to subdistant, broad in front becoming narrower toward the stipe, yellow, often forked near the base. STIPE 1½–4 in. long, ¼–¾ in. thick, equal or nearly so, glabrous, white, spongy-stuffed. SPORES ochraceous yellow, subglobose, (6) 7–8 (9) × 5.5–7.5 μ, ornamented with warts joined by heavy bands.

Solitary or scattered in mixed woods. July–Sept.

This is a yellow to orange species with ochraceous spores and acrid taste in which the lamellae and stipe do not turn gray on drying. Its edible qualities are not known but it is not recommended because of the acrid taste.

RUSSULA CHAMAELEONTINA Fr. *sensu* Kauffm. — Edible

Figure 107, page 51

PILEUS ¾–2 in. broad, thin, fragile, plane or depressed, variable in color, red to purplish or lilac, fading to yellowish, especially on the disk, glabrous, viscid, pellicle separable, margin even, becoming somewhat tuberculate-striate. FLESH thin, fragile, white, mild. LAMELLAE adnexed to almost free, close to crowded, rather narrow, equal or a few forked, ochraceous yellow. STIPE ¾–2 in. long, ¼–½ in. thick, equal or slightly tapering upward, glabrous or slightly marked with lines, white, spongy-stuffed becoming hollow. SPORES ochraceous, subglobose, 7–9 (10) $\times$ 5.5–7.5 μ, ornamented with rather prominent warts and spines, mostly separate or occasionally confluent or joined by fine lines.

Solitary or gregarious on the ground in coniferous or mixed woods. Aug.–Sept.

Considerable confusion exists in the literature concerning this species and it may be a collective species including several recognizable forms. In general, a small *Russula* with ochraceous spore deposit, mild taste, and variable colors with some tendency to fade to yellowish, would probably be referred here.

RUSSULA DECOLORANS Fr. — Edible

Figure 108, page 51

PILEUS 2–5 in. broad, firm, at first globose, becoming plane to slightly depressed, orange-red, light red, or salmon colored, the disk usually ochre, glabrous, slightly viscid, the pellicle partly separable, margin even, becoming slightly striate when old. FLESH at first firm, becoming fragile in age, white, becoming ashy in age or when wounded, mild. LAMELLAE adnexed, close, moderately broad, at first white, becoming pale yellowish, ochraceous, becoming ashy gray on drying, some forked at the base, equal or with a few short ones. STIPE 2–4½ in. long, ½–1 in. thick, equal or nearly so, smooth or somewhat wrinkled with fine lines, white, becoming ashy in age or when bruised, solid or spongy. SPORES subglobose, pale ochraceous yellow, 10–13 $\times$ 8–10 μ, ornamented with high, separate warts and spines, some joined by fine lines.

Solitary or scattered on the ground in coniferous or mixed woods. July-Sept.

This is a large orange-red species particularly characterized by the lamellae and stipe changing to ashy gray when dried or in age.

RUSSULA DELICA Fr. — Edible

Figure 68, page 28

PILEUS 3–6 in. broad, fleshy, firm, at first convex and umbilicate, becoming deeply depressed to infundibuliform, dull white or with rusty-brown stains, glabrous or very finely hairy, dry, margin at first involute, becoming arched, not striate. FLESH compact, firm, white or whitish, not changing color when bruised, mild to slowly and slightly acrid. LAMELLAE adnate-decurrent, subdistant, alternating long and short, few forked, white or whitish, sometimes greenish on edge. STIPE ¾–2 in. long, ½–¾ in. thick, short, stout, equal or tapering downward, white, not turning blackish when bruised, usually with a pale green zone at the apex, glabrous to subtomentose. SPORES subglobose, white in mass, rough 8.5–11 × 7–9 μ, ornamented with rather coarse warts, mostly joined by fine lines or forming short ridges, partly reticulate.

Gregarious on the ground in conifer or frondose woods. July–Oct.

At first sight this species suggests a *Lactarius*, but it has no latex. It is fairly common and is often found pushing up earth or old leaves and partly concealed by them. The greenish zone at the apex of the stipe, though often inconspicuous, can be observed by turning the fruit body in the light and it makes a good field character. Collectors on the west coast should beware of a species resembling *R. delica* but smaller and with the lamellae equal, not alternating long and short. This is *R. vesicatoria* Burl. and it is excruciatingly acrid and may cause blistering of the lips and tongue if tasted.

Another characteristic *Russula* of the west coast is *R. crassotunicata* Singer which is found growing under devil's-club and is a white species that stains brown. It is slightly acrid and its edible qualities appear to be unknown.

RUSSULA DENSIFOLIA (Secr.) Gill. — Edible

Figures 93, 94, page 49

PILEUS 2–4 in. broad, convex-umbilicate becoming depressed to subinfundibuliform, firm and rigid, dull whitish at first, becoming smoky brown, slightly viscid, glabrous, margin even. FLESH thick, firm, white, slowly turning reddish then black when cut or bruised, mild to slowly and slightly acrid. LAMELLAE adnate to slightly decurrent, close to crowded, rather narrow, whitish or grayish, when bruised becoming reddish then black. STIPE 1¾–2½ in. long, ½–1 in. thick, equal or tapering below, glabrous, whitish, becoming reddish then black when bruised, solid. SPORES broadly ellipsoid, white, 7–9 (10) × 5.5–8.5 μ, ornamented with fine warts which are joined by a network of fine lines.

Usually solitary or gregarious on ground in woods from July–Sept.

This species is distinguished from *R. nigricans* principally by the close to crowded lamellae and from *R. sordida* by the appearance of red color in the wounded flesh before it becomes black. The color change is sometimes slow and the red color is transitory so it must be watched for carefully.

It is said to be edible but is unattractive in appearance.

RUSSULA EMETICA (Schaeff. ex Fr.) Pers. ex S. F. Gray — Suspected

Figures 95, 96, page 49

PILEUS 2–4 in. broad, fleshy, firm at first, soon becoming fragile, convex, becoming plane or slightly depressed, rosy red to blood-red, sometimes fading to white, viscid when wet, glabrous, shining, pellicle separable, margin strongly tuberculate-striate. FLESH white, red under the pellicle, very acrid. LAMELLAE slightly adnexed to free, close to subdistant, rather broad, narrowed behind, white. STIPE 1½–3 in. long, ½–¾ in. thick, nearly equal, smooth, white or tinged red, spongy-stuffed. SPORES white, subglobose, (7) 8–10 × (6) 7–8.5 (9) μ, ornamented with fairly high warts and spines, more or less joined by fine lines to form a partial reticulum.

Scattered to gregarious on the ground or on very rotten wood. July–Oct.

This species is very close to *R. fragilis* and some authors consider these to be varieties or subspecies. *R. emetica* is usually a little larger, the flesh is red under the cuticle rather than white, and the spores seem to be slightly larger with slightly higher ornamentation.

The taste is very acrid and although some have claimed that this taste disappears on cooking and that the species is edible, it is not recommended.

RUSSULA FALLAX *sensu* Kauffm. — Suspected

Figure 97, page 49

PILEUS 1–2½ in. broad, thin, fragile, plane or slightly depressed, usually rose or flesh colored on the margin, with an olivaceous zone surrounding the disk which is usually much darker and purplish, viscid, glabrous, pellicle separable, the margin striate. FLESH white, or tinged like the pellicle under the surface, acrid. LAMELLAE adnexed, subdistant, narrow, white. STIPE 1–2 in. long, ¼–½ in. thick, equal, cylindrical or somewhat compressed and with minute longitudinal wrinkles, white, spongy-stuffed to hollow. SPORES white, subglobose, 6–8 × (4.5) 5–7 μ, ornamented with warts that are more or less joined by lines and ridges to form a nearly complete reticulum.

Solitary or gregarious on mossy ground in the woods, or among sphagnum. July–Sept.

This little species with its characteristically colored pileus, white spores, and very acrid taste is fairly common. It is related to the *fragilis-emetica* complex. It is probably not the true *R. fallax* Cke. but is the species Kauffman described under that name.

It is not recommended for eating because of the acrid taste.

RUSSULA FLAVA Rom. — Edible

Figure 98, page 49

PILEUS 2–3 in. broad, rather fragile, at first convex, becoming plane or slightly depressed, dull yellow or golden yellow, sometimes becoming ashy in age, viscid when moist, glabrous, pellicle separable, margin even or slightly striate when old. FLESH white, becoming ashy gray when old, mild. LAMELLAE adnexed, close, moderately broad, narrowed behind, at first white, becoming pale yellow and finally ashy gray in age or in drying. STIPE 2–3 in. long, ½–¾ in. thick, equal or nearly so, smooth or with a fine network of lines, at first white, becoming ashy gray in age, spongy-stuffed. SPORES pale yellow, subglobose, 8–10 (11) × 7.5–9 μ, ornamented with rather fine warts that are more or less joined by fine lines forming a partial reticulum.

Solitary or scattered on the ground in conifer or mixed woods. July–Sept.

The most characteristic feature of this species is the change to ashy gray of the flesh and lamellae in age or on drying. The bright yellow color, mild taste and pale yellow spore deposit are also distinguishing characters. It is probably the same as *R. claroflava* W.B. Grove, and, if so, this would be the correct name since it was published earlier.

RUSSULA FOETENS Pers. ex Fr. — Not edible

Figure 99, page 49

PILEUS 2½–5 in. broad, at first firm, becoming fragile, at first nearly globose, expanding and becoming plane to slightly depressed, yellowish or dingy ochraceous, glabrous, viscid, pellicle separable part way to the disk, margin widely and coarsely tuberculate-striate. FLESH thin, rather fragile, dingy white, yellowish under the pellicle, taste acrid, odor strong, resembling bitter almonds, then fetid. LAMELLAE adnexed, rather close, broad, at first whitish, becoming yellowish with age and dingy when bruised, exuding drops of water when young, some forked. STIPE 1–3 in. long, ½–1 in. thick, equal or nearly so, smooth, white or dingy brown in age or when bruised, stuffed, becoming hollow. SPORES white, subglobose, 8.5–10 × 8–9 μ, ornamented with coarse, separate spines.

Gregarious on the ground in mixed woods. July–Sept.

The dingy colors and unpleasant odor make this an extremely unattractive fungus. It is not recommended as an edible species, but is not likely to be eaten anyway, because of its unpleasant taste.

There is a group of species somewhat similar to *R. foetens*. *R. foetentula* Peck has pale yellow spores and red stains on the stipe. *R. pectinata* Fr. has different spore ornamentation. *R. granulata* Peck is smaller, lacks the odor and has granules on the pileus, and *R. pectinatoides* Peck lacks the granules, is mild or slightly acrid and has whitish spores. *R. ventricosipes* Peck has pronounced red stains on the stipe and grows in sand. In this species the taste is slowly acrid and the spores are pale ochraceous.

RUSSULA FRAGILIS (Pers. ex Fr.) Fr. — Suspected

Figure 100, page 49

PILEUS 1–2 in. broad, thin, fragile, at first convex, becoming plane or slightly depressed, rosy red to pale red, fading to whitish, glabrous, viscid, pellicle separable, margin tuberculate-striate. FLESH white, not red under the pellicle, thin, fragile, very acrid. LAMELLAE adnexed, close to crowded, ventricose, white. STIPE 1–2 in. long, ¼–½ in. thick, equal, smooth, white, spongy-stuffed to hollow, fragile. SPORES white, subglobose, 7–9 × 6–8 μ, ornamented with moderately high warts, more or less joined by lines and ridges to form a partial reticulum.

Scattered on the ground in woods. July–Sept. Common.

See the notes on *R. emetica* for a comparison with this species.

RUSSULA INTEGRA (L. ex Vitt.) Fr. — Edible

Figure 109, page 51

PILEUS 2–5 in. broad, at first firm, soon becoming fragile, convex becoming plane or slightly depressed, color variable, rather dingy or sordid, from buff to reddish brown to dark, dull red, fading, glabrous, viscid when wet, pellicle separable, margin becoming coarsely tuberculate-striate. FLESH white, mild. LAMELLAE adnexed, nearly free, distant, broad, white becoming pale yellow. STIPE 1½–2½ in. long, ½–¾ in. thick, tapering upward to nearly equal, sometimes ventricose, smooth, white, spongy-stuffed, fragile. SPORES pale yellow, subglobose, 7–9 × 5.5–7 μ, ornamented with low to medium separate warts, some more or less confluent or joined by fine lines.

Gregarious on the ground in woods. Aug.-Sept.

The exact identity of *R. integra* appears to be somewhat in doubt but this is believed to be the same fungus that was described by Kauffman under this name. It is a medium-sized, dull red species and has a pale yellow spore deposit and mild taste. The stipe is never red.

RUSSULA LUTEA (Huds. ex Fr.) S. F. Gray — Edible

Figure 101, page 49

PILEUS 1–2½ in. broad, thin, fragile, plane or slightly depressed, bright yellow to golden yellow, glabrous, viscid, pellicle separable, margin even, becoming slightly striate when old. FLESH white, thin, fragile, mild. LAMELLAE free, subdistant, rather narrow, broader at the front, bright ochraceous. STIPE 1–2 in. long, about ¼–½ in. thick, equal or slightly tapering upward, smooth, white, spongy-stuffed becoming hollow. SPORES ochraceous, globose, 8.5–10 × 7.5–9 μ, ornamented with moderately coarse warts, mostly separate or some confluent forming short ridges.

Usually solitary on the ground in mixed woods. Aug.–Sept.

This species is fairly easily recognized with its bright yellow pileus and deep ochraceous lamellae and spores.

Figures 116-125

116. *Russula paludosa.*
117. *R. paludosa.*
118. *R. variata.*
119. *R. variata.*
120. *R. vesca.*
121. *R. vesca.*
122. *R. tenuiceps.*
123. *Amanita flavoconia.*
124. *A. brunnescens.*
125. *A. citrina.*

116
117
118
119
120
121
122
123
124
125

Figures 126-128. *Lepiota americana.* 126, mature fruiting body; 127, young fruiting body; 128, section of young fruiting body.

Figures 129-138

129. *Amanita frostiana.*
130. *A. frostiana.*
131. *A. gemmata.*
132. *A. gemmata.*
133. *A. muscaria.*
134. *A. muscaria.*
135. *A. porphyria.*
136. *A. porphyria.*
137. *A. rubescens.*
138. *A. rubescens.*

129
130
131
132
133
134
135
136
137
138

Figure 139. *Lepiota molybdites.*

RUSSULA MARIAE Peck

Edible

Figure 110, page 51

PILEUS 1–3 in. broad, firm, convex, becoming plane or slightly depressed, dark reddish purple, dark crimson or maroon colored, sometimes paler to slate violet or yellowish when shaded, dry, pruinose-velvety, margin usually even, sometimes becoming striate when old. FLESH white or reddish under the pellicle, becoming sticky when cut or handled, mild or very slightly acrid. LAMELLAE adnate, close to subdistant, rather narrow, somewhat broader in front, white, becoming yellowish in age. STIPE 1½–3 in. long, ½–1 in. thick, equal or tapering downward, pruinose, usually white at each end and rosy red to purplish red in the middle, occasionally entirely white, spongy-stuffed. SPORES whitish to faintly yellowish, subglobose, 7–9 (10) $\times$ (5.5) 6–7.5 (8.5) μ, ornamented with a more or less complete network of bands and ridges, and a few separate warts.

Gregarious on the ground in woods. July–Oct.

This is a beautiful species with purplish red pilei and stipes, cream-colored spore deposit, and mild taste. The pileus is dry and minutely velvety.

RUSSULA NIGRICANS (Bull.) Fr.

Edible

Figure 102, page 49

PILEUS 3–6 in. broad, firm, at first convex and umbilicate, becoming expanded and infundibuliform, whitish, becoming smoky umber to blackish, glabrous, slightly viscid when moist, margin incurved at first, not striate. FLESH firm, whitish, changing to reddish and then black when wounded, mild to slowly acrid. LAMELLAE adnexed, rounded or narrowed behind, subdistant to distant, broad, alternating long and short, whitish, changing to reddish then black when bruised. STIPE ¾–2½ in. long, ½–1 in. thick, stout, equal, glabrous, whitish at first, becoming smoky brown with age, changing to reddish then black when bruised, solid. SPORES white, subglobose 7–9 $\times$ 6–8 μ, ornamented with a network of fine lines joining low warts.

Gregarious or solitary on the ground in conifer or frondose woods. July–Sept.

The pronounced blackening of the flesh and lamellae of this species make it rather unattractive but it is said to be edible. The reddening of the flesh when wounded must be watched for carefully. It is distinguished from *R. densifolia* by the broad, subdistant lamellae.

RUSSULA PALUDOSA Britz.

Figures 116, 117, page 69

PILEUS 3–5 in. broad or sometimes larger, convex, becoming expanded and depressed in the center, deep blood-red to red-purple or red-orange sometimes fading to yellowish, somewhat shining, glabrous, viscid, margin slightly striate-tuberculate. FLESH white, rather fragile, acrid. LAMELLAE white to creamy yellow, broad, subdistant, nearly free. STIPE 3–4 in. long, ½–1 in. thick, nearly equal or slightly narrowed above, white or washed with pinkish, spongy. SPORES pale yellow, subglobose, 9–12 × 8–10 μ, ornamented with prominent warts and spines, a few joined by fine lines, not reticulate.

Singly or gregarious in wet places or among sphagnum. July–Sept.

A large and showy species differing from *R. emetica* in the pale yellow spore deposit and less acrid taste. Some authors say the taste is mild and there is some doubt as to whether the forms with mild taste might be considered a distinct species, *R. rubrotincta* (Pk.) Burl. Specimens labeled as having a mild taste seem to be identical microscopically with the acrid forms. The edges of the lamellae are said to be sometimes red near the margin. Its edible qualities are not known.

RUSSULA PUELLARIS Fr.

PILEUS 1–1½ in. broad, thin, fragile, convex, becoming plane or slightly depressed, variable in color, bluish black, purplish, or yellowish, usually darker or brownish in the center, viscid, glabrous, pellicle separable, margin tuberculate-striate. FLESH white or becoming watery translucent, tinged like the surface under the pellicle, fragile, mild. LAMELLAE narrowly adnate to adnexed, close, narrowed toward the stipe, white, becoming pale yellow. STIPE 1–2 in. long, ¼–½ in. thick, equal or tapering upward, glabrous, white, usually with yellowish stains toward the base especially when old, stuffed or hollow, soft and fragile. SPORES pale yellow, subglobose, rough, (7) 8–10 (11) × (6) 7.5–9 μ, ornamented with medium to high warts and spines, some joined by fine lines or confluent forming short ridges, partly reticulate.

Gregarious on the ground in moist woods. Aug.-Oct.

This species is distinguished from *R. abietina* and *R. chamaeleontina* by the pale yellow spore deposit. The yellowish stains that develop in the stipe are also a distinguishing character. It is probably edible but is a small, fragile species not likely to be gathered for food.

RUSSULA SORDIDA Peck — Edible

Figure 79, page 46

PILEUS 2–6 in. broad, firm, convex, depressed in the center, whitish becoming smoky with age, dry, glabrous, margin at first incurved, not striate. FLESH firm, whitish, when bruised quickly becoming blackish without first turning reddish, mild to slightly and slowly acrid. LAMELLAE adnate to slightly decurrent, close, rather narrow, alternating long and short, white becoming blackish in age or on drying. STIPE 1–2 in. long, ½–1 in. thick, equal, white becoming black when bruised, solid. SPORES white, oblong-ellipsoid, (7) 8–10 (11) × 5.5–7.5 μ, ornamented with fine, low, separate warts, nearly smooth.

Solitary or gregarious on the ground. Said to be associated with hemlocks. July–Aug.

Because of the pronounced blackening of the flesh this is an unattractive species but it is reported to be edible. It is distinguished from *R. densifolia* and *R. nigricans* by the direct change of the wounded flesh to black without any reddening.

R. sordida appears to be very close to *R. albonigra* (Krombh.) Fr. of Europe and may be the same species.

RUSSULA TENUICEPS Kauffm. — Not edible

Figure 122, page 69

PILEUS 2½–4½ in. broad, fragile, convex, becoming plane to slightly depressed, rosy red to blood-red, sometimes uniformly colored, usually whitish spotted or with orange blotches, viscid, glabrous, sometimes with minute wrinkles, pellicle separable, margin striate. FLESH white, red under the pellicle, very fragile in mature plants, acrid, sometimes slowly. LAMELLAE adnexed to free, close to crowded, narrow, white, becoming yellow-ochraceous. STIPE 2–3½ in. long, ¾–1 in. thick, nearly equal or ventricose, smooth or marked with fine lines, white or rosy tinged, spongy-stuffed. SPORES yellow-ochraceous, subglobose, rough (6) 6.5–8.5 (9) × 5.5–8 μ, ornamented with medium high to low warts usually separate or sometimes confluent into short ridges or with a few fine lines, very slightly reticulate.

Gregarious on the ground in mixed woods. July–Sept.

This is a red species with ochraceous spores and acrid taste. There does not appear to be any information concerning its edibility but it is considered doubtful because of the acrid taste.

RUSSULA VARIATA Banning — Edible

Figures 118, 119, page 69

PILEUS 2–5½ in. broad, fleshy, firm, at first convex, expanding and becoming depressed to nearly infundibuliform, reddish purple or brownish purple, often mixed with olive or green, or sometimes entirely greenish, viscid, glabrous, pellicle separable on the margin, margin not striate, sometimes with the pellicle cracking. FLESH white or grayish under the pellicle, mild to slowly acrid. LAMELLAE adnate to slightly decurrent, close to crowded, narrow, forking two or three times, white. STIPE 1½–3 in. long, ½–1 in. thick, equal or nearly so, smooth, white, solid. SPORES white, subglobose, (7) 8–10 (11) × (6) 7–8.5 (9) μ, ornamented with low, separate warts and a few fine lines.

Gregarious on the ground in woods. July–Sept.

This species is close to *R. cyanoxantha* Fr. and is sometimes regarded as a variety of it. It differs in the forked lamellae and slightly acrid taste.

RUSSULA VESCA Fr.

Figures 120, 121, page 69

PILEUS 1¾–4 in. broad, convex-umbilicate becoming rather deeply depressed, sometimes plane, rather firm, brownish red with a tinge of purplish, drying to dull olive-green toward margin and brownish red in center, viscid, glabrous or slightly pruinose, margin slightly striate-tuberculate. FLESH firm, whitish or tinged brownish or violet under cuticle, mild. LAMELLAE white to cream, close to crowded, rather narrow to moderately broad, some forked, adnate. STIPE 1½–3 in. long, ½–1¼ in. thick, nearly equal, glabrous, slightly wrinkled, white, staining yellowish to brownish at the base, solid. SPORES oblong, white, 6–8 (9) × 5–6 (7) μ, ornamented with low, fine, separate warts and some fine lines.

On the ground in mixed woods. July–Sept.

This is a rather dull reddish brown species sometimes drying greenish, with lamellae close to crowded, mild taste and white spore deposit. *R. brunneola* Burl. is probably the same species. The mild taste would suggest that it is probably edible but definite information about it appears to be lacking.

RUSSULA VETERNOSA Fr. *sensu* Kauffm. — Not edible

PILEUS 2–3 in. broad, fleshy, convex, becoming plane or slightly depressed, rosy red or flesh colored, fading to whitish or yellow on the disk, viscid when wet, glabrous, pellicle separable only on the margin, margin even or slightly striate. FLESH white, reddish under the pellicle, very acrid. LAMELLAE adnate, close, narrow, some short ones present, at first white becoming straw colored. STIPE 2–3 in. long, ½–¾ in. thick, equal, smooth, white, fragile, stuffed or hollow. SPORES pale yellowish, subglobose, 7–9 (10) × 6–8 μ, ornamented with moderately high warts and spines, separate or more or less joined by lines and ridges but scarcely reticulate.

Scattered or gregarious on the ground in frondose woods. Aug.–Sept.

Different authors seem to have different ideas about this species and there is considerable doubt as to what *R. veternosa* really is. This is believed to be the same fungus as that described by Kauffman under this name.

It is a species with the pileus red, paler on the disk, acrid taste and pale yellow spore deposit. The stipe is never red. Definite information concerning its edibility appears to be lacking but it is not recommended because of the acrid taste.

RUSSULA XERAMPELINA (Schaeff. ex Secr.) Fr. — Suspected

Figure 111, page 51

PILEUS 2–4 in. broad, firm, convex, becoming plane or slightly depressed, colors variable, more or less reddish purple on the margin to olive green in the center, or colors mixed with brownish purple or olivaceous, dry, glabrous to pruinose, pellicle scarcely separable, margin even. FLESH white, pinkish under the pellicle, taste mild, odor disagreeable, somewhat fishy in age or on drying. LAMELLAE adnexed, close to subdistant, narrowed behind, some forked, whitish to cream colored. STIPE 1½–3 in. long, ½–1 in. thick, equal or nearly so, smooth or slightly wrinkled, white or reddish becoming dingy olivaceous yellow when handled or in age, solid or somewhat spongy. SPORES pale yellow, subglobose, (7) 8–10 (11) × (5.5) 6–8.5 (9) μ, ornamented with rather prominent, separate warts and spines.

Scattered on the ground in woods. Aug.–Oct.

The principal distinguishing characters of this species are the unpleasant fishy odor and the pronounced graying of the lamellae on drying. The odor may not be detected in fresh specimens until they have begun to dry. The colors of the pileus are variable and the species is easier to identify from dried material than fresh because of the characteristic graying of the lamellae.

R. squalida Peck has a similar odor and the spores are very similar. The lamellae also become gray on drying but the pileus dries to an olivaceous or greenish color rather than the vinaceous red of *R. xerampelina*. *R. serissima* Peck seems to be very similar to *R. squalida*, mainly differing in having larger spores.

AMANITA

For anyone interested in eating mushrooms, *Amanita* is the most important of all the genera because it is here that the deadly poisonous species belong. It is absolutely essential for the beginner to learn the diagnostic characters of this genus and avoid it. It is true that some *Amanita* species are edible, but until these species are known beyond any possibility of doubt, no *Amanita* should be eaten.

Amanita is characterized by the combination of three principal characters: white spore deposit and presence of both annulus and volva. The lamellae are typically free from the stipe but in a few species they may be narrowly attached or attached by a line. The stipe separates readily from the pileus.

The annulus is formed from a layer of tissue extending from the stipe to the margin of the pileus and enclosing the lamellae during the button stage. As the pileus expands, this layer of tissue tears apart around the margin of the pileus and remains adhering to the stipe as a more or less definite ring. If this layer of tissue is delicate, the ring may be poorly formed and easily rubbed off or evanescent, hence great care should be taken before deciding that an annulus is absent.

The volva is, perhaps, the most important character and it is also the one most easily missed by careless collecting. It is a layer of tissue completely enclosing the young button, which at this stage may resemble a puffball. However, if it is cut open the outline of the young mushroom can be seen (Figure

91, p. 48). The true puffballs are homogeneous within. As the pileus expands, the volva becomes torn. This may occur in different ways. In some species it tears across the top and the mushroom emerges leaving the volva as a loose membranous sheath enclosing the base of the stipe (Figure 90, p. 48). In other species it may tear around the margin of the pileus and part of it remains on the surface of the pileus as warts or patches whereas the remainder more or less tightly encloses the base of the stipe, sometimes forming a series of irregular rings and patches or sometimes appearing as a more or less evident collar (Figure 103, p. 50). In some species the volva is friable and powdery and tends to disappear. The volva is often buried in the ground and in order to identify them correctly it is necessary to collect *Amanita* species very carefully, making certain to get the base of the stipe.

In spite of the interest of many students in this genus, the identity of a number of the species is still in doubt. It is thus advisable to avoid all species of the genus when collecting for the table.

Key

1.	Pileus white or whitish	2
1.	Pileus not predominantly white	4
2.	Volva forming ragged scales and rings on the stipe; spores nonamyloid	white form of *A. muscaria*
2.	Volva not as above; spores amyloid	3
3.	Fruiting body entirely pure white; volva forming a loose sac-like sheath enclosing the bulb	*A. virosa*
3.	Pileus tinged greenish yellow; volva usually adnate to the bulb but separable at the margin	*A. citrina*
4.	Pileus brown, reddish brown, or yellow-brown	5
4.	Pileus not predominantly brown	10
5.	Wounds in flesh and stipe staining brown or reddish	6
5.	Wounds not conspicuously staining	8
6.	Bulb abrupt, expanded cup-shaped, depressed-marginate, firm, often splitting vertically; spores globose	*A. brunnescens*
6.	Bulb clavate or ovoid, not as above, spores ellipsoid	7
7.	Yellow tones present in pileus, warts, or annulus	*A. flavorubescens*
7.	No yellow tones present; entire fruit body reddening	*A. rubescens*
8.	Annulus gray; bulb not collared; spores amyloid	*A. porphyria*
8.	Annulus white; bulb with a collar; spores nonamyloid	9
9.	Pileus with creamy margin, brownish disk, very large	*A. velatipes*
9.	Pileus yellowish to dark brown with white warts, usually rather small	*A. pantherina*
10.	Pileus orange or red, at least on the disk	11
10.	Pileus yellow or paler, lacking orange or red tones	14
11.	Lamellae clear yellow; volva loose, sac-like	*A. caesarea*
11.	Lamellae white or cream; volva not sac-like	12

12. Volva powdery; pileus nonstriate; spores amyloid .... *A. flavoconia*
12. Volva not powdery; pileus striate; spores nonamyloid 13

13. Pileus large; volva ragged, in rings and scales on the stipe .... *A. muscaria*
13. Pileus usually small; volva a small white boot with a slight collar .... *A. frostiana*

14. Pileus not striate at margin 15
14. Pileus conspicuously striate at the margin 16

15. Pileus pale greenish yellow; bulb abruptly globose .... *A. citrina*
15. Pileus deeper yellow to yellow-brown; bulb clavate or tapering down .... *A. flavorubescens*

16. Lamellae clear yellow; volva loose, sac-like .... *A. caesarea*
16. Lamellae white or cream; volva not loose and sac-like 17

17. Volva forming ragged rings and scales on the stipe; pileus color lacking brown tones .... *A. russuloides*
17. Volva closely enclosing bulb and with a more or less free, collared margin 18

18. Pileus creamy yellowish or buff, lacking brown tones, usually small .... *A. gemmata*
18. Pileus with brown tones 19

19. Pileus with creamy margin, brownish disk, very large, fragile .... *A. velatipes*
19. Pileus yellowish to dark brown with white warts, usually rather small .... *A. pantherina*

AMANITA BRUNNESCENS Atk. — Poisonous

Figure 124, page 69

PILEUS 1½–4 (5) in. broad, convex, becoming expanded, often with a broad obtuse umbo, dark brown, smoky brown, or olive-brown, paler on the margin, usually somewhat streaked with innate fibrils, viscid, decorated with whitish or pallid brownish, floccose warts or patches which may disappear, sometimes faintly striate on the margin. FLESH thin except on the margin, white, tending to stain reddish brown, odor faint. LAMELLAE free or almost so, creamy white, close, rather broad, narrowing toward the stipe, with many short lamellulae interspersed. STIPE 3–6 in. long, ¼–¾ in. thick, with an abrupt, hard, marginate bulb whose margin splits longitudinally in a very characteristic manner, equal or tapering upward above the bulb, stuffed with a pith, subglabrous or minutely scurfy, white, staining reddish brown from the base upward. ANNULUS large, membranous, collapsing against the stipe, white or pallid, staining reddish brown. VOLVA dingy white to pallid brownish, breaking up into membranous-floccose fragments, some of which may cling to the pileus or bulb margin, usually leaving no trace on the bulb. SPORES amyloid, smooth, white, globose, 7–9 (10) μ.

In groups or scattered, on the ground in woods. July–Sept.

Bruises and wounds of the pileus and stipe stain reddish brown. The marginate bulb and globose spores distinguish this species from *A. rubescens.* However, it is very easy to confuse these two species, and, as one is poisonous and the other edible, a mistake might be serious. *A. brunnescens* var. *pallida* Krieger is a whitish form which Singer has called *Amanita aestivalis.*

AMANITA CAESAREA (Scop. ex Fr.) Pers. ex Schw. Edible

Frontispiece; Figures 90, 91, 92, page 48

PILEUS 2–4 in. broad or larger, fleshy, hemispherical, bell-shaped, or convex, expanding to nearly plane, sometimes obtusely umbonate, deep reddish orange on the disk, shading to bright yellow on the margin, or entirely clear yellow with disk slightly deeper in color, smooth, viscid, glabrous, prominently striate on the margin. FLESH thin, white or tinged yellowish, odor faint. LAMELLAE free or attached by a line, close to crowded, moderately broad, yellow. STIPE 3–8 in. long, ¼–¾ in. thick, scarcely bulbous, subequal or tapering toward the apex, hollow, glabrous or slightly floccose-scaly below the annulus, yellow. ANNULUS soft, membranous, yellow to orange, hanging down loosely around the stipe. VOLVA thick, white, membranous, at first encasing the entire fruit body to form a structure the size and shape of a hen's egg, splitting open at the apex as the stipe elongates and the pileus pushes up through, and finally encasing the base of the stipe as a loose, free-margined sac. SPORES nonamyloid, smooth, white, oval, 8–9.5 (11.5) $\times$ 6–7.5 (8) μ.

Solitary or in groups or large fairy rings on the ground in woods. July–Oct.

This mushroom is southern in distribution and seems to be rare as far north as Canada. There are specimens in the herbarium at Ottawa from Elgin County in Ontario and from Kentville, Nova Scotia. It is also said to occur in abundance around Quebec City.

A. caesarea is said to be a fine edible mushroom, but anyone collecting it for food should first be very certain of its correct identification. The poisonous, *A. muscaria* has similar colors on the pileus but can readily be distinguished if a careful examination is made of the volva and other characters such as the color of the lamellae, stipe, and annulus.

AMANITA CITRINA (Schaeff.) S. F. Gray Suspected

Figure 125, page 69

PILEUS 1.5–3.5 in. broad, convex becoming expanded, pale lemon-yellow or almost white, viscid when moist, decorated with appressed, friable, whitish or dingy buff, volval patches which may disappear, nonstriate on the margin. FLESH rather thin, white. LAMELLAE free or slightly attached, close, moderately broad, creamy white, floccose on the edges. STIPE 3–5 in. long, ¼–½ in. thick, equal or tapering slightly toward the apex, whitish, entirely glabrous or slightly scurfy at the base, stuffed or hollow, with a rather soft, globose bulb. ANNULUS creamy yellow or whitish, fragile, membranous, collapsing against the stipe. VOLVA whitish to dingy buff, membranous, more or less adnate to the soft bulb but usually separable at the margin, occasionally exceeding the bulb margin slightly. SPORES amyloid, smooth, white, globose, 7–9 μ.

Solitary or scattered on the ground in woods. Aug.-Oct. Fairly common, at least in Eastern Canada.

This mushroom has been known in North America as *A. mappa*, but the correct name is *A. citrina*. Bruises on the pileus, lamellae, and stipe tend to stain reddish brown. It has been reported as nonpoisonous but is better avoided.

AMANITA FLAVOCONIA Atk. — Possibly poisonous

Figure 123, page 69

PILEUS 1–3 in. broad, convex, expanding to almost plane, sometimes broadly umbonate, brilliant orange to bright yellow, the entire pileus more or less unicolorous, usually paling slightly toward the margin, smooth, viscid, with scattered, very friable, floccose or granular-powdery patches of the bright yellow volva, soon becoming glabrous, nonstriate on the margin, occasionally becoming faintly short-striate in places in age. FLESH thin except on the disk, white or tinged yellow next to the pellicle, odor not distinctive. LAMELLAE free or almost so, white or creamy, often yellow on the edges, close, moderately broad. STIPE 2–5 in. long, ¼–½ in. thick, equal or tapering upward from the oval to clavate-bulbous base, stuffed, then hollow, white or pale yellow, pruinose above the annulus, minutely scurfy below, usually with a few friable, yellow, volval fragments clinging to it. ANNULUS membranous, creamy to yellow. VOLVA entirely bright yellow, very friable, floccose to granular-powdery present on the pileus in scattered friable warts, a few fragments usually clinging to the stipe but most remaining in the soil when the fruit body is collected. SPORES smooth, white, ovoid, amyloid, 7–9 (10) × (4.5) 5–6 (7) μ.

Fairly common at least in Eastern Canada, scattered or in groups on the ground in woods. June–Sept.

The two orange-yellow species most likely to be confused with this one, *A. muscaria* and *A. frostiana*, are both conspicuously striate, and both have nonamyloid spores. *A. flavoconia* has much more brilliant coloring than *A. flavorubescens* and the base of the stipe does not stain reddish.

AMANITA FLAVORUBESCENS Atk. — Suspected

Figure 142, page 89

PILEUS fleshy, 1½–4 in. broad, convex to campanulate or broadly gibbous, expanding, light yellow to deep, dull yellow, or tinged olive-yellow, often dingy yellow-brown to umber on the disk, bruising brown to reddish brown, smooth, slightly viscid, decorated with small, very floccose warts which vary in color from brilliant ochre-yellow to dingy buff, nonstriate, or in places faintly striate on the margin. FLESH thin except on the disk, white or tinged yellow, bruising reddish. LAMELLAE moderately broad to rather narrow, free or attached by a line, close, creamy white. STIPE 2–5½ in. long, ¼–1 in. thick, equal or tapering upward, slightly swollen at the base to form an oval-clavate bulb, occasionally tapering below the bulb, stuffed or hollow, white or tinged yellow above, staining reddish toward the base, pruinose or minutely floccose, sometimes bearing a few bright yellow volval fragments. ANNULUS membranous, fragile, yellowish on the lower surface, creamy white with a yellow margin on the upper surface, striate. VOLVA on the bulb thick, membranous, sordid buff, staining reddish, usually forming a neat appressed boot so tightly appressed as to be inconspicuous and overlooked, occasionally leaving rings of dingy tissue on the stipe above the bulb, present on the pileus in dingy buff, floccose warts, or (especially in young buttons) in bright yellow, floccose fragments, a few of which may cling to the stipe or annulus but which seem to be lacking from the bulb except perhaps around its margin. SPORES smooth, white, amyloid, ellipsoid, 7.5–9 (10) × 5.5–6.5 μ.

Solitary or in groups on the ground in woods. June–Sept.

This seems to be a rather rare *Amanita*, although it may not be uncommon in certain localities. The combination of reddening stipe base, lack of any orange-red coloring in the pileus, and amyloid, ellipsoid spores will distinguish it from other yellowish species. The warts frequently disappear from the pileus, leaving it glabrous. The thick white flesh of the bulb may pull apart into slight scales or rings as the mushroom grows.

AMANITA FROSTIANA Peck — Not edible

Figures 129, 130, page 71

PILEUS 1–2½ in. broad, convex, expanding to plane, deep orange or reddish orange on the disk, often changing to clear yellow on the margin, conspicuously striate, viscid, bearing scattered, friable-floccose warts which are usually yellow or more rarely whitish with a few bright yellow fragments clinging to them. FLESH thin, white, tinged yellow beneath the cuticle, odor not distinctive. LAMELLAE free, moderately broad, close, white or tinged creamy yellow, at times yellow-marginate. STIPE 2–4½ in. long, about ¼ in. thick, subequal or tapering upward above the small, oval or subglobose bulb, stuffed, becoming hollow, white or pale yellow, subglabrous. ANNULUS fragile, membranous, often yellow. VOLVA on the bulb forming a little white boot with a free collar at the margin, usually with few to many bright yellow friable fragments clinging to the bulb and stipe base, on the pileus either entirely yellow and friable, or at times white-floccose with a few bright yellow friable fragments adhering. SPORES nonamyloid, smooth, white, subglobose to globose, apiculate, 7.5–9.5 × 7–9 μ.

Solitary or in groups on the ground in mixed woods. July–Sept. Rather rare.

This species is easily confused with both *A. muscaria* and *A. flavoconia*, although it differs from both in its globose spores and the collared white boot on the bulb. From *A. muscaria* it also differs in its small stature and the yellow friable warts on the pileus. The nonamyloid spores and the prominent striations on the pileus margin are further characters that separate it from *A. flavoconia*.

It is said to be nonpoisonous but the danger of confusing it with *A. muscaria* is too great and it should be avoided at all times.

AMANITA GEMMATA (Fr.) Gill. Doubtful

Figures 131, 132, page 71

PILEUS ¾–2¼ in. broad, at first ovoid to convex, expanding to plane, slightly depressed in center, occasionally subumbonate, smooth, viscid, glabrous or with floccose-membranous, whitish warts on the disk, pale creamy yellow on the margin, more dingy yellowish to buff on the disk, margin conspicuously striate. FLESH soft, fragile, thin, white or tinged creamy yellow, odor not distinctive. LAMELLAE free, close, broad in comparison to the flesh, creamy white, edge minutely flocculose. STIPE 2–4 in. long, ⅛–⅜ in. thick, nearly equal or tapering upward slightly, with a small, round to oval bulb about ¼–¾ in. diam. at the base, whitish, finely pruinose at the apex, subglabrous to appressed-floccose or appressed-fibrillose below, stuffed becoming hollow. ANNULUS white, membranous, fragile, evanescent or sometimes remaining attached to the margin of the pileus. VOLVA adnate to the bulb, at first with a slight free margin forming a collar at the top of the bulb, this later disappearing and leaving a slightly torn bulb margin. SPORES white, subglobose to globose, nonamyloid, 7–10 (11) $\times$ 7–9.5 μ.

Singly or gregarious on the ground in mixed woods. June–Sept.

This is a rather small, creamy yellowish to buff colored *Amanita*. The annulus is very fragile and may disappear very soon so that one would be inclined to look for the species in *Amanitopsis*. It might be confused with *A. russuloides*.

Konrad and Maublanc assert that they have eaten this species and that it is edible; however it cannot be recommended.

AMANITA MUSCARIA Fr. Deadly poisonous

Figures 103, 104, page 50; 133, 134, page 71

Fly Agaric

PILEUS 3–8 in. broad, hemispherical, becoming convex, then expanded, viscid when fresh, striate on the margin, blood-red to scarlet, orange, or yellow, or sometimes white, darkest on the disk, adorned with thick floccose to pyramidal warts which are whitish or tinged buff or straw-yellow. FLESH white or creamy, moderately thick on the disk, thinning to a line at the margin, tinged yellow under the cuticle, odor not distinctive. LAMELLAE free but reaching the stipe, close to crowded, rather broad, white or creamy, often minutely floccose on the edges. STIPE 4–8 in. long, ½–¾ in. thick, equal or tapering upward above the clavate bulb, stuffed, whitish or tinged yellow, subglabrous above, lacerate-scaly toward the base from the remains of the torn volva. ANNULUS large, membranous, white to yellowish. VOLVA whitish or tinged buff or straw color, broken up into rings of shaggy scaliness on bulb and base of stipe, in thick warts on the pileus. SPORES nonamyloid, smooth, white, ellipsoid, 8–11 $\times$ 6–8 μ.

On the ground in scattered groups or large colonies, sometimes in fairy rings, along roadsides and in open woods. July–Oct.

This poisonous *Amanita* is fairly common and widely distributed. The ragged volval scales on the stipe are very characteristic but the scales on the pileus may become washed off by rains. On the west coast of North America the form with reddish pileus seems to be the common one; while in the East the pileus tends to be yellow to orange. A white form is sometimes found. For a comparison with *A. frostiana* and *A. flavoconia*, see the notes on those species. See also the notes on *A. velatipes*.

AMANITA PANTHERINA (DC. ex Fr.) Secr. — Deadly poisonous

PILEUS 2–4 in. broad, convex at first, becoming expanded to nearly plane, viscid, surface covered with whitish, pyramidal warts which later may fall off or be washed off, typically smoky brown in color but varying to yellowish brown or olive-brown, sometimes quite yellowish, margin striate. FLESH whitish, thick in center to thin at margin, odor not distinctive. LAMELLAE white, free or attached by a line at first, close to crowded, with 1–2 tiers of lamellulae. STIPE 2½–4 in. long, ⅜–¾ in. thick, swollen at the base, white, silky above the annulus, fibrillose below, stuffed. ANNULUS median or superior, floccose-membranous, fibrillose below, margin yellowish to grayish brown. VOLVA closely adhering to the bulb as a sheath with a free collar, sometimes leaving a few concentric rings of volval tissue on the stipe, forming whitish warts on the pileus. SPORES white, smooth, nonamyloid, broadly ellipsoid to ovoid, (8) 9–11 (12) $\times$ 6.5–8 μ.

Under conifers apparently only along the west coast. May–Nov.

This is a very poisonous mushroom and according to Smith (1949) it has caused more deaths in Europe than *A. muscaria*, which usually receives more publicity. All of the records in our herbarium are from British Columbia except one, which is from the Yukon. *A. pantherina* may not occur in eastern North America but it has been included because of its very poisonous properties.

It is reported to be variable in color, typically brown but sometimes varying to yellow. Yellow forms may be confused with *A. muscaria* but the collar-like margin of the volva should distinguish it. *A. velatipes* is paler, larger, and more fragile.

AMANITA PORPHYRIA (A. & S. ex Fr.) Secr. — Suspected

Figures 135, 136, page 71

PILEUS 1–2½ in. broad, convex, becoming expanded, sometimes broadly subumbonate, brown to gray-brown or muddy brown, smooth, viscid, usually bearing a few remnants of the friable, gray volva, nonstriate, tending to remain for a long time decurved on the margin. FLESH thin, white, odor not distinctive. LAMELLAE free, close, moderately broad, creamy white. STIPE 2–4½ in. long, ¼–½ in. thick, equal or tapering upward above the rather soft, subglobose bulb, often patterned with innate gray flecking on a white background. ANNULUS thin, membranous, ashy gray, collapsing against the stipe. VOLVA pallid or grayish on the bulb, usually separable at the margin, often leaving a few friable, gray patches on the pileus, and around the stipe base. SPORES amyloid, smooth, white, globose, 7–9 μ.

Solitary or in groups of several on the ground in woods. Aug.–Oct. Infrequent.

The brown pileus, ashy gray annulus and soft, globose bulb are the distinguishing marks of this species. *A. tomentella* Krombh. is said to differ in the densely powdery, gray coating of the pileus and stipe. A form answering to the description of *A. tomentella* is occasionally collected and seems distinct from the usually glabrous *A. porphyria*. Whether or not this is a variation of *A. porphyria* is a question. *A. porphyria* often bears a few fragments of gray pulverulence on the pileus and occasionally it has a rather large powdery volval patch. *A. spreta* Peck is another brown to umber species in which the stipe is equal throughout and not bulbous at the base.

AMANITA RUBESCENS (Pers. ex Fr.) Gray — Edible

Figures 137, 138, page 71

PILEUS 2–6 in. broad, at first ovoid, expanding to convex or with a broad obtuse umbo, variable in color, usually dingy reddish or dull reddish brown, often with muddy brown or olive-umber shades present, slightly viscid, adorned with numerous, floccose, grayish or dirty pinkish scales which are readily washed off, nonstriate or the extreme margin indistinctly striate. FLESH thin, soft, white, staining reddish, odor not distinctive. LAMELLAE free or scarcely attached, close to crowded, moderately broad, narrowing toward the stipe, dingy white, staining reddish. STIPE stout, sometimes slightly excentric, 3–8 in. long, ¼–¾ in. thick, swollen at the base, subequal or tapering upward, stuffed, subglabrous to minutely fibrillose, staining dingy pink to reddish. ANNULUS large, membranous, fragile, collapsing against the stipe, dingy white or pale greenish yellow, staining pinkish. VOLVA fragile, gray, tinged sordid reddish, breaking up into scales on the pileus, usually lacking or almost so on the stipe base as most of the fragments remain in the soil. SPORES amyloid, smooth, white, ellipsoid, 7–9 (10) × 5–7 μ.

Solitary or scattered, on the ground in woods. July–Sept.

This is one of the edible species of *Amanita*, but the danger of confusing it with the poisonous *A. brunnescens* is great. Wounds and bruises in *A. rubescens* stain a sordid reddish color, while in *A. brunnescens* the stains are more reddish brown. *A. brunnescens* has a marginate bulb and globose spores. *A. flavorubescens* also stains reddish, but the dull yellow coloring, especially in the pileus margin, should distinguish it.

AMANITA RUSSULOIDES Peck — Poisonous

Figures 140, 141, page 89

PILEUS 1–2½ in. broad, convex, expanding to plane, prominently striate on the margin, smooth, viscid, pale straw-yellow to yellowish buff, paler on the margin, occasionally bearing a few whitish volval fragments. FLESH thin, white. LAMELLAE free or almost so, close to crowded, not broad, white or creamy. STIPE 2–4 (5) in. long, ⅛–¼ in. thick, tapering upward slightly above the clavate bulb, white or pallid. ANNULUS membranous, collapsed against the stipe, sometimes disappearing. VOLVA white or pallid, in ragged rings and fragments up the stipe, resembling the volva of *A. muscaria*. SPORES smooth, white, nonamyloid, ellipsoid, 8–10.5 (11.5) × 5.5–7.5 μ.

Scattered, on the ground in open places. July–Sept. Infrequent to rare.

This Peck species, which is infrequently collected, has been relegated by most authors to the long list of synonymy with *A. junquillea* and *A. gemmata*. However, it does not belong in that group. As Peck himself says, "The bulb is ovate and the volva fragile and easily broken into fragments. Its nearest relationship is with *A.* [*Agaricus*] *muscarius*. . . . "*A. russuloides* can be distinguished by its small stature, its straw colored, striate pileus, its muscaria-type volva, and its nonamyloid, ellipsoid spores.

A. russuloides may be confused with *A. gemmata* but the latter is more creamy yellow in color and it has globose spores, an evanescent annulus and an inconspicuous volva that is often marked only by a slight torn line around the margin of the bulb. It lacks the ragged volval fragments on the stipe that are characteristic of *A. russuloides*.

Definite information regarding its edibility appears to be lacking but it should be avoided.

AMANITA VELATIPES Atk. — Doubtful

Figure 143, page 89

PILEUS large, 3–5½ in. broad, fleshy, ovoid in the button stage, later convex and finally broadly expanded, striate on the margin, in age grooved-striate or somewhat tuberculate-striate, smooth on the disk, viscid, creamy yellow, usually darkening to brownish or umber on the disk and paling to cream on the margin (or, according to Atkinson, at times entirely hair-brown), decorated with thick, dingy white, floccose warts. FLESH moderately thick on the disk, thinning to a line at the margin, white, tinged yellowish beneath the cuticle, odor faint. LAMELLAE free, broad, elliptical, crowded, creamy white, slightly floccose on the edges, with many shorter, truncate lamellulae of varying lengths present. STIPE stout, 5–8 in. long, ⅜–1 in. thick, subequal or tapering upward slightly, clavate-bulbous, stuffed with a pith or becoming hollow within, dry, creamy white, minutely pruinose at the apex. ANNULUS large, membranous, creamy white, median, collapsed against the stipe. VOLVA thick, membranous, dingy white, tightly booting the clavate bulb but with a more or less free, thick margin, sometimes leaving one or two rings of tissue on the stipe above the bulb, scattered in thick, dingy white, floccose warts on the pileus. SPORES nonamyloid, smooth, white, ellipsoid, 8–10 × 6–7 μ.

Scattered or in groups on the ground in open woods, or in grassy places at the edge of woods. July–Sept.

This large *Amanita*, with its pale yellowish color, has been mistaken at first glance for a faded *A. muscaria* by more than one collector. Plate 6 in Güssow and Odell labeled *A. muscaria* is undoubtedly *A. velatipes*. However, the complete lack of any orange-red coloring, even in buttons, and the neat volval boot on the stipe will distinguish it. The booted stipe has caused some people to call this mushroom *A. cothurnata* Atk., and certainly these two are very closely related. The type specimens of both these Atkinson species have been examined. *A. cothurnata* is a small slender mushroom, pure white or with a very faint tinge of color on the disk, and with slightly broader spores which are thus more broadly ovoid than ellipsoid in shape.

Amanita glabriceps Peck, which has ellipsoid and not globose spores as stated by Peck, is a later synonym.

There is no information concerning the edibility of this species but it should be left alone.

AMANITA VIROSA Lam. ex Secr. — Deadly poisonous

Figures 115, page 52; 291, page 193

Destroying Angel

PILEUS 2–5 in. broad, at first ovoid, becoming convex and finally expanded, nonstriate, pure white, at times becoming faintly discolored on the disk, viscid, glabrous, rarely bearing a patch of volval membrane. FLESH soft, white, thin except on the disk, developing a disagreeably sweet odor. LAMELLAE free or reaching the stipe by a line, close to crowded, fairly broad, white, often minutely flocculose on the edges. STIPE 3–8 in. long, ¼–¾ in. thick, swollen at the base to a clavate or oval bulb, equal above or tapering upward, smooth or appressed-fibrillose, sometime floccose, white, stuffed within. ANNULUS large, white, membranous, hanging skirt-like near the apex of the stipe or clinging in shreds to the pileus margin. VOLVA white, membranous, sac-like, loosely encasing the bulb and stipe base, usually extending up the stipe well beyond the bulb. SPORES amyloid, smooth, white, subglobose varying to ovoid or ellipsoid, (8) 9–10.5 (12) × (7) 8–9.5 (10) μ.

Solitary or scattered on the ground in woods, sometimes on lawns beneath trees. July–Oct. Frequent.

This is the most deadly poisonous of our wild mushrooms. Everyone collecting mushrooms should learn to recognize it. The distinguishing marks are the pure white color of all parts, the presence of annulus and sac-like volva, the free lamellae and the white spore print. The volva may be left in the ground and overlooked if the mushroom is pulled carelessly. The danger of mistaking *A. virosa* for the edible *Lepiota naucina* is great, and a mistake could be fatal.

The folly of applying simple popular 'tests' to determine the edibility of a mushroom is well illustrated here. The pileus of *A. virosa* peels readily, and yet many people use this peeling test as a proof of edibility.

A. verna (Lam. ex Fr.) Pers. ex Vitt. is said to differ in having truly ellipsoid spores. In the collections of *A. virosa* in the herbarium at Ottawa, spores vary from globose to ellipsoid. Most specimens have at least some ellipsoid spores and a few have a considerable number of truly ellipsoid spores, but always mixed with globose ones. It does not seem possible in our collections to separate two species on the basis of spore shape or other characters. Since globose spores predominate, it seems best to use the name *A. virosa* for our species.

A. bisporigera Atk. seems to be identical with *A. virosa* except for the 2-spored basidia. Size is not a distinguishing feature as *A. bisporigera* can vary from very large to very small.

A. virosa has been mistakenly called *A. phalloides* in North America. The true *A. phalloides*, however, is a greenish species native to Europe, which has not yet been found in Canada but may possibly occur in California.

Amanita aestivalis Singer is not a synonym of *A. virosa* but of *A. brunnescens* var. *pallida* Krieger. It would key to this place in the above key but it has a marginate bulb which bruises reddish brown, and its volva is not loose and sac-like.

AMANITOPSIS

Species of *Amanitopsis* are white-spored. They have the lamellae free from the stipe, and possess a volva but no annulus. They are very similar to species of *Amanita* in appearance and stature, differing only in the absence of an annulus, and some authors do not regard them as generically distinct from *Amanita*. The name *Amanitopsis* has been officially conserved against the earlier name *Vaginata*.

It is not a large genus and, except for the varieties of *Amanitopsis vaginata*, the species are not commonly collected. None are known to be poisonous but because of their great similarity to *Amanita* and the danger of mistaking an *Amanita* that may have lost its annulus for an *Amanitopsis*, the amateur should avoid using these species as food until he is thoroughly familiar with them.

Figures 140-149

140. *Amanita russuloides.*
141. *A. russuloides.*
142. *A. flavorubescens.*
143. *A. velatipes.*
144. *Amanitopsis vaginata.*
145. *Armillaria imperialis.*
146. *A. mellea.*
147. *Limacella glischra.*
148. *L. illinita.*
149. *Lepiota acutaesquamosa.*

140
141
142
143
144
145
146
147
148
149

Figure 150. *Lepiota naucina.* Note that the base of the stipe is somewhat bulbous but no volva is present, also note the rather stiff annulus which stands out from the stipe.

Figures 151-160

151. *Lepiota brunnea.*
152. *L. clypeolaria.*
153. *L. cristata.*
154. *Cystoderma cinnabarinum.*
155. *Pleurotus applicatus.*
156. *P. porrigens.*
157. *P. subpalmatus.*
158. *Clitocybe aurantiaca.*
159. *C. clavipes.*
160. *C. decora.*

151
152
153
154
155
156
157
158
159
160

Figures 161-163. *Lepiota procera*. 161, two immature fruiting bodies; 162, mature fruiting body; 163, young fruiting body after the partial veil has torn and formed the annulus. Note the furfuraceous stipe.

AMANITOPSIS VAGINATA Fr. — Edible

Figure 144, page 89

PILEUS 2–4 in. broad, soft and fleshy, at first ovate, becoming campanulate or convex, then plane, umbonate, occurring in three color forms — white, (var. *alba* Sacc.), tawny, (var. *fulva* Sacc.), or grayish, (var. *livida* Pk.), surface glabrous or occasionally bearing fragments of the volva, margin grooved-striate. FLESH white. LAMELLAE free, not close, moderately broad, whitish. STIPE 3–7 in. long, about ¼ in. thick, tapering slightly towards the apex, base not bulbous, extending some distance into the ground, stuffed, becoming hollow, surface glabrous or somewhat mealy, whitish. ANNULUS lacking. VOLVA ample, white, membranous, mostly underground, encasing the base of the stipe and usually collapsed against it. SPORES smooth, white, globose, 8–10 μ.

Solitary or scattered, on the ground in woods. June–Sept. Fairly common.

Edible but not to be confused with specimens of *Amanita* from which the annulus has disappeared, especially the poisonous *Amanita spreta* Pk. whose stipe also lacks a bulb.

It seems probable that the three color varieties mentioned above are good autonomous species. They seem to be entirely distinct without intermediate forms. *Amanitopsis inaurata* (Secr.) Fayod has a gray volva and large spores.

LIMACELLA

Limacella is a small genus and most of the species are rather rare. They were formerly placed in *Lepiota*, but constitute a fairly well-marked, related group that seems worthy of generic rank. The fundamental character separating them from *Lepiota* is the microscopic structure of the trama of the lamellae, but they can usually be recognized in the field by the viscid pileus; the stipe, too, is often viscid. The lamellae are free or in one species slightly attached and an annulus is present. The North American species have been studied by Helen V. Smith (1945) and she recognized twelve species. Two species may be found fairly often in the Ottawa district.

There does not appear to be much information on the edibility of *Limacella* species, but since the genus is generally considered to be closely related to *Amanita*, they are probably best avoided or tried very cautiously.

LIMACELLA GLISCHRA (Morg.) Murr.

Figure 147, page 89

PILEUS ¾–2½ in. broad, convex or subumbonate, slimy-viscid, yellow-brown to reddish brown. FLESH white, thick, soft. LAMELLAE free, close to crowded, white, broad. STIPE 2–3 in. long, ⅛–¼ in. thick, equal or nearly so, glutinous, colored like the pileus, cottony at the base, solid. ANNULUS slight, evanescent, glutinous-fibrillose. SPORES white, subglobose, 4–6 × 4–5 μ.

Usually singly or scattered on ground in woods. Aug.–Oct.

The rather bright colored, thick gluten covering both pileus and stipe, together with the white spores and free lamellae characterize this species. *L. glioderma* (Fr.) Earle is also brown but is darker and the stipe is not viscid.

LIMACELLA ILLINITA (Fr.) Earle

Figure 148, page 89

PILEUS 1¾–2½ in. broad, at first ovoid becoming campanulate to plane or subumbonate, glutinous, viscid, glabrous, white, sometimes yellowish on the disk. FLESH white, thin, soft, no taste. LAMELLAE free, close, moderately broad, white. STIPE 2–3½ in. long, ⅛–¼ in. thick, equal or tapering upward slightly, white, glutinous below annulus, silky above, stuffed to hollow. ANNULUS fibrillose, evanescent. SPORES white, subglobose to broadly ellipsoid, smooth, 5–6 × 4–5.5 μ.

Singly to gregarious on the ground in woods. Sept.–Oct.

The very glutinous white pileus and stipe, and the free lamellae distinguish this species. Sometimes the gluten is so copious that it drips from the pileus. A form is sometimes found in which the gluten on the stipe becomes pinkish or red and this has been described as *L. illinita* var. *rubescens* H. V. Smith.

LEPIOTA

The genus *Lepiota* contains a great many species including some of the largest and most important edible species. It has white spores and lamellae free from the stipe. An annulus is present, but the volva is lacking. The stipe is a different texture from the flesh of the pileus and separates readily from it. An exception to the spores being white is found in *Lepiota molybdites* where they are greenish.

It is an important genus for those who are interested in mushrooms as food. *L. procera* is one of the finest edible species we have and *L. brunnea* and *L. americana* are also very good. However, the occurrence of the poisonous *L. molybdites*, which might be mistaken for either *L. procera* or *L. brunnea* if spore prints are not taken, makes it important that the characters of the species be examined closely.

Careful identification is particularly important in the case of *L. naucina*. This beautiful white species, which occurs commonly in lawns and meadows, is edible itself but has probably been indirectly responsible for many deaths from mushroom poisoning because of people mistaking the deadly poisonous

Amanita virosa for it. For this reason no one should eat *L. naucina* or any pure white mushroom until he is absolutely certain he knows and can recognize *A. virosa.*

As with most of the older genera of agarics, modern investigators tend to split *Lepiota* into a number of smaller genera composed of groups of closely related species. Two of these segregated genera have been recognized here, *Limacella* including the viscid species formerly placed in *Lepiota*, and *Cystoderma* including the species with granulose covering of the pilei and with the lamellae attached to the stipe.

Key

1. Spore print green .... *L. molybdites*
1. Spore print white 2

2. Pileus glabrous, white; lamellae faintly pinkish in age .... *L. naucina*
2. Pileus more or less scaly 3

3. Annulus movable, fruiting bodies very large 4
3. Annulus not movable, fruiting bodies small to medium 5

4. Stipe glabrous, striate, turning reddish when wounded; fruiting body about as broad or broader than tall .... *L. brunnea*
4. Stipe furfuraceous, not reddening when wounded; fruiting body much taller than broad .... *L. procera*

5. Stipe glabrous 6
5. Stipe clothed with a floccose or filamentous sheath, sometimes scaly 7

6. Stipe equal, mostly less than ¼ in. thick .... *L. cristata*
6. Stipe bulbous or ventricose, mostly ¼ in. or more thick; whole fruiting body staining reddish when bruised or in age .... *L. americana*

7. Scales of pileus erect, wart-like; spores 7-9 μ long .... *L. acutaesquamosa*
7. Scales of pileus more or less appressed, patch-like; spores 10-6 μ long .... *L. clypeolaria*

LEPIOTA ACUTAESQUAMOSA (Weinm.) Kummer — Edible

Figure 149, page 89

PILEUS 2–5 in. broad, convex expanding to almost plane, obtuse or broadly umbonate, bright tawny, tomentose, covered with erect, beaked or squarrose, tawny brown scales which gradually fall off and leave tawny, fibrillose, scar-like patches with paler flesh showing between. FLESH not very thick, soft, white. LAMELLAE free, very crowded, unforked, moderately broad, white, pruinose on the edge, minutely saw-toothed. STIPE 2–4 in. long, about ¼ in. thick, bulbous at base, equal or tapering slightly to the apex, dingy white, covered with a dense cottony-fibrillose sheath, stuffed or hollow. ANNULUS white, with brownish scales on the lower surface, membranous, hanging loosely around the stipe, sometimes disappearing. SPORES smooth, white, long-ellipsoid, 7–9 × 2.5–3 μ.

In groups, on the ground or on decayed wood in woods, gardens, greenhouses. Aug.–Sept.

It can be recognized by its bright tawny color and spiny scales. *L. friesii* Lasch is similar but with forking lamellae.

LEPIOTA AMERICANA Peck — Edible

Figures 126, 127, 128, page 70

PILEUS 1–4 in. or more broad, at first subglobose or ovoid, becoming conic-expanded, convex, or broadly expanded, with more or less striate margin, dull reddish brown, breaking up into large scales except on the umbo and exposing the whitish flesh beneath. FLESH thin, white, reddening where bruised, gradually turning pinkish brown with age, taste mild. LAMELLAE free, close, moderately broad, narrowing toward the stipe, white, turning red where bruised, in dried specimens a smoky cocoa-brown color. STIPE central, separable, 2–5 in. long, often with a decided swelling at the base or just above the base, varying above from moderately stout to very slender, stuffed, glabrous, white, bruising reddish. ANNULUS fairly ample, sometimes disappearing. SPORES smooth, white, broadly oval, 7–9 (11) $\times$ 5.5–6 (7) μ, many slightly inequilateral.

Solitary or in clusters, on the ground in grassy places. Rare. Aug.–Oct.

The tendency of the entire fruiting body to redden where handled or in age is characteristic. In dried specimens the entire fruiting body is a smoky cocoa-brown color. Another distinctive character is that the stipe is usually broader above the base and tapers both up and down.

LEPIOTA BRUNNEA Farlow & Burt — Edible

Figure 151, page 91

PILEUS 3–7 in. or more broad, thick, soft, ovoid, expanding to convex, then plane, brown to smoky brown, cracking (except on disk) into concentric rings of large, coarse, persistent scales whose outer edges become reflexed. FLESH beneath cuticle whitish, darkening to smoky color or reddish on exposure to air. LAMELLAE free but not remote from the stipe, broad, crowded, dull whitish, darkening on drying. STIPE stout, central, 2–6 in. or more long, up to 1 in. thick, with a large underground bulb at the base, silky-striate to fibrillose-striate, smoky brown, paler above the annulus, becoming reddish where wounded, hollow, easily separable from pileus. ANNULUS large, thick, flaring, persistent, smoky brown on the lower surface, whitish on the upper surface, fixed becoming free and movable. SPORES smooth, white, variable in size and shape, ellipsoid to subglobose, inequilateral, obliquely apiculate, often truncate, (7.5) 9.5–11 $\times$ (4.5) 5.5–7 μ.

Solitary or in small clusters on the ground in open grassy places, laneways, etc. Sept.–Oct.

L. brunnea is distinguished from *L. procera* by its stout stature, dingy coloring, striate stipe, and thick recurving scales as well as by the spores.

The poisonous *L. molybdites* is very similar in size and stature. If *L. brunnea* is being collected for food, a spore print should be obtained to make certain the spores are white. The spores of *L. molybdites* are green.

L. brunnea is very close to *L. rachodes* (Vitt.) Quél. (sometimes spelled *rhacodes* or *racodes* by different authors). The characters by which *L. brunnea* is said to differ are the darker brown color, striate stipe, and less remote lamellae. The most important of these characters would seem to be the striate stipe, and most European illustrations of *L. rachodes* do not show a striate stipe, although there is a suggestion of it in Cooke's Plate 22 in *Illustrations of British Fungi*. There are two European specimens labeled *L. rachodes* from England and Norway in the herbarium at Ottawa and in these the stipes appear to be identical with those of dried specimens of *L. brunnea* from Canada in which the stipe was known to be striate when fresh. On the other hand, A. H. Smith's photograph in *Mushrooms in their Natural Habitats*, Reel 21, No. 142, shows specimens with smooth nonstriate stipes. Since the species commonly collected around Ottawa does have a striate stipe we are referring it to *L. brunnea* until more information is available. From the standpoint of edibility the problem is of no significance because both *L. brunnea* and *L. rachodes* are edible but it is very important to distinguish the poisonous *L. molybdites*.

LEPIOTA CLYPEOLARIA (Bull. ex Fr.) Kummer — Suspected

Figure 152, page 91

PILEUS 1–2 in. broad, at first ovate or acorn-shaped, coated with a thin layer of yellowish-buff or brownish tissue, expanding to campanulate-convex, the outer tissue being drawn apart into scales which range in color from creamy white to ochraceous or brown and which vary from appressed or floccose patches to somewhat squarrose, brown-tipped scales, the exposed flesh between the scales creamy white, fibrillose, the disk umbonate or obtuse, smooth, brownish, in age the pileus becoming nearly plane, the scaliness partly or almost entirely disappearing, margin often ragged with fragments of veil, sometimes striate. FLESH thin, soft, white. LAMELLAE free, close, moderately broad, white, edges somewhat floccose. STIPE 1½–4 in. long, about ⅛ in. thick, tapering slightly upward, hollow, whitish, silky-fibrillose, sheathed with white or creamy yellow, cottony fibrils which may partly disappear. ANNULUS white, floccose, disappearing. SPORES smooth, white, variable in size and shape, subfusiform to ellipsoid, often slightly beaked or curved at one end, 10–16 (18) $\times$ 4–6 μ.

In groups, on the ground in open woods or fields. Aug.–Oct.

Suspected of being poisonous.

L. cristata may be somewhat similar in size and coloring but it has a glabrous stipe and quite different spores.

LEPIOTA CRISTATA (A. & S. ex Fr.) Kummer — Edible

Figure 153, page 91

PILEUS ½–2 in. broad, at first ovate, becoming campanulate-convex, then expanded, umbonate, cuticle reddish tan, intact on the umbo, elsewhere broken up into scales which become finer toward the margin and tend to disappear, leaving exposed the white surface beneath, margin sometimes striate. FLESH thin, fleshy, white, odor rather unpleasant, taste mild. LAMELLAE free, close to crowded, rather narrow, white, edges finely crenulate. STIPE slender, equal, 1–2 in. long, about ⅛ in. thick, glabrous or slightly fibrillose, sometimes striate, stuffed or hollow, whitish or tinged lilac or pinkish brown. ANNULUS soft, white, tending to disappear. SPORES white, irregular and variable, elliptical, wedge-shaped, or angular, 5.5–7 (8) $\times$ 3–4 μ.

On the ground, usually in groups, in open grassy places. June–Oct.

The glabrous stipe and the small wedge-shaped spores easily separate this species from *L. clypeolaria.* The odor of *L. cristata* was described as 'fishy' by Krieger (1936). Apparently the odor varies from strong to weak or absent under different conditions.

LEPIOTA MOLYBDITES (G. Meyer ex Fr.) Sacc. — Poisonous

Figure 139, page 72

PILEUS 3–11 in. broad, sometimes even larger, at first subglobose, expanding to convex, sometimes slightly umbonate, buff colored, soon breaking up (except on disk) to form irregular patchy scales which tend to disappear, whitish between the scales. FLESH thick, firm, white. LAMELLAE free, remote from the stipe, close, broad, at first white, becoming dull green. STIPE 4–8 in. or more long, up to 1½ in. thick at the swollen base, tapering somewhat toward the apex, fibrous-stuffed, stout and firm, glabrous, grayish white or tinged with brown. ANNULUS large, thick, firm, somewhat floccose, movable. SPORES smooth, subelliptical, bright to dull green in mass, 9–12 $\times$ 6–8 μ.

In large colonies, sometimes in fairy rings. On the ground in grassy places and open woods. Aug.–Sept.

L. molybdites is poisonous, at least to some people, and should be avoided as food. The danger of mistaking it for *L. brunnea* or *L. rachodes* is great. A spore print should be obtained first, if either of the latter two species is to be eaten. The green spore print of *L. molybdites* will distinguish it. The flesh of *L. molybdites* is said to redden slightly on bruising. This species tends to be southerly in distribution but has been collected near Ottawa. It has been generally known under the name of *L. morgani* Peck. Some authors place it in a separate genus, *Chlorophyllum.*

LEPIOTA NAUCINA (Fr.) Kummer — Edible: Use caution

Figure 150, page 90

PILEUS 2–4 in. broad, soft, fleshy, at first subglobose, becoming hemispherical, then expanded-convex, white or somewhat smoky, in age darkening to buff or leather color, smooth, dry, glabrous, or occasionally with the cuticle cracking into scales. FLESH soft, white, thick, thinning toward the margin, odor and taste not distinctive. LAMELLAE free, close, rather broad, often rounded behind, white, gradually taking on a pinkish tinge with age, finally darkening to pinkish brown. STIPE stout, 2–4 in. long, up to ½ in. in diameter, subequal or tapering upward from the swollen base, smooth, glabrous, pruinose above the ring, white, stuffed to hollow, easily separable. ANNULUS median to superior, thick, white, rolling back upon itself to form a stiff, collar-like ring on the stipe, persistent, becoming movable in age. SPORES smooth, white, oval to ellipsoid, slightly inequilateral, 7–9 × 5–6 μ.

In scattered colonies on the ground in grassy places, common. Aug.–Oct.

This species is edible, but is not recommended for food because of the danger of confusing it with the deadly *Amanita virosa*. In *A. virosa* the annulus is pendent and skirt-like. In *Lepiota naucina* the annulus is rolled and collar-like. If *Amanita virosa* is carelessly collected, the volva may be left behind and its presence overlooked, thus increasing the danger of mistaking it for a *Lepiota*. The two species can be distinguished with certainty by the spores.

LEPIOTA PROCERA (Fr.) S. F. Gray — Edible

Figures 161, 162, 163, page 92

Parasol Mushroom

PILEUS 3–9 in. broad, or sometimes larger, at first subglobose or egg-shaped, becoming campanulate, then plane, umbonate, cuticle reddish tan except on the disk, soon breaking up into more or less concentric rings of scales which are larger and more scattered toward the margin and tend to disappear, exposing the finely fibrillose, white surface beneath. FLESH soft, white, thick, thinning toward the margin. LAMELLAE free, remote from the stipe, close to crowded, broad, ventricose, white, floccose on the edge. STIPE tall and slender, 6–12 in. or more long, tapering upward from a bulbous base, ¼–½ in. thick at the apex, white, silky-fibrillose, covered with fine, brown, floccose or fibrillose scales, sometimes with several brown rings of scaliness near the annulus, hollow, easily separable from the pileus. ANNULUS large and flaring, thick, soft, movable. SPORES smooth, white, oval, 14–18 × 9–12 μ.

Solitary or in groups, on the ground in grassy places and open woods. July–Sept.

L. procera is taller and more slender in stature than either *L. brunnea* or *L. molybdites*. Its nonstriate, floccose stipe will separate it from *L. brunnea* and its white spore print from *L. molybdites*.

L. procera is one of the largest of our mushrooms and is frequently called the Parasol Mushroom. As long as care is taken to distinguish it from the green-spored *L. molybdites*, it is not likely to be confused with any other poisonous species and it is one of the most desirable mushrooms for the table.

CYSTODERMA

Cystoderma is characterized by having white spores, lamellae attached to the stipe, a more or less distinct annulus, and a granulose covering of the pileus. *Cystoderma* species suggest *Lepiota* in general appearance and were formerly placed in that genus but are now separated from it because the lamellae are not free. *Cystoderma* is distinct from *Armillaria* in the granulose covering of the pileus and it forms a natural group that is easily recognized in the field.

They are mostly small species unlikely to interest the mycophagist but they are usually attractive in appearance. As far as we know, none are poisonous. The species are distinguished to a large extent by microscopic characters. Smith & Singer (1945) have published a good monograph on the genus in which they recognized fourteen species and a few more have been added since.

CYSTODERMA CINNABARINUM (Alb. & Schw. ex Secr.) Fayod — Edible

Figure 154, page 91

PILEUS 1–2½ in. broad, at first ovoid, then expanded-convex to plane, finely granular-scaly with bright cinnamon to brownish orange or rusty brown particles, pallid between the scales, darkest on the disk, margin incurved at first, sometimes with fragments of the annulus clinging to it. FLESH thin, whitish or stained rusty near the surface. LAMELLAE at first adnate to the stipe but later separating from it, white or creamy, close to crowded, not broad. STIPE short and stout, 1–2 in. long, about ¼-in. thick, equal or slightly thickened at the base, coated up to the annulus with mealy-granular, cinnamon particles, above the annulus paler and glabrous. ANNULUS slight, disappearing. SPORES minute, ellipsoid, smooth, white, nonamyloid, 3.5–5 × 2.5–3 μ. CYSTIDIA with spear-shaped tips.

Singly or in small groups on the ground in woods. Sept.–Oct.

This is one of the largest species of the genus and is a very attractive and beautiful mushroom. It is reported to be edible. *C. granulosum* (Batsch ex Fr.) Fayod and *C. amianthinum* (Scop. ex Fr.) Fayod are also fairly common species that are somewhat similar in appearance and a microscope is required to identify them with certainty. *C. amianthinum* has amyloid spores and *C. granulosum* lacks cystidia on the lamellae.

ARMILLARIA

Armillaria is generally used to include the species that have an annulus, white spores, and lamellae attached to the stipe. No volva is present. It is generally agreed among taxonomists that the genus as so understood includes several groups of unrelated species but there is not agreement as to the most suitable way of splitting the genus in order to bring out the relationships. It therefore seems preferable to use it in the broad sense at present.

ARMILLARIA IMPERIALIS (Fries in Lund) Quél.

Figure 145, page 89

PILEUS 3–8 in. broad, very large and firm, at first convex, becoming expanded, whitish to smoky gray with innate darker fibrils near margin, glabrous, slightly viscid, sometimes becoming cracked on the disk, margin decurved, strongly inrolled at first. FLESH white, thick, no odor, strong taste. LAMELLAE decurrent, white to yellowish white, drying brownish, close, rather narrow. STIPE 2–4 in. long, ½–1 in. thick, equal, concolorous with pileus or more yellowish, floccose to scaly, solid. ANNULUS double, the outer ring membranous, concolorous with pileus, the inner ring more filamentous, whitish, somewhat evanescent. SPORES hyaline, smooth, oblong-ellipsoid, 11–15 × 5–6 (7) μ.

On ground under conifers, single or gregarious. Aug.–Sept.

This is a large, massive mushroom whose fruit bodies develop slowly and persist for a long time. It is rather rare but is a very striking fungus when found.

A. ventricosa (Peck) Peck is another large species with a double annulus. It is perhaps a little smaller than *A. imperialis*, whiter in color, and has smaller spores 9–12 (15) × 4.0–5.5 μ.

There does not seem to be any information regarding the edibility of these species but a somewhat similar large species is said to be used extensively for food by the Japanese on the west coast of the United States. This is *Armillaria ponderosa* (Peck) Sacc. and is as large as *A. imperialis* but paler in color, with smaller spores, and only a single annulus. This species is also known to occur in the East but is less common than *A. imperialis.*

The species with oblong-ellipsoid spores, double veil, and decurrent lamellae have been placed by some authors in a separate genus, *Catathelasma.*

ARMILLARIA MELLEA (Fr.) Kummer — Edible

Figure 146, page 89

Honey Mushroom

PILEUS 1¼–4 in. broad, at first acorn-shaped to hemispherical with inrolled margin, then convex or expanded, sometimes subumbonate, yellow-brown, yellow-buff or rusty tinged, minutely scaly, especially on the disk, with buff to brown or blackish tufts of fibrils, becoming striate on the margin. FLESH thin except on the disk, white to rusty tinged, odor mild to slightly unpleasant, taste mild or slightly acrid and unpleasant. LAMELLAE adnate or subdecurrent, fairly close to subdistant, moderately broad, white or creamy, staining rusty brown. STIPE 2–6 in. long, ¼–¾ in. thick, subequal or broadening below into a clavate base, stuffed, then hollow, finely fibrillose to fibrillose-scaly, paler than the pileus, becoming rusty stained, paler at apex. ANNULUS white or tinged brown, fibrillose-membranous, subpersistent or evanescent. SPORES smooth, white, broadly oval, obliquely apiculate, 7.5–9.5 × 5–6.5 μ.

In dense clusters around the base of living trees and old stumps, common. July–Oct.

This is perhaps the only really common species of *Armillaria*. It is widely distributed and often occurs in considerable abundance. It is rather variable and may confuse the amateur at first, but it has a characteristic appearance that is difficult to describe though soon recognized. The colors may vary considerably and the annulus may be somewhat evanescent.

Another feature of this fungus is the presence of tough, black strands of mycelium which may be found under the bark of the tree or stump from which it is growing. These strands, sometimes called 'shoe strings' are more properly termed rhizomorphs.

The unpleasant taste of this fungus when raw disappears on cooking and the species is widely used as food. Smith recommends it as being very good but suggests that younger stages should be selected.

PLEUROTUS

The genus *Pleurotus* has been used to include those species with excentric or lateral stipes, or lacking stipes altogether, and having a white spore deposit. However, in *P. sapidus* the spore deposit is lilac tinged and in *P. subpalmatus* it is pinkish. The genus corresponds to *Claudopus* in the pink-spored group and *Crepidotus* in the brown-spored group.

Most of the species occur on decaying wood and they vary in size from very minute to very large. Some species might be mistaken for *Clitocybe* but the most closely related forms are to be found in *Panus* and *Lentinus*. It is now generally recognized that *Pleurotus*, *Panus*, and *Lentinus* are all artificial genera, and modern authors tend to redistribute the species of all three among a number of other genera. *Pleurotus ostreatus* (Jacq. ex Fr.) Kummer, which may be the same as *P. sapidus* Kalchbr., is regarded as the type species of *Pleurotus*.

No poisonous species are known in *Pleurotus*.

Key

1. Pileus with excentric to central stipe 2
1. Pileus sessile or with lateral stipe 3

2. Pileus at first whitish, becoming tan, smooth ..... *P. ulmarius*
2. Pileus reddish tan to pinkish, veined ..... *P. subpalmatus*

3. Pileus tiny, usually much less than 1 inch across 4
3. Pileus larger, usually more than 1 inch across 5

4. Pileus pure white ..... *P. candidissimus*
4. Pileus gray to blackish ..... *P. applicatus*

5. Pileus olivaceous to yellow-brown ..... *P. serotinus*
5. Pileus white or whitish 6

6. Pileus thin, fragile, sessile .... *P. porrigens*
6. Pileus thick, fleshy; lamellae decurrent on the lateral stipe .... *P. sapidus*

PLEUROTUS APPLICATUS (Fr.) Kummer

Figure 155, page 91

PILEUS less than 1/4 in. broad, sand-color to pinkish gray, darkening to nearly black, growing from the underside of logs, etc., sessile, resupinate, at first almost cylindrical, expanding to deep cup-shaped, then saucer-shaped, somewhat irregular in outline because of the excentric to lateral attachment, coarsely pruinose, margin inrolled. FLESH thin, gelatinous. LAMELLAE radiating from a central point, subdistant, rounded behind, moderately broad, thick, with bluntly rounded edges, sand-color to dark gray, densely pruinose, alternate lamellae short. STIPE lacking, or sometimes with the thickened flesh at the point of attachment prolonged into a stubby, stipe-like base, densely pruinose to white-mycelioid at the base. SPORES smooth, white, subglobose, 4–5 μ in diameter.

In groups on decaying wood. Aug.–Oct.

This tiny mushroom is not common and will often be overlooked because of its small size and its occurrence on the underside of logs and planks. When dry it tends to fold up and appears as a small blackish spot on the wood, which at first glance, would not be taken for a mushroom. However, when moistened and expanded, the numerous fruiting bodies with their radiating lamellae form rather a pretty sight. It is obviously not closely related to other species that have been placed in *Pleurotus* and is not likely to be confused with any of them. *Trogia crispa* Fr. may also be found on the underside of branches, growing resupinate at times, but it is larger and the upper surface is reddish tan to yellowish in color.

PLEUROTUS CANDIDISSIMUS B. & C.

Figure 174, page 110

PILEUS pure white, thin, soft, 1/8–3/4 in. broad, laterally attached, sessile or almost so but never resupinate, white-mycelioid at the point of attachment, semicircular to shell-shaped or fan-shaped in outline, at first convex with inrolled margin, expanding to nearly plane, with a soft powdery appearance to the surface, somewhat radiately wrinkled in dried specimens. FLESH thin, white, membranous. STIPE absent or insignificant, lateral, minutely tomentose, whitish. LAMELLAE reaching the point of attachment or (if stipe is present) subdecurrent, distant or subdistant, broad, narrowing toward each end, creamy white, with edges fimbriate. SPORES white, smooth, globose, 4–6 μ in diameter.

In scattered colonies on decaying wood. July–Sept.

Several small white species of *Pleurotus* have been described, of which this one appears to be the most common. It has a somewhat chalky appearance and is very delicate, soon becoming shriveled.

PLEUROTUS PORRIGENS (Fr.) Kummer

Figure 156, page 91

PILEUS sessile, laterally attached, elongated, ½–3 in. long, up to 2 in. broad, at first resupinate with inrolled margin, expanding to almost plane, or depressed toward the base, narrowing toward the base, variable in shape, mostly fan-shaped to ear-shaped, white, watery and slightly striate on the margin when moist, varying from almost glabrous at the margin to densely tomentose at the base, sometimes lobed on the margin. FLESH thin, white, fragile, odor and taste mild. LAMELLAE mostly reaching the point of attachment, close, rather narrow, linear, white or cream, in some specimens showing a certain amount of forking near the base. STIPE lacking, base somewhat white-mycelioid. SPORES smooth, white, subglobose to broadly ovoid, 5–7.5 × 5–6 μ.

In overlapping, shelving clusters, on decaying wood of conifers. Sept.–Nov.

A number of whitish species of *Pleurotus* that may key out here are known. From our herbarium records, *P. porrigens* appears to be the most common, but the others are often difficult to distinguish and some can be determined only by microscopic characters.

The fruiting bodies of one group of species, including *P. porrigens*, are resupinate when very young but soon become reflexed, whereas those of another group are never resupinate. Among the resupinate forms, *P. albolanatus* Pk. in Kauffm. differs in having a gelatinous layer of tissue in the pileus, and is more hairy. *P. porrigens* appear slightly hairy when dry, especially toward the base, but *P. albolanatus* is decidedly hairy all over and is a larger, firmer plant. Among the nonresupinate forms, *P. petaloides* Fr. is a brownish mushroom with slightly smaller spores and also has cystidia on the lamellae. *P. porrigens* lacks cystidia. *P. spathulatus* (Fr.) Peck has ovoid-ellipsoid spores. *Panus angustatus* Berk. is another species that might be confused with this group. It is tougher in consistency, has cystidia on the lamellae, and has a gelatinous layer of tissue in the pileus.

PLEUROTUS SAPIDUS Kalchbr. — Edible

Figure 186, page 112

Oyster Mushroom

PILEUS firm to pliant, fleshy, white to ashy or brownish, 2–8 in. broad or sometimes larger, fan-shaped to shell-shaped or elongated, usually marginate behind, sometimes more or less circular and nearly centrally stipitate, convex, sometimes depressed toward the stipe, smooth, moist, glabrous or minutely tomentose toward the stipe, margin thin, inrolled, faintly striate when moist, sometimes lobed and wavy. FLESH thick, white, soft when young, becoming tougher with age, odor and taste agreeable. LAMELLAE broad, white or whitish, close to subdistant, decurrent, extending down the stipe in vein-like lines with varying amounts of converging and branching at the base. STIPE very short, usually lateral or almost lacking, occasionally excentric to nearly central, stout, firm and solid, sometimes hairy at the base. SPORES smooth, tinged lilac in heavy deposits, oblong, 7–10 (12) × 3–4 μ.

Usually growing in overlapping shelving clusters on wood of deciduous trees. May–Oct. Common.

The name 'oyster mushroom' has been applied to this species because of the shape of the pileus, which often suggests a shell. It is edible, but authors disagree as to its quality. The manner of cooking is important if this species is to be fully appreciated. It is recommended that this mushroom be cut in pieces, which are then dipped in seasoned beaten egg, rolled in bread or cracker crumbs and fried in hot fat.

Many authors have tried to distinguish *Pleurotus sapidus* from *P. ostreatus* (Fr.) Kummer on the basis of the color of the spore deposit. The spore print of *P. sapidus* was said to be lilac-tinted and that of *P. ostreatus* white. However, we have been unable to find any specimen in which the lilac color did not show up in a good spore deposit. This has also been the experience of other collectors in North America and, although it is possible that a similar white-spored species occurs in Europe, it is either absent or very rare in North America. P. *subareolatus* Peck is somewhat similar in appearance and has a white spore deposit but larger spores. If it should prove to be universally true that all the forms with small spores have a lilac-tinted spore deposit, *P. sapidus* will become a synonym of *P. ostreatus*, which is the older name.

The amount of development of the stipe is variable and might lead to possible confusion with some forms of *Pleurotus ulmarius*, but the decurrent lamellae and cylindric spores will distinguish it readily.

It has been claimed that successive crops of fruit bodies may be obtained by watering a log on which they are known to occur.

PLEUROTUS SEROTINUS (Fr.) Kummer Edible

Figure 112, page 51; Figure 413, page 296

PILEUS compact, 1–3 in. broad, typically more or less semicircular in outline, varying to kidney-shaped, convex with inrolled margin, expanding somewhat, varying from densely tomentose to almost glabrous, slimy-viscid when moist, muddy olivaceous or yellow-green to dull shades of yellow-brown or reddish. FLESH thick, white, firm, odor and taste not distinctive. LAMELLAE narrowly adnate, often showing a sharp line of demarcation between the end of the lamellae and beginning of tomentum on stipe, thin, close, narrowing in front and behind, whitish to yellowish tan. STIPE solid, stubby, ¼–¾ in. long, up to ⅜ in. thick, laterally attached, continuous with pellicle of the pileus on the upper surface, densely tomentose below or partially dotted with minute, dark brown scales, yellowish. SPORES smooth, white, narrowly oblong, some slightly curved, 4–6 × 1–1.5 μ. CYSTIDIA up to 28μ long, about 12 μ broad at widest point, narrowing slightly toward the apex and considerably toward the base, sac-like.

Solitary, or more often in overlapping clusters, on wood of deciduous trees. Aug.–Nov. Fairly common.

It is most likely to be confused with *Phyllotopsis nidulans*, but the greenish or olive tints in the pileus distinguish it. In addition, the spore print of *P. serotinus* is white whereas in *P. nidulans* it is pinkish. It is sometimes found late in the fall. It is said to be edible but seems rather tough.

PLEUROTUS SUBPALMATUS (Fr.) Gill.

Figure 157, page 91

PILEUS 1–2¾ in. broad, fleshy, convex to plane, flesh-colored to brick-red, glabrous, the cuticle gelatinous and forming coarse reticulations on the surface. FLESH reddish, fairly thick. LAMELLAE adnate, close, rather broad, sometimes forked, pinkish. STIPE ½–1½ in. long, ⅛–¼ in. thick, excentric, usually curved, equal, fibrillose, reddish. SPORES pinkish in mass, subglobose, echinulate, 5–7 × 4.5–6.5 μ.

Usually on fallen logs, occasionally on standing trees, singly or gregarious, rare. June–Sept.

There is no information concerning the edible qualities of this mushroom but it is so rare that it is not of any importance as an edible species in any case. It is not a good *Pleurotus*. The pinkish, spiny spores and the peculiar, gelatinous, reticulated cuticle separate it sharply from other species of this genus. It has been made the type of a new genus, *Rhodotus* by Maire.

PLEUROTUS ULMARIUS (Fr.) Kummer — Edible

Figure 187, page 112

PILEUS firm, compact, 2–6 in. broad, convex with inrolled margin, expanding to nearly plane, varying from fairly regular in shape with near-central stipe to irregular with strongly excentric stipe, white to dull buff becoming darker with age, sometimes tinged with yellow or reddish brown shades, moist, glabrous. FLESH thick, white, odor and taste mild. LAMELLAE sinuate-adnexed becoming rounded or notched at the stipe, close to subdistant, broad, white or whitish. STIPE firm, stout, solid, 1–3 in. long, up to ¾ in. thick, sometimes swollen at the base, often curving to bring the pileus into upright position, whitish, varying from glabrous to densely tomentose. SPORES white, smooth, broadly ovoid to subglobose, 5.5–8 × 4.5–6 μ.

Solitary or in clusters on wood of deciduous trees, especially elm. Sept.–Nov.

According to Kauffman, the pileus may at times be somewhat tomentose. In robust specimens the surface sometimes cracks to form a network pattern.

This is the large white mushroom commonly seen in late fall, often high up in a tree, arising from a wound or branch stub. The fruiting bodies do not decay very quickly and sometimes persist until after snowfall. *P. ulmarius* is edible but inclined to be tough, especially in older specimens, and requires careful cooking. According to Singer (1951) the European *P. ulmarius* is a different fungus and the North American species is really *P. tessulatus* (Bull. ex Fr.) Gill.

Considerable variation may occur in the size and position of the stipe: from central to decidedly excentric or almost lateral. Plants of the latter type might be confused with *P. ostreatus*, but the attachment of the lamellae will distinguish them and, if a microscope is available, the size and shape of the spores will provide a sure diagnostic character. Another species was described by Peck as *Pleurotus elongatipes*, which differs from *P. ulmarius* chiefly in having a stuffed to hollow stipe and slightly smaller spores. It seems to be rare but has probably been confused with *P. ulmarius*. This makes no difference from the standpoint of edibility.

CLITOCYBE

Species of *Clitocybe* are mostly white-spored, with decurrent lamellae, and lack a volva and annulus. In some species the spores are pinkish buff or pale yellowish in a good deposit, but these species would not likely be looked for in the yellow-spored group. The stipe is fibrous, more or less similar to the pileus in texture and not separating readily from it. In this respect as well as in the attachment of the lamellae they differ from *Collybia*, in which the stipe is more cartilaginous in texture than the pileus and separates readily.

It is sometimes difficult to distinguish between *Clitocybe* and *Tricholoma* and the attachment of the lamellae is the principal distinguishing character. In *Clitocybe* the lamellae are typically decurrent to adnate whereas in *Tricholoma* they are more or less sinuate to emarginate or adnexed, but since the attachment may vary to some extent at different stages of maturity and in individual fruiting bodies, a clear-cut distinction is not always easy.

Other genera that might be confused with *Clitocybe* are *Cantharellus*, *Laccaria*, *Leucopaxillus* and *Omphalina*. *Cantharellus* differs in having blunt-edged, more or less fold-like, forking lamellae, although species like *Cantharellus umbonatus* and *Clitocybe aurantiaca* make sharp separation difficult. *Laccaria* has globose, spiny spores and lamellae of waxy consistency. Some species of *Leucopaxillus* have the form and stature of a *Clitocybe* but differ in having rough-walled, amyloid spores. *Omphalina* includes a group of small, umbilicate species with decurrent lamellae and cartilaginous stipes. The size of the fruiting body and the texture of the stipe seem to be the principal characters distinguishing them from *Clitocybe* but it seems almost impossible to make a clear-cut separation. No species of *Omphalina* are described here.

Clitocybe is a fairly large genus and many of the species, especially the small whitish forms are difficult to identify. Only a few of the commoner and more striking species are described here. Most of the species appear to be edible but information is lacking concerning several species and at least two, *C. illudens* and *C. dealbata*, are known to be poisonous.

Key

1. Fruiting bodies with strong anise odor;
   pileus greenish, bluish, or white .... *C. odora*
1. Fruiting bodies not with odor of anise 2

2. Pileus and lamellae yellow or yellowish 3
2. Pileus and lamellae not yellow 6

3. Pileus glabrous, growing in dense clusters .... *C. illudens*
3. Pileus not glabrous 4

4. Pileus deeply depressed to infundibuliform;
   lamellae subdistant .... *C. ectypoides*
4. Pileus convex to slightly depressed; lamellae crowded 5

5. Pileus minutely scaly; lamellae yellow .... *C. decora*
5. Pileus fibrillose to subtomentose; lamellae orange .... *C. aurantiaca*

6. Fruiting bodies white or whitish 7
6. Fruiting bodies not white 9

7. Fruiting bodies in notable cespitose clusters .... *C. multiceps*
7. Fruiting bodies occurring singly or gregarious, small 8

8. Fruiting bodies convex, or slightly depressed, grayish white .... *C. dealbata*
8. Fruiting bodies deeply depressed to
   infundibuliform, buff-white .... *C. adirondackensis*

9. Pileus and lamellae gray .... *C. cyathiforme*
9. Pileus and lamellae not gray 10

10. Pileus obtuse, grayish brown; stipe clavate .... *C. clavipes*
10. Pileus deeply depressed to infundibuliform 11

11. Pileus buff-white, usually less than 2 in. broad .... *C. adirondackensis*
11. Pileus reddish tan fading to dingy white,
    usually more than 2 in. broad .... *C. infundibuliformis*

Figures 164-173

164. *Clitocybe illudens.*
165. *C. illudens.*
166. *C. ectypoides.*
167. *C. infundibuliformis.*
168. *C. odora.*
169. *Tricholoma aurantium.*
170. *Leucopaxillus giganteus.*
171. *L. giganteus.*
172. *L. laterarius.*
173. *L. laterarius.*

164
165
166
167
168
169
170
171
172
173

Figure 174. *Pleurotus candidissimus*. Figure 175. *Panus operculatus*.

Figures 176-185

176. *Tricholoma flavovirens*.
177. *T. flavovirens*.
178. *T. irinum*.
179. *T. personatum*.
180. *T. resplendens*.
181. *T. rutilans*.
182. *T. saponaceum*.
183. *T. sejunctum*.
184. *T. vaccinum*.
185. *T. terreum*.

176
177
178
179
180
181
182
183
184
185

Figure 186. *Pleurotus sapidus.*

Figure 187. *Pleurotus ulmarius.*

Figures 188-197

188. *Melanoleuca alboflavida.*
189. *M. melaleuca.*
190. *Hygrophorus borealis.*
191. *H. cantharellus.*
192. *H. chrysodon.*
193. *H. chrysodon*
194. *H. conicus.*
195. *H. flavescens.*
196. *H. marginatus.*
197. *H. miniatus.*

188
189
190
191
192
193
194
195
196
197

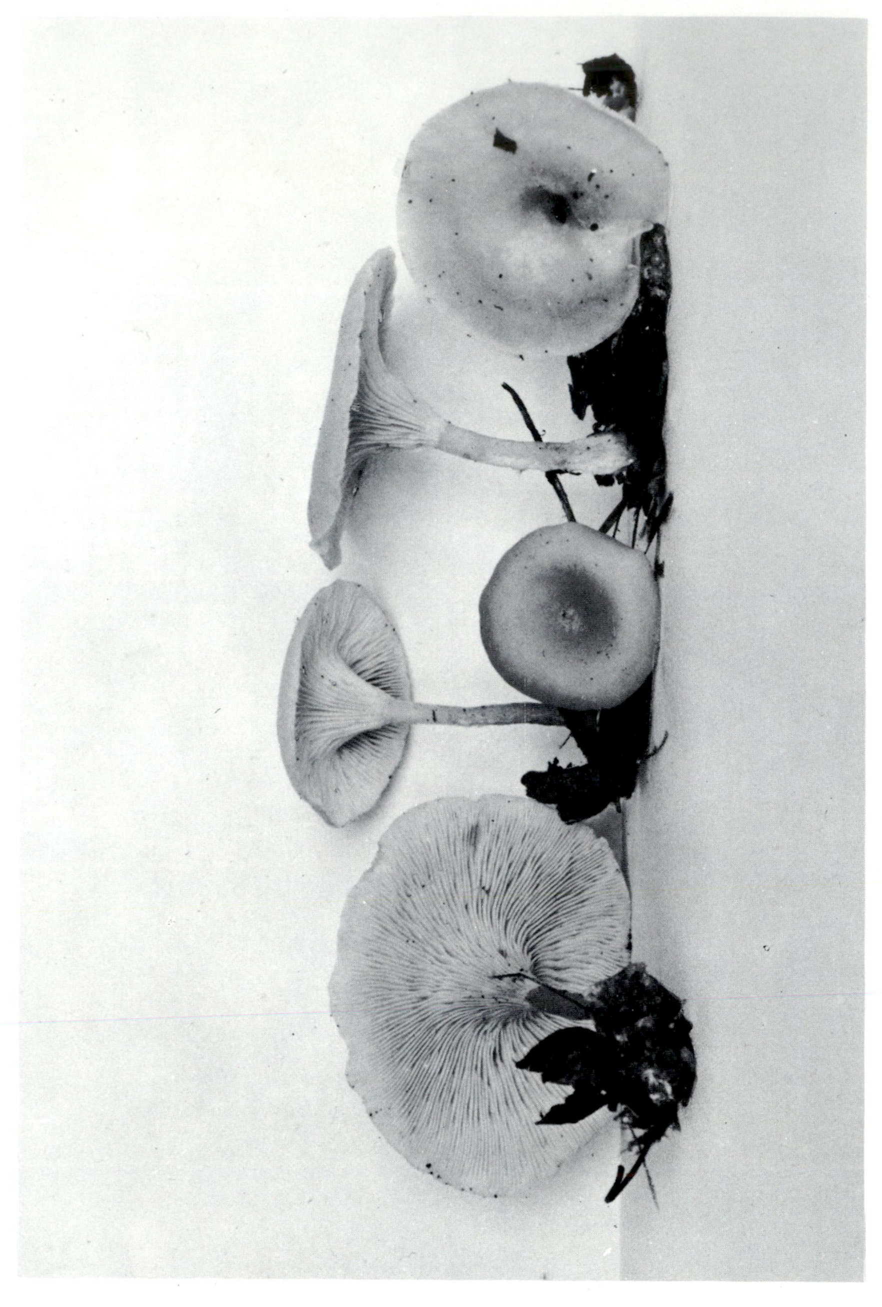

Figure 198. *Clitocybe adirondackensis.*

Figures 199-208

199. *Hygrophorus nitidus.*
200. *H. nitidus.*
201. *H. olivaceoalbus.*
202. *H. olivaceoalbus.*
203. *H. pratensis.*
204. *H. psittacinus.*
205. *H. pudorinus.*
206. *H. speciosus.*
207. *H. puniceus.*
208. *H. puniceus.*

199
200
201
202
203
204
205
206
207
208

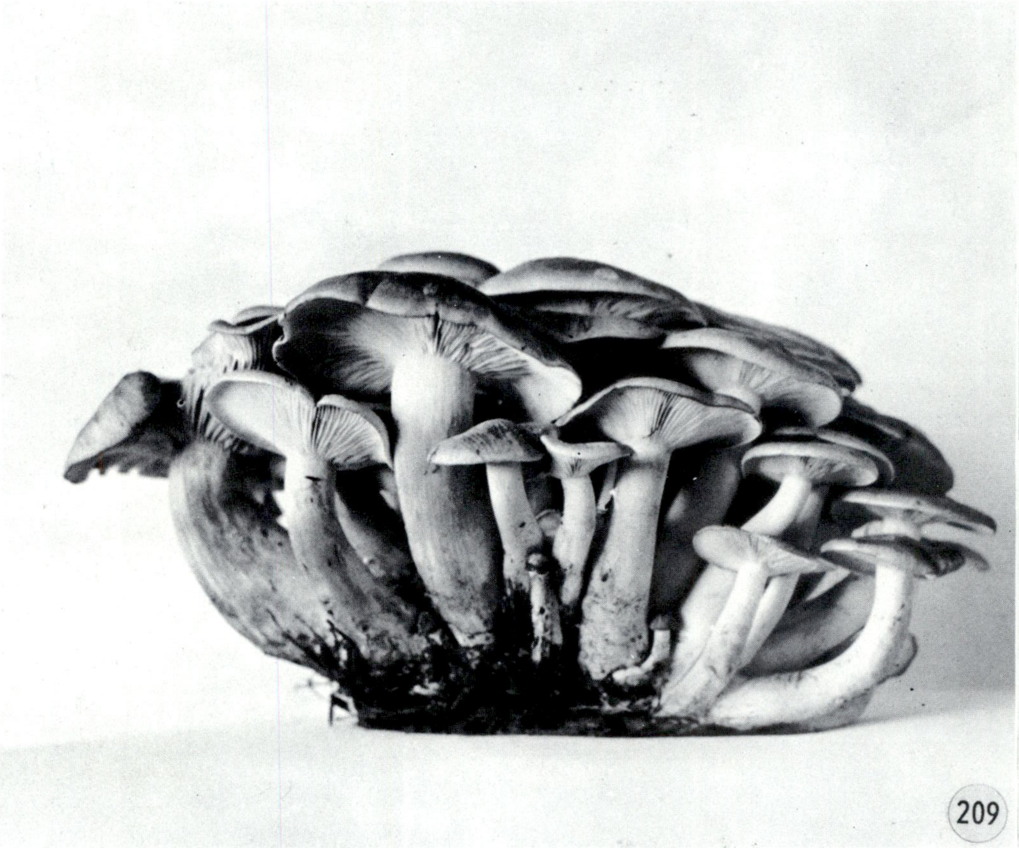

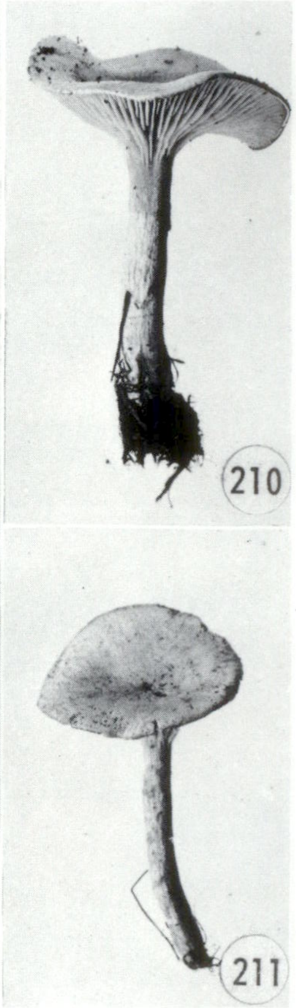

Figure 209. *Clitocybe multiceps.*

Figures 210-211. *Clitocybe dealbata.*

CLITOCYBE ADIRONDACKENSIS (Pk.) Sacc.

Figure 198, page 114

PILEUS 1–2 in. broad, thin, pliant, somewhat funnel-shaped with depressed disk and elevated, decurved margin, dull whitish at first, then tinged buff, sometimes darkening to a dingy buff-brown, glabrous, not hygrophanous, slippery-smooth when moist, margin even. FLESH thin, white. LAMELLAE decurrent, close to crowded, very narrow and tapering toward each end, whitish. STIPE 1–1 ¾ in. long, about ⅛ in. thick, straight, equal or slightly swollen at the base, pruinose, concolorous with the pileus or paler, stuffed, becoming hollow. SPORES smooth, white, oval, apiculate, 4–6 × 2.5–3.5 μ.

Scattered, on the ground among debris. Aug.–Oct.

The pallid color, rubbery-pliant consistency, and slippery, smooth pileus are characteristic. It is a small species, not likely to be collected for food, and its edible qualities are not known.

CLITOCYBE AURANTIACA (Fr.) Studer — Doubtful

Figure 158, page 91

PILEUS 1–3 in. broad, fleshy, pliant, convex to plane, becoming depressed in age, orange-yellow to brownish orange, fibrillose to subtomentose, sometimes nearly smooth, margin at first involute, then elevated and somewhat wavy. FLESH soft, thick in the center, thin on the margin, yellowish, odor and taste mild. LAMELLAE decurrent, crowded, forked, thin, narrow, bright orange to salmon-tinged. STIPE 1–2 in. long, ¼–½ in. thick, tapering upward, minutely tomentose, pale orange, varying to brownish or pale yellowish, spongy within, sometimes hollow. SPORES white, elliptical, smooth, 5–7 × 3–4 μ.

Gregarious on the ground or rotten wood, in both coniferous and frondose woods. July–Oct.

This species has long been known as *Cantharellus aurantiacus*. The forked lamellae suggest *Cantharellus* but in other respects it is not closely related to other *Cantharellus* species. Some authors consider it better placed in *Clitocybe* although Singer (1951) considers it to be more closely related to *Paxillus* and treats it and one other southern species in the genus *Hygrophoropsis*.

It varies considerably in color from pale yellow to dark brown, but the crowded, forked, orange lamellae are distinctive.

There have been conflicting reports in the literature concerning its edibility, some maintaining that it is poisonous and others that they have eaten it quite safely. In view of this doubt it is not recommended or should be tried only with caution, although Singer (1951) states that its edibility has been established.

CLITOCYBE CLAVIPES (Pers. ex Fr.) Kummer — Edible

Figure 159, page 91; Figure 414, page 296

PILEUS ¾–2¾ in. broad, fleshy, convex becoming expanded, often obtusely umbonate, sometimes centrally depressed, drab grayish brown, smooth and glabrous, extreme margin tending to remain incurved for some time. FLESH white, thick at the disk, odor and taste mild. LAMELLAE decurrent, subdistant, white or yellowish, broadest in the center, narrowing toward the ends. STIPE stout, ¾–2½ in. long, ¼–⅜ in. thick at the apex, broadening downward into a clavate-bulbous base ½–1 in. thick, concolorous with the pileus, white and somewhat spongy within. SPORES smooth, white, ellipsoid, 6–8 × 3.5–5 μ.

In groups or occasionally in clusters of two or three on the ground in woods, often under conifers. July–Oct. Fairly common.

The broadly clavate, gray-brown stipe and the subdistant lamellae are the distinguishing marks of this *Clitocybe*.

Clitocybe nebularis (Fr.) Kummer is a large cloudy-gray species with crowded lamellae. It is apparently more common on the west coast than in the East and collectors in the West are likely to find it. It may reach 6 inches in diameter. Kauffman reported it to be edible although he noted that some European authors had considered it dangerous. There are no collections of this species from Eastern Canada in the herbarium.

CLITOCYBE CYATHIFORMIS (Bull. ex Fr.) Kummer

PILEUS ¾–2 in. broad, at first convex, soon umbilicate-depressed to infundibuliform, smoky brown when moist becoming more grayish when dry, hygrophanous, glabrous or innately fibrillose, margin even, inrolled. FLESH thin, grayish, rather watery, taste mild. LAMELLAE decurrent, narrow, close to subdistant, grayish brown. STIPE 1–2 in. long, ⅛–¼ in. thick, equal or tapering upward, brownish or grayish, fibrillose, tomentose at base, spongy-stuffed. SPORES smooth, white, elliptical-ovate, 7.5–10.5 × 5.0–6.5 μ.

Singly or gregarious on rotten wood. Aug.–Sept.

The gray lamellae are the most striking character of this species, and a spore print is required to make certain of the color of the spores. *Clitopilus noveboracensis* Pk. is somewhat similar in appearance but has pink spores, a farinaceous odor, and a bitter taste. The spores of *C. cyathiformis* are amyloid and Singer has on this account removed it from *Clitocybe* and placed it in the genus *Cantharellula* along with *Cantharellus umbonatus* and *Clitocybe ectypoides*.

CLITOCYBE DEALBATA (Sow. ex Fr.) Kummer — Poisonous

Figures 210, 211, page 116

PILEUS ½–1½ in. broad, rarely larger, thin, hygrophanous, convex, obtuse, expanding to plane, sometimes depressed, pallid, whitish when dry, pale grayish brown when moist, smooth, glabrous, margin tending to remain incurved. FLESH thin, whitish, odor none, taste mild or slightly astringent. LAMELLAE adnate-decurrent, fairly close, narrow, broadest in the center, narrowing to each end, whitish to pallid. STIPE short, ¾–1½ in. long, about ⅛ in. thick, sometimes excentric, equal or nearly so, straight or curved, tough, subfibrillose, pruinose, concolorous with pileus, solid. SPORES smooth, white, ellipsoid, 4–5 × 2.5–3 μ.

Solitary or scattered in grass on lawns, etc. Aug.–Oct.

This poisonous little mushroom is dangerous because of its habit of growing in or near rings of *Marasmius oreades*, which is often collected for food. The color of the pileus is somewhat similar in the two species although the *Clitocybe* is whiter, but a glance at the lamellae will quickly distinguish the two.

Clitocybe morbifera Pk. and *C. sudorifica* Pk. appear to be synonyms of *C. dealbata*. *C. rivulosa* (Fr.) Kummer is a closely related species more pinkish in color and it is also said to be poisonous.

CLITOCYBE DECORA (Fr.) Gillet — Edible

Figure 160, page 91

PILEUS 1–2½ in. across, convex becoming expanded, finally plane or depressed at disk, surface moist, yellow to ochre or brownish ochre, sometimes with a slight olive cast, covered with very minute gray to dark brown fibrillose scales which are dense on the disk and more scattered toward the margin, margin thin, incurved, tomentose. FLESH rather thin, firm, yellow, odor and taste mild. LAMELLAE mostly decurrent with a narrow tooth, sometimes merely adnate or adnexed, seceding readily, close, moderately broad except for decurrent portion on stipe, yellow, often with alternate lamellae short. STIPE 1½–2½ in. long, ⅛–⅜ in. thick, subequal, yellow, subglabrous to slightly fibrillose-scaly, becoming hollow. SPORES smooth, white, broadly oval, 6–7.5 × 4–5 μ.

Single, in small clusters of several, or in groups, on decaying coniferous wood. June–Oct.

The attachment of the lamellae is not typical for the genus *Clitocybe*, and the color of the lamellae might lead one to look for it in *Flammula*. A spore print should be obtained.

This species is considered to be closely related to *Tricholoma rutilans* (Fr.) Kummer, and has been called *Tricholoma decorum* by Quélet. Singer places both these species in a new genus, *Tricholomopsis*.

CLITOCYBE ECTYPOIDES Pk.

Figure 166, page 109

PILEUS ¾–2 in. broad, rather thin, broadly umbilicate, approaching funnel-form, watery, gray-buff to yellow-buff, virgate with brown or dark brown fibrils, usually slightly punctate with minute dark scales, margin even. FLESH rather thin, tinged the color of the pileus. LAMELLAE decurrent, occasionally forking, subdistant, narrow, tapering at each end, yellowish. STIPE ¾–2 in. long, about ⅛ in. thick, equal or tapering slightly upward or downward, concolorous with pileus or paler, densely tomentose at base, less tomentose to subglabrous upward, solid when young, often becoming hollowed by grubs. SPORES smooth, white, ellipsoid, 7–9 × 4–5 μ.

In groups or small clusters on decaying logs. July–Sept.

The subdistant, forking lamellae in this species might lead one to look for it in *Cantharellus*. The spores are amyloid, and Singer has placed this species in the genus *Cantharellula* along with *Cantharellus umbonatus*. The minute, dark scales in the center of the pileus are a valuable diagnostic character.

CLITOCYBE ILLUDENS Schw. — Poisonous

Figures 164, 165, page 109

Jack-o'-lantern

PILEUS 2–4½ in. broad (large specimens reaching 8 inches), sometimes irregularly lobed especially in crowded situations, at first shallow-convex with umbonate disk and inrolled margin, becoming expanded-convex with depressed disk, the umbo sometimes persisting within the depression, bright orange-yellow, smooth, at first moist, becoming dry, more or less innately fibrous-streaked, pellicle rather tough, separable, margin at length elevated and wavy but with the extreme edge remaining incurved. FLESH very thin except on the disk, fibrous-pithy, tinged orange, drying whitish, continuous with the context of the stipe, odor strong, sweetish, pleasant. LAMELLAE unequally long-decurrent, close, rather narrow for so large a mushroom, narrowing at each end, occasionally forking, brittle, orange-yellow, either brighter or paler than the pileus, phosphorescent in the dark. STIPE stout, tough, 3–8 in. long, ⅜–¾ in. thick, tapering at the base, often twisted and contorted, central or slightly excentric, solid, surface finely white-pruinose at first, smooth and dry, fibrillose, tinged pinkish orange. SPORES creamy white, smooth, globose, 4–5 μ.

In dense clusters at base of old stumps, sometimes from buried wood. July–Sept.

This mushroom is remarkable for its phosphorescent properties. Although apparently not common in Canada, its bright colors and its habit of growing in large clusters make it very conspicuous when it is present. Over 100 fruiting bodies may grow in a single cluster. The odor apparently varies, as it is reported as unpleasant by some authors.

It is sometimes mistaken for *Armillaria mellea* or *Cantharellus cibarius*. The much brighter colors and lack of an annulus should distinguish it easily from *A. mellea* and the crowded narrow lamellae and the habit of growing in dense clusters separate it from *C. cibarius*. Neither *A. mellea* nor *C. cibarius* is phosphorescent, but old or dried-out specimens of *Clitocybe illudens* will not show this character either, so that it cannot be relied on by itself as a certain means of identifying *C. illudens*.

It is not considered to be a deadly poisonous mushroom but will certainly cause more or less severe illness in most people if it is eaten. The flavor is said to be good so that this is no guide to its poisonous properties.

CLITOCYBE INFUNDIBULIFORMIS (Schaeff. ex. Fr.) Quél. Edible

Figure 167, page 109

PILEUS 1¾–2¾ in. broad, at first convex with subumbonate disk, later becoming depressed on the disk and finally funnel-shaped, reddish tan, becoming faded, finely silky, margin thin. FLESH white, thin on the margin, thicker on the disk, odor and taste not distinctive. LAMELLAE decurrent, close, not very broad, tapering to each end, thin, whitish. STIPE 1½–3 in. long, ⅛–¼ in. thick, often slightly larger at the base, elsewhere subequal, white-mycelioid at the base, glabrous upward, concolorous with pileus or pallid. SPORES smooth, white, ovoid to ellipsoid or subpyriform, obliquely apiculate, 5–8 × 3–4 μ, with many smaller immature spores present.

Solitary or in groups on the ground in woods. July–Oct. Fairy common.

This is the type species of *Clitocybe*, and illustrates well the characteristics of the genus. It is a fairly common species occurring throughout a long period of the growing season. The color may fade to nearly white. *C. gibba* (Fr.) Kummer is probably the correct name for this species.

CLITOCYBE MULTICEPS Pk. Edible

Figure 209, page 116

PILEUS 1–3 in. broad or larger, convex, sometimes irregular in outline when growing in a crowded cluster, whitish, often tinged with gray or buff, moist, glabrous, margin thin. FLESH white, thick on disk. LAMELLAE adnate to short-decurrent, or slightly sinuate, close, moderately broad in center, narrowing to each end, whitish. STIPE 2–5 in. long, ¼–½ in. thick, stout, subequal or tapering upward slightly, pruinose at apex, elsewhere subglabrous to fibrillose-scaly, whitish, solid within, central or slightly excentric in crowded clusters. SPORES smooth, white, globose, 5–7 μ broad.

In clusters, often densely crowded, on the ground in grassy places or open woods. July–Oct.

Kauffman reports this mushroom tò be edible but does not recommend it very highly. Güssow and Odell also describe it as rather insipid.

Clitocybe cartilaginea Bres. is somewhat similar in habit but is darker in color, gray to brown, and has a cartilaginous cuticle.

CLITOCYBE ODORA (Bull. ex Fr.) Kummer — Edible

Figure 168, page 109

PILEUS 1–3 in. broad, convex with margin incurved at first, becoming expanded, varying in color from bluish green or grayish green to whitish, tinged green or entirely lacking the green tints, smooth and glabrous. FLESH white, thin toward the margin, odor sweet, fragrant, sometimes faint. LAMELLAE broadly adnate to subdecurrent or short-decurrent, close, moderately broad, white to creamy yellowish or tinged green. STIPE 1–3 in. long, ⅛–¼ in. thick, equal or slightly thickened at the base, whitish to pallid, concolorous, pruinose above, white-mycelioid at base, becoming hollow. SPORES smooth, white, oval, 6–8 $\times$ 4–5 μ.

Solitary or in clusters of 2 or 3, on the ground in woods, often attached to leaves and debris. July–Oct.

The greenish colors and the fragrant odor of anise are the distinguishing marks of this species, but the greenish color is sometimes entirely lacking and the fruiting bodies may then be white. A smaller and thinner species *C. fragrans* (Sow. ex Fr.) Kummer has a similar odor.

LEUCOPAXILLUS

Leucopaxillus is rather difficult to define in such a way that the amateur collector can readily recognize the genus by the gross appearance, but it is fairly easy to determine by microscopic characters. It includes *Clitocybe*-like or *Tricholoma*-like species with rough-walled spores that turn blue in iodine. These species are mostly whitish or dull colored, with fairly large to large pilei and fleshy stipes. The attachment of the lamellae varies from decurrent to sinuate. Unless a microscope is available the beginner will have some difficulty at first in recognizing a *Leucopaxillus* and will be inclined to look in *Clitocybe* or *Tricholoma*.

Singer & Smith (1943) published a monograph on the genus and recognized twelve species. Some of the species they included in *Leucopaxillus* were formerly known as *Clitocybe gigantea* (Fr.) Quél., *Tricholoma laterarium* (Pk.) Sacc., *T. tricolor* Peck, *Clitocybe albissima* (Peck) Sacc., and as varieties of the latter, *C. piceina* Peck, *C. subhirta* Peck, and *Tricholoma lentum* (Post in Romell) Sacc.

LEUCOPAXILLUS ALBISSIMUS (Pk.) Sing.
var. **PICEINUS** (Peck) Singer & Smith — Edible

Figure 222, page 134

PILEUS 2–4 in. broad, sometimes larger, convex becoming plane or nearly so, dry, glabrous to slightly fibrillose, especially toward the margin, whitish to cream, or pale tan on the disk, margin inrolled at first, sometimes irregular and ribbed. FLESH white, firm, rather thick, taste bitter and disagreeable, odor unpleasant. LAMELLAE short-decurrent with anastomosing lines at apex of stipe, narrow, close to subdistant, whitish to yellowish in age, separating readily from the context. STIPE 1¼–3¼ in. long. ¼–1 in. thick, at first bulbous and tapering upward, then elongating and becoming nearly equal, white or tinged buff, glabrous or fibrillose to strigose toward the base, solid. SPORES white, ellipsoid, rough, amyloid, 5.5–8 × 4.5–5 μ.

Usually gregarious to single on beds of needles in conifer woods. Aug.–Oct.

The bitter taste, anastomosing ridges at the apex of the stipe and the development of yellowish colors distinguish this variety. *L. albissimus* var. *albissimus* is pure white. The fruit bodies persist a long time without decaying. It has been reported edible in spite of the bitter taste.

LEUCOPAXILLUS GIGANTEUS (Fr.) Sing. — Edible

Figures 170, 171, page 109

PILEUS 4–12 in. or more broad, at first convex to plane, then becoming depressed and finally infundibuliform, dry or slightly moist, glabrous to slightly pubescent at the margin, whitish to buff or tan, margin at first inrolled, then spreading and becoming ribbed, sometimes splitting. FLESH white or whitish, thick, firm becoming softer, taste mild. LAMELLAE slightly decurrent, crowded, at first whitish, darkening with age, narrow to moderately broad, separable from the context. STIPE 1¾–3 in. long, ¾–2 in. thick, short, stout, swollen at the base, glabrous, white or colored like the pileus, solid. SPORES white, ellipsoid, nearly smooth, slightly amyloid, 5.5–8 × 3–5.5 μ.

Singly or in small groups on the ground in woods or open places. Aug.–Oct.

This species is remarkable because of the large size it sometimes reaches but smaller specimens may be recognized by the color, crowded lamellae and short stipe.

L. tricolor (Pk.) Kühner is another species that reaches considerable size, sometimes exceeding 12 inches in diameter. It is yellowish to pale tan, dry and unpolished, sometimes appearing matted-fibrillose, and usually more or less ribbed or grooved on the margin. The lamellae are close to crowded and separate readily from the pileus. They are whitish to yellowish when fresh but on drying change to vinaceous or purplish. Kauffman called this fungus *Clitocybe maxima*. Its edible qualities appear to be unknown.

LEUCOPAXILLUS LATERARIUS (Peck) Singer & Smith — Not edible

Figures 172, 173, page 109

PILEUS 2–4 in. broad or sometimes larger, convex to plane, sometimes umbonate, dry, slightly fibrillose to scurfy, white to faintly pinkish, sometimes yellowish on the disk, margin inrolled, slightly ridged. FLESH white, thick, firm, taste very bitter, odor farinaceous to disagreeable. LAMELLAE adnate to sinuate, decurrent by lines, narrow, crowded, white to pale cream. STIPE 1½–4 in. long, ¼–¾ in. thick, nearly equal or enlarged at the base, white, at first pruinose to finely tomentose, becoming fibrillose, solid. SPORES white, subglobose to globose, slightly rough, amyloid, 3.5–5.5 × 3.5–4.5 μ.

Gregarious to subcespitose on the ground in frondose woods. June–Oct.

The bitter taste, non-anastomosing, close, narrow lamellae, and ridged or ribbed margin are characteristic of this species. It is fairly common, and will attract the collector because of its size, but it has a very unpleasant taste.

TRICHOLOMA

Tricholoma is a large genus and the species are often difficult to identify. Usually they are fairly large forms growing on the ground and often appearing late in the season. The genus is characterized by the white spores, fleshy stipes, lamellae adnexed to sinuate and frequently notched at the stipe, and the lack of a volva or annulus. The type species of the genus is *T. flavovirens* (Fr.) Lundell. Some of the species included here, such as *T. personatum* and *T. irinum*, have a creamy to dirty pinkish spore deposit and most authors now put these in the genus *Lepista.*

Only some of the more common and easily recognized species are included here. Many of the species are edible and highly prized, but some are disagreeable in flavor and a few are known to be poisonous. Therefore, only those species that can be definitely identified and are known to be edible should be used.

Key

1. Pileus viscid 2
1. Pileus not viscid 6

2. Fruiting body white .... *T. resplendens*
2. Fruiting body not white 3

3. Pileus yellow or yellowish 4
3. Pileus without yellow in the coloration 5

4. Lamellae yellow; pileus yellowish, usually reddish on disk .... *T. flavovirens*
4. Lamellae white; pileus yellowish with innate black fibrils .... *T. sejunctum*

5. Stipe peronate with reddish orange scales .... *T. aurantium*
5. Stipe smooth to minutely fibrillose .... *T. pessundatum*

6. Lamellae yellow; pileus tomentose-scaly, purple-red .... *T. rutilans*
6. Lamellae white or becoming reddish stained 7

7. Lamellae stained with reddish spots; pileus fibrillose-scaly, dark reddish brown .... *T. vaccinum*
7. Lamellae not reddish spotted 8

8. Pileus with prominent, acute umbo, gray .... *T. subacutum*
8. Pileus not umbonate 9

9. Pileus fibrillose to scaly, gray .... *T. terreum*
9. Pileus glabrous 10

10. Spore deposit white; pileus pale gray usually tinged olive; flesh turning pinkish; taste disagreeable, soapy .... *T. saponaceum*
10. Spore deposit creamy to dirty pinkish 11

11. Pileus and lamellae more or less tinged with blue or lavender .... *T. personatum*
11. Pileus whitish to buff, no blue or lavender shades .... *T. irinum*

TRICHOLOMA AURANTIA (Schaeff. ex Fr.) Ricken

Figure 169, page 109

PILEUS 1–3 in. broad, convex, becoming expanded, slightly umbonate, reddish ochraceous to orange-red, rather bright colored, viscid, soon becoming appressed-scaly, margin floccose, glutinous, inrolled at first. FLESH white, thick on disk, thin on margin, odor farinaceous. LAMELLAE adnexed, close, white, becoming spotted with rusty brown, a few forked. STIPE 1½–2½ in. long, ¼–½ in. thick, equal or narrowed at the base, more or less covered with rings of scales the same color as the pileus up to an obscure annular zone, white at apex and between the scales, solid. SPORES white, broadly ellipsoid to ovoid, 4.5–6 × 3–4 μ.

Usually gregarious on the ground. Aug.–Oct.

Because of the suggestion of an annulus in this species one might be inclined to look for it in *Armillaria* and some authors have placed it in this genus. However, the structure of the trama of the lamellae indicates that it is more closely related to *Tricholoma* than to *Armillaria*. It is a rather bright colored species and can be recognized by the characteristic scaliness on the stipe, the viscid pileus and the lamellae staining brownish. Its edible qualities are not known.

TRICHOLOMA FLAVOVIRENS (Fr.) Lundell — Edible

Figures 176, 177, page 111

PILEUS 2–4 in. broad, compact, fleshy, convex, expanding, sometimes obtuse on the disk, pale to bright yellow, usually stained brownish or reddish on the disk, viscid, glabrous or faintly scaly on the disk, incurved on the margin at first. FLESH white or tinged yellow, odor not distinctive, taste slightly unpleasant. LAMELLAE free or almost so, rounded behind, rather broad, close to crowded, sulphur-yellow. STIPE stout, 1–2½ in. long, ¼–¾ in. thick, equal or slightly thickened at the base, pale yellow or white, solid, smooth or slightly scaly. SPORES smooth, white, ellipsoid, 6–7 × 4–4.5 μ.

In groups, on the ground in conifer woods. Sept.–Oct.

This species has been well known by the name *Tricholoma equestre* (Fr.) Kummer, but under the International Code of Botanical Nomenclature, *T.*

flavovirens is the correct name. It is a striking species with its bright yellow pileus and lamellae. It might be confused with *T. sejunctum* but the latter has usually dark radiating lines on the pileus whereas *T. flavovirens* tends to become stained reddish or brownish on the disk and is usually a more robust species. The lamellae of *T. sejunctum* are usually white or whitish but may show some yellow. *T. sulphureum* (Fr.) Kummer, another yellow species, is not viscid and is characterized by a disagreeable odor resembling coal tar.

TRICHOLOMA IRINUM (Fr.) Kummer — Edible

Figure 178, page 111

PILEUS 1½–6 in. broad, fleshy, convex becoming expanded-plane, glabrous, not viscid, pale alutaceous, tinged flesh color to nearly white, margin at first inrolled, then spreading. FLESH thick, firm, whitish, taste mild. LAMELLAE sinuate to adnexed, crowded, whitish or nearly the same color as the pileus. STIPE ¾–2 in. long, ¼–¾ in. thick, equal or bulbous at the base, fibrillose-striate, whitish, solid. SPORES ellipsoid, smooth or some minutely rough, pale cream in mass, 7–9 × 3.5–4.5 μ.

Usually gregarious in troops on the ground in woods or open places. Sept.–Oct.

This species has the appearance and stature of *T. personatum* but entirely lacks the violet or lilac colors. According to Singer the true *T. irinum* has smooth white spores but the species described above is evidently the fungus described and illustrated by Lange and other European authors as *T. irinum.* The spores appear smooth under ordinary magnifications but under oil immersion a few seem to be minutely roughened.

T. irinum may be found growing in large fairy rings in the woods. I have seen such rings 18, 24 feet, and one even 50 feet in diameter, forming almost perfect circles and containing hundreds of fruiting bodies. When a tree is in the way the mycelium appears to grow around each side of the tree and join up again on the other side.

TRICHOLOMA PERSONATUM (Fr. ex Fr.) Kummer — Edible

Figure 179, page 111

PILEUS 2–5 in. broad, sometimes larger, at first broadly convex, becoming plane or slightly umbonate, glabrous, moist, becoming subviscid and water-soaked in wet weather, grayish to brownish, tinged more or less with lilac, fading to buff or whitish, margin at first inrolled, pruinose, then spreading and often wavy and irregular. FLESH whitish, tinged lavender, becoming water-soaked in wet weather, taste mild. LAMELLAE sinuate to adnexed, close to crowded, rather broad, at first blue, becoming buff-lilac to grayish buff, usually with tinge of lilac, with two or three tiers of shorter ones interspersed. STIPE 1–3 in. long, ½–1 in. thick, equal or often somewhat bulbous at base, solid, pale lilac or bluish, fading to pallid whitish, fibrillose-pruinose, becoming glabrous, sometimes striate. SPORES elliptical, minutely rough, pale dirty flesh color, 7–8 × 4–5 μ.

Singly, in groups, or in clusters of several, on the ground, often under old leaves, in the woods or on decaying vegetable matter. Sept.-Oct.

This species is rather common and is one of the better edible species, although it may have a disagreeable flavor if the pilei are old and water-soaked. It varies considerably in color but there is always some lilac.

According to European accounts, *T. personatum* exhibits the blue color only in the stipe, and specimens with blue pilei and lamellae are referred to *T. nudum* (Bull. ex Fr.) Kummer. We have referred a smaller species with much deeper color and slightly differently colored spore print to *T. nudum*. The exact identity of these species is somewhat in doubt but they are all edible. This group of species in which the spore print is not pure white and at least some of the spores are minutely roughened has been separated from *Tricholoma* as the genus *Lepista*.

TRICHOLOMA PESSUNDATUM (Fr.) Quélet — Edible

Figure 113, page 51

PILEUS 2–4 in. broad, at first convex becoming expanded, reddish bay to reddish brown or rufous tan, paler toward the margin to a whitish flesh color, viscid, glabrous, margin inrolled at first. FLESH white, tinged reddish, firm, odor and taste farinaceous. LAMELLAE sinuate to adnate or decurrent by a tooth, crowded, white staining reddish. STIPE 1 ¼–3 in. long, ¼–1 in. thick, equal to slightly bulbous and narrowed below the bulb, glabrous or with a few fibrils, whitish or becoming stained reddish brown, solid. SPORES white, ovoid-ellipsoid, 4.5–6 × 2.5–4 μ.

Gregarious, in troops under conifers. Sept.–Oct.

This species is included as the representative of a group of somewhat similar species occurring with conifers, especially pines, and with reddish brown pilei and the lamellae staining reddish. From our herbarium records this appears to be the commonest species.

T. flavobrunneum (Fr.) Kummer also has a farinaceous odor but the stipe is sulphur-yellow within and the lamellae are pale sulphur-yellow. *T. albobrunneum* (Fr.) Kummer has a faint farinaceous odor, the pileus is minutely striate with innate fibrils, and the stipe is somewhat squamulose above. *T. transmutans* Peck has a farinaceous odor but the taste of the surface of the pileus is bitter. *T. ustale* (Fr.) Kummer lacks the farinaceous odor and the pileus is glabrous and viscid, whereas *T. imbricatum* (Fr.) Kummer, which also lacks the farinaceous odor, is dry and the surface of the pileus breaks up into rather coarse, more or less imbricate scales.

TRICHOLOMA RESPLENDENS (Fr.) Quél. — Edible

Figure 180, page 111

PILEUS 1½–3 in. broad, convex to plane, white, viscid, glabrous. FLESH white, rather soft, odor and taste mild. LAMELLAE adnexed, emarginate, close, white, fairly broad. STIPE 1½–3 (5) in. long, ¼–⅝ in. thick, equal or tapering downward, glabrous, dry, white, solid or becoming hollow. SPORES white, elliptical, smooth 5–7 × 3.5–5 μ.

Singly or gregarious on the ground, usually in hardwood forests. Aug.–Oct.

This species is similar in stature to *T. sejunctum* but is pure white.

TRICHOLOMA RUTILANS (Schaeff. ex Fr.) Kummer — Edible

Figure 181, page 111

PILEUS 1½–3 in. broad, campanulate-convex, expanding to nearly plane, sometimes broadly umbonate, dry, covered with a dense brick-red to wine-colored tomentum which separates into tomentose scales exposing yellowish flesh between, incurved on the margin at first. FLESH yellow, thin at the margin, thick at the disk, taste mild. LAMELLAE adnate, becoming rounded at the stipe, crowded, rather narrow to moderately broad, clear yellow, floccose on the edges. STIPE 2–4 in. long, ¼–½ in. thick, equal or nearly so, stuffed in the center, becoming hollow, often curved, yellow, dotted with minute wine-colored tomentose scales which may partly or almost entirely disappear, yellow within. SPORES smooth, white, broadly ovoid, 6–7 × 3.5–5 μ.

Solitary or slightly clustered, on conifer wood, sometimes apparently on the ground. June–Sept.

It can be easily distinguished by its purple-red scales, yellow flesh and lamellae, and by its habitat on wood. It is not a typical *Tricholoma* in many ways and has been made the type of a new genus *Tricholomopsis* by Singer.

It might be confused with *Clitocybe decora* (Fr.) Gill. also growing on wood and having yellow lamellae, but the scales of the latter are blackish and the attachment of the lamellae is different. *C. decora* has also been placed in *Tricholomopsis* and these two species are undoubtedly closely related.

TRICHOLOMA SAPONACEUM (Fr.) Kummer — Not edible

Figure 182, page 111

PILEUS ¾–3 in. broad, convex becoming expanded, variable in color, pale gray or pale brown, usually more or less tinged olive or greenish, darker on disk, glabrous or becoming cracked, not viscid, margin incurved. FLESH white becoming pinkish, thick, firm, odor and taste rather soapy, disagreeable. LAMELLAE adnate, emarginate, with a decurrent tooth, subdistant, rather broad, whitish. STIPE 1½–3 in. long, ¼–¾ in. thick, equal to ventricose, white, becoming pinkish within, glabrous to minutely floccose, solid. SPORES white, elliptical to ovoid, 5.5–7 × 3.5–5 μ.

Singly or gregarious on the ground in mixed woods. Aug.–Oct.

The gray-green pilei, the flesh staining pink, and the very disagreeable odor and taste are the distinguishing characters of this species. Sometimes the odor is not very pronounced.

TRICHOLOMA SEJUNCTUM (Sow. ex Fr.) Quél. Edible

Figure 183, page 111

PILEUS 1½–3 in. broad, fleshy, convex becoming expanded, umbonate, often somewhat irregular in outline, smooth, slightly viscid, white to yellowish, streaked with minute, radiating, innate, blackish fibrils, in dark specimens with the entire disk umber-brown to blackish. FLESH white or tinged yellow, taste bitter to nauseous. LAMELLAE broad, white, adnate, becoming notched at the stipe, close to moderately distant. STIPE stout, 2–4 in. long, ¼–½ in. thick, more or less equal, often curved, usually solid, smooth, white to straw-colored. SPORES smooth, white, broadly ovoid, 6–7 × 4–5.5 μ.

In groups, on the ground in woods. Sept.–Oct.

T. sejunctum differs from *T. flavovirens* in its fibrous-streaked pileus and white lamellae. Specimens with a more or less yellow tinge in the lamellae have been found and these might be *T. intermedium* Peck but they were growing along with typical forms of *T. sejunctum* and it seems more likely that this yellow color in the lamellae represents a variation rather than a distinct species.

TRICHOLOMA SUBACUTUM Pk. Doubtful

Figure 233, page 136

PILEUS 1½–3 in. broad, at first conic-campanulate with incurved margin, expanding to broadly convex with a conspicuous acute umbo, varying in color from slate-gray or pale gray to ashy or grayish brown, sometimes blackish on the umbo, fading toward the margin, streaked with minute, radiating, dark fibrils, dry, glabrous or slightly fibrillose-scaly. FLESH thin except at the umbo, white, odor not distinctive, taste somewhat acrid. LAMELLAE adnexed, close, broad, white. STIPE 2–4 in. long, ¼–½ in. thick, equal, smooth or slightly fibrillose-scaly, solid, white. SPORES smooth, white, broadly ovoid, 6–7.5 × 4.5–5 μ.

In groups, on the ground in conifer woods. Sept.–Oct.

Apparently there are several closely related forms similar to *T. subacutum*. There is some disagreement as to their edibility but none are known to be dangerously poisonous. The gray color and prominent, acute umbo are the most distinctive characters. *T. virgatum* (Fr.) Kummer is probably the same fungus and, if so, this would be the correct name for it.

TRICHOLOMA TERREUM (Schaeff. ex Fr.) Kummer Edible

Figure 185, page 111

PILEUS 1–2½ in. broad, convex to expanded, subumbonate, gray or mouse-colored, dry, fibrillose, becoming fibrillose-scaly to slightly floccose, with the scales concolorous or dark gray to sooty. FLESH thin, fragile, white, grayish beneath the cuticle, odor and taste mild. LAMELLAE adnexed, close, moderately broad, white to dingy. STIPE short, 1–2 in. long, ⅛–¼ in. thick, equal, solid or stuffed, white or grayish. SPORES smooth, white, broadly ovoid, 6–7.5 × 3.5–5.5 μ.

In groups, or clusters of several, on the ground, in open woods. July–Oct.

There are a number of closely related forms in the *T. terreum* group that are puzzling to distinguish from one another. Just how many species or forms are involved in this complex is uncertain. The species described above is fairly common and is characterized by the gray color and fibrillose to scaly pilei. *T. myomyces* (Pers.) Lange is similar in appearance but has a farinaceous odor and taste and slightly smaller spores.

TRICHOLOMA VACCINUM (Pers. ex Fr.) Kummer — Suspected

Figure 184, page 111

PILEUS 1½–3 in. broad, convex to campanulate, sometimes subumbonate, becoming expanded, dry, covered with cinnamon-brown to dark reddish brown, appressed scales, margin tomentose, incurved at first. FLESH thin except at the disk, white, staining faintly reddish, taste slightly disagreeable. LAMELLAE adnexed or almost adnate, becoming sinuate, close, moderately broad, dingy white, staining reddish brown. STIPE 2–3 in. long, ¼–½ in. thick, subequal, hollow, pale reddish brown, fading to whitish at the apex, fibrillose-scaly with reddish brown scales. SPORES smooth, white, broadly ovoid to subglobose, 5.5–6 × 4 μ.

In groups or slightly clustered on the ground under conifers. Sept.–Oct.

T. vaccinum is a fairly common species, well characterized by the reddish, scaly pileus, hollow stipe, and by the lamellae staining reddish brown. Other species in which the lamellae also stain reddish brown are *T. imbricatum* (Fr.) Kummer, which has a less scaly pileus and solid stipe, and *T. transmutans* Peck, which is viscid with the surface tasting bitter. Both the latter species are reported edible but there seems to be some doubt regarding *T. vaccinum*.

MELANOLEUCA

This genus includes a group of species formerly placed in *Tricholoma*. The characters used to separate it from *Tricholoma* are mainly microscopic, being the rough-walled, amyloid spores and the harpoon-shaped cystidia on the edges of the lamellae. However, we can usually recognize a *Melanoleuca* in the field by its stiff stature and the texture of the stipe, which is almost cartilaginous as in *Collybia*.

The type of the genus, *M. melaleuca* (Pers. ex Fr.) Murr., is a fairly common and widely distributed species. As far as is known, all the species are edible but, as with all mushrooms, they should be tried cautiously at first.

MELANOLEUCA ALBOFLAVIDA (Pk.) Murr. Edible

Figure 188, page 113

PILEUS 2–4½ in. broad, at first somewhat campanulate, expanding to almost plane with disk often slightly obtuse or slightly depressed and margin tending to remain decurved for a long time, smooth, moist, glabrous, dingy yellowish brown at first, becoming dingy yellowish buff to whitish, darkest on disk. FLESH white, odor and taste not distinctive. LAMELLAE thin and crowded, moderately narrow, sinuate-adnexed, white to dingy. STIPE rather tall and straight, giving the plant a stiff, rigid appearance, 3–7 in. long, ¼–½ in. thick, equal, subbulbous, solid within, with a cartilaginous rind, glabrous, fibrillose-striate, whitish or tinged the color of the pileus. SPORES ovoid, thick-walled, minutely punctate, strongly amyloid, white, 7–10 × 4.5–5.5 μ. CYSTIDIA lanceolate, often encrusted at the apex into a harpoon-like tip.

Solitary or in small groups on the ground in woods. June–Sept. Fairly common.

The general appearance and stature of this species and especially the subcartilaginous stipe would lead one to look for it in the genus *Collybia*. It was placed in *Collybia* by Kauffman, although it was described by Peck as a *Tricholoma*, and its closest relatives appear to be in the *T. melaleucum* group. Since this group is now separated from *Tricholoma* as a distinct genus, *Melanoleuca*, this seems to be the proper place for this species. It is larger than *M. melaleuca* and paler in color.

MELANOLEUCA MELALEUCA (Pers. ex Fr.) Murr. Probably edible

Figure 189, page 113

PILEUS 1–3 in. broad, convex, subumbonate, expanding to almost plane, moist, hygrophanous, smoky brown, drying much paler, smooth and glabrous, sometimes wavy on the margin. FLESH thin, whitish. LAMELLAE adnexed, notched at the stipe, close, moderately broad, white or whitish. STIPE 1–3 in. long, ⅛–¼ in. thick, equal or slightly swollen at the base, whitish, marked with darker fibrils, centrally stuffed. SPORES white, ellipsoid, rough-walled, strongly amyloid, 6–8 × 4–5 μ. CYSTIDIA lanceolate, encrusted at apex forming a harpoon-like tip.

Solitary or scattered, on the ground, in woods and open places. Sept.–Oct.

This species is distinguished by its rather stiff stature and rigid stipe, the strongly hygrophanous, fading pileus, and the rough-walled amyloid spores. *M. brevipes* (Bull. ex Fr.) Pat. is very similar but has a short stipe, less than the diameter of the pileus in height.

HYGROPHORUS

Hygrophorus is a large and important genus that includes some of our most beautiful mushrooms and several fine edible species. The spore deposit is white and the chief distinguishing character of the genus is the texture of the lamellae. This texture is difficult to describe but is usually termed waxy, and is fairly readily recognizable with a little field experience. The lamellae are as a rule rather thick, more or less triangular in shape, and usually subdistant to distant. The attachment to the stipe varies considerably in different species from adnexed to decurrent.

Some of the species are brilliantly colored, scarlet to orange, yellow, or in one species bright green. Others are duller colored, brown, gray, dull violaceous, or white. They may be moist or dry, or both the pileus and stipe may be viscid, or only the pileus viscid.

The genus *Hygrophorus* is divided into three sections based on the structure of the trama of the lamellae. This is a character that can be determined only by making thin transverse sections of the lamellae and examining them under the microscope. In this treatment not much emphasis is laid on the structure of the trama but from a scientific standpoint it is such a valuable character in making identifications when a microscope is available that it has been included in the descriptions. However, the key and descriptions are so arranged that it should be possible to identify the species described here without knowing the structure of the trama.

Three types of structure may be found. In the first the trama is said to be 'divergent' and in this type the hyphae form a more or less definite core in the center of the lamella and from it they curve out obliquely and rather loosely to the hymenium. The other two types are said to be 'parallel' or 'interwoven' and sometimes difficulty may be encountered in deciding whether a particular specimen has parallel trama or interwoven trama.

In general there is a correlation between the structure of the trama and the diameter of the hyphae. If the tramal hyphae are consistently more than 7μ broad the species will usually be classified in the group with parallel hyphae, and if they are consistently less than $7\ \mu$ broad, it will be found in the group with interwoven hyphae. Those species with divergent trama are placed in the subgenus *Limacium*, those with interwoven trama in *Camarophyllus*, and those with parallel trama in *Hygrocybe*.

Although some of the species are small, many of them are among our best edible mushrooms. Only *H. conicus* is considered to be dangerous and it is easily recognized by its conical shape and blackening flesh.

Laccaria laccata is likely to be mistaken for a *Hygrophorus* but it has spiny spores.

Figures 212-221

212. *Hygrophorus russula.*
213. *Mycena pura.*
214. *Laccaria laccata.*
215. *L. ochropurpurea.*
216. *Xeromphalina campanella.*
217. *X. campanella.*
218. *Mycena leaiana.*
219. *M. leaiana.*
220. *Collybia acervata.*
221. *C. acervata.*

212
213
214
215
216
217
218
219
220
221

Figure 222. *Leucopaxillus albissimus* var. *piceinus*.

Figures 223-232

223. *Collybia confluens*.
224. *C. maculata*.
225. *C. dryophila*.
226. *C. dryophila*.
227. *C. platyphylla*.
228. *C. tuberosa*.
229. *C. velutipes*.
230. *C. velutipes*.
231. *Marasmius oreades*.
232. *M. oreades*.

223
224
225
226
227
228
229
230
231
232

Figure 233. *Tricholoma subacutum*.

Key

1. Pileus viscid 2
1. Pileus not viscid 13

2. Pileus white or whitish 3
2. Pileus not white 5

3. Pileus with yellow granules, at least at the margin and at the apex of the stipe .... *H. chrysodon*
3. Pileus not with yellow granules 4

4. Pileus glutinous .... *H. eburneus*
4. Pileus moist, occasionally subviscid .... *H. borealis*

5. Fruiting body bright green, fading to orange, yellow, or pinkish, very viscid .... *H. psittacinus*
5. Fruiting body not green 6

6. Fruiting body gray-brown with tinge of olive; stipe more or less floccose-scaly .... *H. olivaceoalbus*
6. Not as above 7

7. Blackening when handled; pileus conical .... *H. conicus*
7. Flesh of pileus and stipe not blackening 8

8. Lamellae becoming reddish spotted, close; pileus rosy red, somewhat variegated .... *H. russula*
8. Lamellae not reddish spotted 9

9. Pileus pale tan or pale flesh colored, often flushed with pink .... *H. pudorinus*
9. Pileus yellow to orange or red 10

10. Stipe viscid 11
10. Stipe not viscid 12

11. Pileus ⅜-1½ in. broad, wax-yellow, fading, becoming deeply umbilicate .... *H. nitidus*
11. Pileus 1-3 in. broad, bright reddish orange, fading to yellowish at margin, plane or slightly umbonate .... *H. speciosus*

12. Pileus scarlet to bright red .... *H. puniceus*
12. Pileus yellow to orange, never red .... *H. flavescens*

13. Fruiting body white .... *H. borealis*
13. Fruiting body not white 14

14. Pileus becoming squamulose 15
14. Pileus remaining glabrous or may be somewhat cracked in age 16

15. Lamellae decurrent .... *H. cantharellus*
15. Lamellae adnexed .... *H. miniatus*

16. Pileus bright orange, fading to yellowish or whitish; the lamellae also orange and not fading, brighter than the pileus in age .... *H. marginatus*
16. Pileus reddish brown when young, fading to tawny or buff, often turbinate in shape .... *H. pratensis*

HYGROPHORUS BOREALIS Pk. — Edible

Figure 190, page 113

PILEUS 3/8–1 1/2 in. broad, slightly fleshy, convex to obtusely subumbonate, at length slightly depressed on the disk, surface smooth, glabrous, moist, occasionally slightly subviscid, watery white to white, margin decurved, finally expanded, striatulate when moist. FLESH thickish on the disk, white to watery, odor and taste mild. LAMELLAE not very broad, arcuate, decurrent, subdistant to distant, white, trama of interwoven hyphae. STIPE 1–2 1/2 in. long, approximately 1/8 in. thick, equal or tapering downward, white, smooth, glabrous, stuffed. SPORES smooth, white, ellipsoid, 7–10 × 5–6.5 μ.

In groups, on the ground in woods. Aug.–Oct.

Hygrophorus niveus Fr. is said to differ from *H. borealis* in its thin submembranous pileus, which is decidedly viscid. *H. eburneus* Fr. is usually larger and is also viscid, and has the trama of the lamellae of divergent hyphae.

HYGROPHORUS CANTHARELLUS Schw. — Edible

Figure 191, page 113

PILEUS 1/4–1 1/4 in. broad, convex, with disk at first plane or slightly depressed, usually becoming decidedly depressed to umbilicate as pileus expands, dry, glabrous at first but soon becoming minutely scurfy to squamulose, bright scarlet, fading to orange or yellowish, margin often regularly crisped. FLESH thin, concolorous with or paler than the surface, odor and taste not distinctive. LAMELLAE decurrent to long-decurrent, subdistant, broad, yellowish or tinged orange, paler than the pileus, trama of parallel hyphae. STIPE 1 1/4–3 1/2 in. long, about 1/8 in. thick, equal or almost so, stuffed at first, becoming hollow, glabrous, more or less concolorous with pileus, paler at the base. SPORES smooth, white, broadly oval, apiculate, 8–10 × 4–6 μ.

In groups on the ground in moist woods or bogs. June–Oct. Fairly common.

Some authors consider this to be a variety of *H. miniatus* and the two species are very close but *H. cantharellus* has a longer, more slender stipe, decidedly decurrent lamellae, and is usually not as broad. It is a very attractive little mushroom and is reported to be edible.

HYGROPHORUS CHRYSODON Fr. — Edible

Figures 192, 193, page 113

PILEUS 1–3 in. broad, at first convex with margin incurved, becoming expanded, disk sometimes remaining obtusely subumbonate, white, viscid when fresh, sprinkled with minute, golden yellow granules. FLESH soft, white, thick at the disk, odor and taste mild. LAMELLAE decurrent, broad, subdistant to distant, white, trama of divergent hyphae. STIPE 1 1/2–4 in. long, 1/4–1/2 in.

thick, equal or tapering downward, stuffed, viscid when fresh, white, sprinkled with minute, golden yellow granules, especially toward the apex where they sometimes form a yellowish annular zone. SPORES smooth, white, ellipsoid, apiculate, 7–10 × 4–5 μ.

In groups on the ground in woods. Sept.–Oct.

The yellow granules on the pileus and at the apex of the stipe provide an easy means of recognizing this species. It is apparently more common on the west coast than in the east. It is reported to be edible although Smith and Hesler (1939) report unfavorably on the flavor.

HYGROPHORUS CONICUS Fr. — Suspected

Figure 194, page 113

PILEUS 1–2 in. broad, acutely conic to obtusely conic, remaining unexpanded, orange-red, orange-yellow, or yellowish, often tinged with olive to blackish streaks, blackening when bruised or in age, glabrous, sometimes obscurely fibrous-streaked, viscid when wet, becoming dry, margin often splitting as the pileus expands, sometimes lobed. FLESH thin, tinged orange, odor and taste not distinctive. LAMELLAE almost free, fairly close, moderately broad, broadest in center, pallid yellowish, trama of parallel hyphae. STIPE 1½–3½ in. long, ⅛–¼ in. thick, equal, yellowish or orange-tinged, blackening where bruised, moist or dry, becoming hollow, readily splitting longitudinally, fibrillose-striate, the striations sometimes twisting around the stipe. SPORES smooth, white, ovoid to slightly irregular 9–13 × (4.5) 5.5–6.5 (7.5) μ.

In groups or singly on the ground in woods. Fairly common. June–Oct.

The entire fruit body blackens with age or on handling or drying but traces of blackening can be found on nearly any plant, especially at the base of the stipe or on the disk. The bright colors, conical shape, and twisted stipe are characteristic features. *H. cuspidatus* Peck is somewhat similar in color and shape but does not blacken.

HYGROPHORUS EBURNEUS Fr. — Edible

Figure 244, page 154

PILEUS 1–3 in. broad, pure white, glutinous, convex or obtusely subumbonate, becoming expanded, margin at first slightly floccose and incurved, becoming expanded, in age somewhat elevated. FLESH white, rather thick on the disk, odor and taste not distinctive. LAMELLAE subdecurrent, becoming decurrent, subdistant to distant, moderately broad, narrowing toward the margin, pure white, becoming dingy with age, trama of divergent hyphae. STIPE 2–6 in. long, ⅛–⅜ in. thick, subequal or tapering downward, stuffed then hollow, glutinous, pure white becoming dingy, apex dotted with minute white squamules. SPORES smooth, white, ellipsoid, 6–8 × 4–5.5 μ.

In groups on the ground in woods. Sept.–Oct.

This is a fairly common species, distinguished by the very glutinous covering of both pileus and stipe. It is somewhat shiny when dry. It is usually larger than *H. borealis* Pk. and *H. niveus* Fr.

HYGROPHORUS FLAVESCENS (Kauffman) Smith & Hesler — Edible

Figure 195, page 113

PILEUS brittle-fragile, ¾–2½ in. broad, convex to expanded-convex with decurved margin, often slightly irregular in outline, striate on the margin when moist, smooth, glabrous, viscid, shining when dry, at first bright orange, fading in streaks to bright yellow, then paler yellow. FLESH thin, pale yellowish, odor and taste not distinctive. LAMELLAE unevenly attached, mostly adnexed, varying from broad to moderately narrow, close to subdistant, thick and waxy, deep yellow to pale lemon-yellow, many shorter lamellulae present, trama of parallel hyphae. STIPE 1¼–3 in. long, ⅛–½ in. thick, subequal or tapering downward, often compressed or grooved, hollow, waxy to the touch but not viscid, orange to yellow, usually paler than the pileus, whitish at the base. SPORES smooth, white, ellipsoid, 7–8 × 3.5–4.5 μ.

In groups or scattered, on the ground in woods, fairly common. June–Sept.

Kauffman described this fungus as a variety of *H. puniceus* but it is evidently a distinct species. *H. chlorophanus* Fr. is similar but has a viscid stipe. The stipe of *H. flavescens* may feel slippery or subviscid on handling but it is not truly viscid. *H. chlorophanus* is apparently rare but *H. flavescens* is common and probably most specimens identified as *H. chlorophanus* are, in reality, *H. flavescens*.

HYGROPHORUS MARGINATUS Pk. — Not recommended

Figure 196, page 113

PILEUS ½–2 in. broad, at first obtusely conic with incurved margin, becoming more or less convex to broadly expanded, disk often tending to remain obtuse, smooth, glabrous, moist, hygrophanous, bright orange, fading gradually to pale yellowish. FLESH thin, fragile, concolorous with pileus, odor and taste not distinctive. LAMELLAE adnexed, broad, ventricose, subdistant, intervenose, bright orange, retaining this deep color, especially on the edges, after the rest of the plant has faded, trama of subparallel to interwoven hyphae. STIPE 1–3 in. long, up to ¼ in. thick, subequal, often slightly compressed, hollow, smooth, glabrous, moist, concolorous with pileus or paler. SPORES smooth, white, oval, apiculate, 7–9 × 4–6 μ.

In groups on ground in woods. Not common. July–Sept.

The most striking character of this species is the manner in which the lamellae retain their colors after the pileus has faded. Sometimes the edges of the lamellae are brighter colored but not always. Kauffman reported this species as suspected and we have no further information about it, hence it is not recommended for food.

HYGROPHORUS MINIATUS Fr. Edible

Figure 197, page 113

PILEUS ¾–1½ in. broad, slightly convex to expanded, with disk plane or depressed, glabrous when fresh and moist but very soon becoming minutely scurfy as pileus loses its moisture, scarlet at first, fading to orange or yellow, margin at first incurved. FLESH more or less concolorous with the pileus, thin, odor and taste not distinctive. LAMELLAE broad, subdistant, adnate to adnexed, paler than the pileus, fading to yellowish, trama of parallel hyphae. STIPE 1–2 in. long, about ⅛ in. thick, equal, glabrous or almost so, concolorous with the pileus, fading gradually to orange, then yellowish, stuffed at first, becoming hollow. SPORES smooth, white, oval, apiculate, 7–9 $\times$ 4–5 μ.

In groups on the ground or on much-decayed logs in woods, fairly common. June–Sept.

This species is rather variable in appearance and different authors have described several varieties. The surface is not viscid and appears glabrous at first, soon becoming fibrous or scurfy to squamulose. The color is brilliant scarlet at first, soon fading to orange or yellow. The attachment of the lamellae is another variable character in this species.

HYGROPHORUS NITIDUS B. & C.

Figures 199, 200, page 115

PILEUS ⅜–1½ in. broad, flattened-hemispherical to convex, with disk becoming depressed then deeply umbilicate, smooth, glabrous, viscid when moist, clear bright yellow, fading to cream or whitish, margin striate when moist, incurved, gradually becoming elevated but with the extreme margin tending to remain decurved. FLESH thin and fragile, yellowish, fading, odor and taste not distinctive. LAMELLAE arcuate-decurrent, becoming long-decurrent, moderately broad, rather distant, with a soft, waxy appearance, yellow, usually retaining the yellow color after the pileus and stipe have faded to whitish, trama of parallel hyphae. STIPE 1–3 in. long, about ⅛ in. thick, equal, hollow within, glabrous, viscid, concolorous with the pileus, fading. SPORES smooth, white, broadly oval, apiculate, 7–8 $\times$ 4–5 μ.

In groups on the ground in moist woods, fairly common. July–Sept.

This is a fairly common species but is too small and fragile to be of any interest as food. The most distinctive character is the pileus, which fades from bright yellow to whitish while the lamellae retain their yellow color. *H. ceraceus* Fr. is similar in color but does not fade and the pileus is not umbilicate.

HYGROPHORUS OLIVACEOALBUS Fr. Edible

Figures 201, 202, page 115

PILEUS 1–3 in. broad, fleshy, at first convex to campanulate with incurved margin, becoming expanded but disk sometimes remaining obtuse or broadly subumbonate, smooth, viscid, dark gray-brown on the disk, paler toward the margin, streaked with blackish fibrils beneath the viscid layer. FLESH white, thickest on the disk, odor and taste mild. LAMELLAE broad, white, close to subdistant, broadly adnate to subdecurrent, trama of divergent hyphae. STIPE 1½–3½ in. long, ¼–½ in. thick, solid, equal or tapering toward the base or apex, streaked up to an annular zone with dark gray fibrils under a viscid coating, white above this zone, occasionally with the fibrils on the lower part of the stipe arranged in a series of rings or bands. SPORES smooth, white, oval, apiculate, 9–12 × 5–7 μ.

In groups on the ground beneath conifers. Sept.–Oct.

The stipe in this species has a double sheath, an outer glutinous layer and an inner fuscous-fibrillose layer which breaks up into irregular bands as the stipe elongates. *H. paludosus* Peck is very similar but does not have the fuscous inner sheath on the stipe, and in wet weather it develops greenish spots and stains on the lamellae and upper part of the stipe. *H. fuligineus* Frost is darker colored and has smaller spores. *H. tephroleucus* Fr. is a smaller gray species with whitish fibrils on the stipe, which soon become gray.

HYGROPHORUS PRATENSIS Fr. — Edible

Figure 203, page 115

PILEUS ¾–3 in. broad, fleshy, convex or with a broad obtuse umbo, often turbinate, smooth, dry, sometimes cracking around the disk, light reddish brown fading to pinkish tan or pale tan, margin at first incurved, gradually becoming expanded, in age the margin becoming elevated and the disk somewhat depressed. FLESH thick on the disk, tinged the color of the pileus, odor and taste mild. LAMELLAE thick, decurrent, distant, intervenose, rather broad, narrowing toward the margin, flesh colored, trama of interwoven hyphae. STIPE 1½–3 in. long, ¼–½ in. thick, equal or tapering upward or downward, stuffed, dry, pallid or tinged the color of the pileus. SPORES smooth, white, ellipsoid, 6–8 × 4–5 μ.

In groups on the ground in woods and open places, fairly common. July–Oct.

The stipe is usually short and the pileus more or less top-shaped. It often grows in more open and exposed places and sometimes fades to whitish.

HYGROPHORUS PSITTACINUS Fr. — Not edible

Figure 204, page 115

PILEUS ⅜–1⅜ in. broad, conic-campanulate becoming convex or expanded, sometimes persistently umbonate, at first deep olive-green to parrot-green, quickly fading on drying out to salmon color, flesh color, pinkish orange or yellowish, smooth, glabrous, slimy-viscid and very slippery when moist, margin striate when moist. FLESH thin, fragile, brittle, more or less concolorous with the pileus. LAMELLAE adnate, moderately broad, subdistant, thick especially next to the flesh, somewhat intervenose, greenish at first, soon fading to flesh color, orange, or yellowish, trama of parallel hyphae. STIPE 1½–2½ in. long, about ⅛ in. thick, equal, glabrous, slippery-viscid, at first green, drying to flesh color or yellowish, the apex retaining the green color longest, hollow. SPORES smooth, white, oval, obliquely apiculate, 6–9 × 4–5 μ.

In groups on the ground in grassy places and woods. July–Oct.

The bright green color is unusual in mushrooms and this is an attractive fungus for collectors, but is too small and slimy to be of any interest as food. The color fades rather rapidly but traces of it can usually be found around the margin of the pileus and at the apex of the stipe. Faded specimens might be confused with *H. laetus* Fr., which is very variable in color and may be of mixed colors, but it is not bright green and the lamellae are decurrent.

HYGROPHORUS PUDORINUS Fr. — Edible

Figure 205, page 115

The Blushing Hygrophorus

PILEUS 1½–4 in. broad, fleshy, convex to somewhat campanulate with incurved margin, becoming expanded but disk tending to remain obtuse, smooth, glabrous, viscid, pale pinkish tan. FLESH rather thick, firm, white or pinkish, odor and taste mild. LAMELLAE not very broad, adnate to subdecurrent, thick, subdistant, intervenose and tending to fork, whitish to cream or flesh colored, trama of divergent hyphae. STIPE stout, 1½–3½ in. long, ¼–¾ in. thick, equal or tapering downward, solid or stuffed, dry, whitish or tinged the color of the pileus, somewhat fibrillose below, apex dotted with minute white flecks which become reddish as the plant dries. SPORES smooth, white, ellipsoid, apiculate, 7–9 × 4–5.5 μ.

In groups or slightly clustered on the ground in woods. Sept.–Oct.

According to Smith and Hesler (1939) there is some doubt as to whether or not this is the true *H. pudorinus* and there is a western species, *H. fragrans* Murr., which differs in having the base of the stipe ochraceous and tends to stain yellow when bruised, and which in some respects seems closer to the original description of *H. pudorinus*.

HYGROPHORUS PUNICEUS Fr. — Edible

Figures 207, 208, page 115

PILEUS 1–2½ in. broad, at first bluntly conic with incurved margin, expanding to convex or nearly plane, disk often remaining obtusely umbonate, smooth, glabrous, viscid, deep blood-red when fresh, soon becoming streaked with orange, finally fading entirely to orange. FLESH thin, watery, reddish orange to yellowish, odor and taste not distinctive. LAMELLAE adnate to adnexed, broad, subdistant, reddish orange to yellowish, trama of parallel hyphae. STIPE 1½–3½ in. long, ¼–½ in. thick, subequal or tapering at the base, stuffed then hollow, at first reddish above, fading to orange, then yellow, base paler, yellow to white. SPORES smooth, white, ellipsoid, apiculate, (7) 8–10 × 4.5–6 μ.

In groups on the ground in woods. July–Nov.

Young, fresh specimens of this species are among the most brilliantly colored of our mushrooms. The white base of the stipe and the broad, deeply colored lamellae are also distinctive. *H. coccineus* Fr. is said to be similar but not viscid and is apparently very rare in North America.

HYGROPHORUS RUSSULA (Fr.) Quél. — Edible

Figure 212, page 133

PILEUS 2–4½ in. broad, firm, fleshy, convex or with a broad obtuse umbo, viscid when fresh, rosy red to wine color on the disk, paling toward the margin to flesh-pink or whitish, sometimes flecked with wine-colored spots, becoming very minutely areolate, especially on the disk, margin at first incurved and slightly floccose, becoming expanded and finally elevated. FLESH thick, firm, white to pinkish, odor and taste not distinctive. LAMELLAE adnate to decurrent, moderately narrow, close to crowded, white then pinkish, finally spotted with purplish red stains, trama of divergent hyphae. STIPE stout, 1½–3 in. long, ⅜–1 in. thick, equal or tapering downward, dry, white then pinkish, solid. SPORES smooth, white, ellipsoid, 6–8 × 3.5–5 μ.

Scattered or in groups on the ground in hardwoods. Sept.–Oct.

This species is rather unusual in the genus because of the close to crowded lamellae and some authors have placed it in *Tricholoma*. However, it now seems to be generally agreed that it properly belongs in *Hygrophorus*. It sometimes occurs late in the autumn under fallen leaves. This is considered to be one of the best edible species.

H. purpurascens Schw. can be distinguished by the presence of an evanescent, fibrillose annulus.

HYGROPHORUS SPECIOSUS Peck — Edible

Figure 206, page 115

PILEUS 1–3 in. broad, at first subconic to campanulate, then expanded, often umbonate, scarlet to orange-red, fading to yellowish near the margin but remaining red on the disk, viscid, glabrous, margin incurved at first, then spreading. FLESH white, tinged orange under the pellicle, soft, odor and taste mild. LAMELLAE decurrent, distant, rather broad, thick, white to yellowish, trama of divergent hyphae. STIPE 1¼–4 in. long, ¼–½ in. thick, equal or slightly compressed, floccose-fibrillose up to an evanescent annular zone, subglabrous above, viscid, solid. SPORES broadly ellipsoid, white, smooth, 8–10 × 5–6 μ.

Usually gregarious in larch swamps. Sept.–Oct.

This species is not common but is a very showy and beautiful fungus and will certainly attract notice when it is found. It is said to be edible.

LACCARIA

Species of *Laccaria* have white or pale lilac spores, usually strongly echinulate and not amyloid. They have rather thick, somewhat waxy-looking lamellae that are usually purplish to flesh colored, and lack annulus and volva. They have been placed in *Clitocybe* but they are not closely related to this genus. They are most likely to be confused with *Hygrophorus* because of the waxy appearance of the lamellae but the echinulate spores will distinguish them. They might even be mistaken for the fruiting bodies of a *Lactarius* in which the latex had dried up, but the iodine reaction of the spores is a sure way of separating them. The spores of all *Lactarius* species turn blue in iodine. *Laccaria* species are edible but are not reputed to be of good flavor.

LACCARIA LACCATA (Fr.) Berk. & Br. — Edible

Figure 214, page 133

PILEUS ¾–2 (3) in. broad, convex becoming plane, sometimes slightly umbilicate, glabrous at first, then scurfy to somewhat scaly, hygrophanous, variable in color, reddish brown to reddish flesh colored, fading to ochraceous or pallid, margin even or wavy to notched. FLESH thin, moist, taste mild. LAMELLAE emarginate to short decurrent, broad, distant to subdistant, thick, tinged flesh color. STIPE 1–4 in. long, ⅛–¼ (⅜) in. thick, equal, tough, fibrous, glabrous to scurfy, sometimes striate, solid to stuffed or becoming hollow, colored like the pileus. SPORES globose, echinulate, white, 8–10 μ diam.

Common in woods or open places. May–Nov.

This is one of our commonest fungi and one of the most puzzling to the beginner. It is very variable and is likely to be collected many times before it is recognized with certainty. The broad, distant, flesh-colored lamellae are, perhaps, the best field mark and if a microscope is available the spiny, non-amyloid spores are characteristic.

L. amethystina (Bolt. ex Fr.) B. & Br. is similar in stature but the whole fruiting body is a beautiful deep violet.

LACCARIA OCHROPURPUREA (Berk.) Peck — Edible

Figure 215, page 133

PILEUS 2–4 in. broad, sometimes larger, convex to nearly plane and depressed in the center, hygrophanous, minutely downy-tomentose at first, becoming glabrous, or sometimes cracking into scale-like areas, purplish brown at first, becoming tawny yellow to grayish leather colored, margin even or wavy. FLESH tough, taste unpleasant. LAMELLAE adnate to slightly decurrent, broad, thick, distant, purple. STIPE 1½–4 in. long, ½–1 in. thick, variable in shape, equal or tapering either upward or downward, sometimes curved or twisted, tough and hard, solid, colored like the pileus or paler. SPORES globose, echinulate, pale lilac in mass, 8–10 μ in diameter.

Scattered to subcespitose in woods, open grassy places or bare soil. Aug.–Oct.

The purple lamellae are characteristic and might suggest a *Cortinarius* but there is no veil and the spores are pale lilac. It is not as common as *L. laccata*.

XEROMPHALINA

The genus *Xeromphalina* includes a small group of species, most of which were formerly put in *Omphalia*. However, the name *Omphalia* is illegitimate under the International Code of Nomenclature and, furthermore, the species formerly placed there are not regarded as a taxonomic unit by modern taxonomists. They have been divided among several genera of which *Xeromphalina* is one. There is some disagreement as to the exact limits of this genus but the type species is *X. campanella* (Fr.) Kühner & Maire.

The species of this genus have white, amyloid spores. The lamellae are adnate to decurrent and the stipes dark brown to blackish, horny in consistency, and with a bright-colored tomentum at the base. There is no annulus or volva. The amyloid spores provide the best character for separating *Xeromphalina* from *Marasmius*. The species are mostly small and of no value as food.

XEROMPHALINA CAMPANELLA (Fr.) Kühner & Maire — Edible

Figures 216, 217, page 133; Figure 415, page 297

PILEUS ¼–¾ in. broad, fragile, convex, expanding, umbilicate, rusty orange-yellow, orange-brown, or reddish-tinged, glabrous or almost so, hygrophanous, fading on drying, margin slightly incurved, tending to remain so, becoming striate. FLESH membranous, yellowish, odor and taste not distinctive. LAMELLAE decurrent, close to subdistant, rather narrow, connected by veins, yellowish. STIPE slender, ½–1½ in. long, 1/16–⅛ in. thick, smooth, cartilaginous, even, straight or curved, hollow, dark reddish brown, yellowish at apex, pruinose, with a hairy, orange, root-like tuft at the base. SPORES smooth, white, ellipsoid to long-ellipsoid, amyloid, 6–7.5 × 2.5–3.5 μ.

In clusters, often densely crowded, on decaying logs and stumps of conifers, common. May–Oct.

This species has been reported as edible but is too small to be of much interest as food although it sometimes occurs in very large clusters. Its abundance, bright colors, and rather graceful stature make it a very attractive little mushroom and it may be found throughout the growing season.

XEROMPHALINA TENUIPES (Schw.) Smith

Figure 255, page 156

PILEUS ½–1½ (2½) in. broad, convex, becoming plane or broadly umbonate, rather variable in color, orange-brown, tinged olive or ochraceous, dry, velvety to slightly granulose when old, margin even or becoming slightly striate in age. FLESH watery brown, pliant, no odor. LAMELLAE adnate or decurrent by a line, whitish soon becoming pale yellow, close, moderately broad. STIPE 1–3 in. long, ⅛–¼ in. thick, equal or slightly enlarged toward base, concolorous with pileus, drying more yellowish, velvety-tomentose. SPORES white, ellipsoid, smooth, amyloid, 7–9 $\times$ 4–5 μ.

Singly or in clusters on hardwood. June–July.

This species appears early in the season and is the largest species in the genus. It somewhat resembles *Collybia velutipes* but differs in the dry, velvety pileus and amyloid spores. There is no information concerning its edibility but its texture seems rather tough.

MYCENA

Mycena is a very large genus of which the species are small and difficult to identify. Smith (1947) published a monograph of the genus in North America and recognized 232 species. With rare exceptions the species can be identified only from microscopic characters and the group is not one for the beginner. The species are all too small to be of any interest as food.

As interpreted by Smith the genus includes white-spored species with cartilaginous, hollow stipes, and usually conical or convex pilei with the margin straight and appressed to the stipe when young, although some forms with incurved margins and decurrent lamellae are included. The fruit bodies are fragile, fleshy or membranous. In delimiting the genus, Smith also placed considerable emphasis on microscopic characters.

Some of the species are rather attractive and brightly colored but many are small, brownish or grayish forms, all looking rather similar. Only a few of the more distinctive ones are included here.

MYCENA ALCALINA (Fr.) Kummer

Figure 266, page 174

PILEUS ¼–1 in. broad, occasionally larger, fragile but pliant, ovoid in the button stage, then obtusely conic, expanding to conic-campanulate with the disk usually remaining obtusely umbonate, becoming long-striate on the margin as the pileus expands, moist, when young dark grayish brown with a pruinose bloom, soon glabrous, drab grayish brown at maturity, pallid on the margin. FLESH thin except on the disk, white to pallid, odor alkaline. LAMELLAE ascending-adnate, moderately broad, uncrowded, white or tinged grayish, with shorter lamellae interspersed. STIPE 2–3½ in. long, up to ⅛ in. thick, equal, hollow, brittle, subconcolorous with the pileus, at first with a pruinose bloom, soon glabrous, usually somewhat white-mycelioid at the base. SPORES smooth, white, amyloid, ovoid, 7.5–10 × 4.5–6μ. CYSTIDIA fusoid-ventricose on the sides of the lamellae, up to 60 μ long, rare to abundant; those on the edges of the lamellae varying from ventricose to clavate, sometimes with one to several fingerlike projections at the apex.

In groups or in clusters of several on decaying conifers, common. May–Sept.

This is one of the commonest species of the genus and is a fairly typical *Mycena*. It may sometimes be found early in the spring. The characteristic alkaline odor is the most distinctive single character of the species, but it may vary from being quite strong to only detectable when the flesh is crushed or even occasionally apparently absent.

MYCENA GALERICULATA (Fr.) S. F. Gray — Edible

Figure 277, page 176

PILEUS ¾–1½ in. broad, at first conic, becoming campanulate to expanded-umbonate, the umbo sometimes disappearing, buff-brown on the margin, darker to umber on disk, fading, glabrous, somewhat slippery, not viscid, margin striate. FLESH grayish white to pallid, cartilaginous, odor and taste slightly farinaceous. LAMELLAE adnexed to adnate or sinuate, close to subdistant, moderately broad, whitish becoming tinged with pale pink, edges even. STIPE 1½–3½ in. long, sometimes longer, $\frac{1}{16}$–⅛ in. thick, equal, glabrous, cartilaginous, smooth or twisted-striate, grayish white, darker below to brownish at base, hollow, rooting. SPORES white, smooth, ellipsoid, amyloid, 8–10 × 5–7 μ.

Usually in clusters or sometimes scattered, on rotten wood. May–Oct.

Although this is a small fragile species it sometimes occurs in such large clusters that it may be of interest as food. Another common species that occurs in clusters on wood is *M. inclinata* (Fr.) Quél. It is more grayish in color, the margin is more or less scalloped, and the stipe has a white fibrillose coating when young that usually leaves flecks or fibrils on the stipe as the fruiting body matures.

MYCENA LEAIANA (Berk.) Sacc.

Figures 218, 219, page 133

PILEUS ¼–1¼ in. broad, tough, pliant, convex, becoming expanded-convex, with a slight depression on the disk, translucent-striate on the margin, smooth and very viscid, bright flaming orange, fading to yellowish orange and finally pale yellow. FLESH very thin, yellowish, odor and taste not distinctive. LAMELLAE adnate, ventricose, moderately broad, close to subdistant, yellowish to pinkish-tinged with the edges bright orange. STIPE ¾–2 in. long, about 1/16 in. thick, gradually tapering downward, hollow, glabrous, viscid, orange, not fading as readily as the pileus. SPORES smooth, white, ellipsoid, 7–9 × 5–6 μ.

In clusters of few to many individuals on old logs and stumps. June–Oct.

The bright orange edges of the lamellae are a distinctive field mark of this species since they retain their color after the pileus has faded to yellow. The brilliant colors of this species are certain to attract attention.

MYCENA PURA (Fr.) Kummer

Figure 213, page 133

PILEUS ½–1½ in. broad, convex to expanded, usually obtusely umbonate, smooth, glabrous, moist, hygrophanous, translucent-striate on the margin, varying in color from rosy red to violet or shades of grayish violet, sometimes nearly white. FLESH moderately thick on the disk, thin on the margin, tinged the color of the pileus or whitish, odor and taste of radish. LAMELLAE adnate to sinuate at the stipe, broad, subdistant to moderately close, interveined, variable in color, white or more often tinged the color of the pileus. STIPE 1½–3 in. long or longer, ⅛–¼ in. thick, equal, hollow, glabrous or pruinose, concolorous with the pileus or paler to whitish, sometimes twisted, fibrillose-striate. SPORES smooth, white, long-ellipsoid, 6–8 × 3.5–4 μ.

Solitary or in groups on the ground in woods. June–Oct. Common.

This is a common and widely distributed species and is attractive because of its beautiful colors. It is one of the larger species of the genus but is too small to be of value as food.

COLLYBIA

The genus *Collybia* has usually been defined as comprising those species with white spores, cartilaginous stipes, the lamellae adnate to adnexed, the margin of the pileus incurved to inrolled at first, and lacking both annulus and volva. However, modern taxonomists seem to be generally agreed that this concept is too broad and brings together many unrelated species. The tendency now is to split the genus into several genera with *Collybia dryophila* as the type of *Collybia* in a more restricted sense. Since some of the divisions are based principally on microscopic characters and there is not yet general agreement as to precise generic limits, for the purpose of this book *Collybia* is retained in the older and broader sense.

It is sometimes difficult to draw a distinction between *Collybia* and *Marasmius* on the one hand and *Collybia* and *Mycena* on the other. The principal difference between *Collybia* and *Marasmius* is in the ability of *Maras-*

mius species to revive after drying but sometimes this distinction is not very clear-cut and *Collybia confluens*, for example, might equally well be placed in *Marasmius*. *Mycena* is usually distinguished by the small, more or less conic to campanulate pilei which do not become expanded, and also by the margin of the pileus which in the young stage lies straight along the stipe rather than being incurved or inrolled as in *Collybia*.

No poisonous species of *Collybia* are known and some of the larger ones are considered to be very good but many of the species are too small to be of any value as food.

Key

1. Fruiting bodies densely cespitose, reddish brown or vinaceous brown; the stipes glabrous above, tomentose below and more or less bound together .... *C. acervata*
1. Fruiting bodies single, gregarious, or subcespitose 2

2. Stipe deeply rooting .... *C. radicata*
2. Stipe not deeply rooting 3

3. Stipe glabrous 4
3. Stipe velvety or tomentose to pruinose 6

4. Fruiting body white becoming spotted or stained rusty brown .... *C. maculata*
4. Fruiting body not white 5

5. Lamellae very broad; fruiting body gray .... *C. platyphylla*
5. Lamellae narrow; fruiting body reddish tan to yellowish fawn .... *C. dryophila*

6. Fruiting body small, usually less than ½ in. broad; stipes arising from a small sclerotium .... *C. tuberosa*
6. Fruiting bodies more than ½ in. broad; stipes not arising from sclerotia 7

7. Stipe velvety, dark brown to black; pileus viscid; on wood .... *C. velutipes*
7. Stipe densely whitish pubescent; pileus not viscid; usually among fallen leaves on the ground .... *C. confluens*

COLLYBIA ACERVATA (Fr.) Kummer — Probably edible

Figures 220, 221, page 133

PILEUS ¾–2 in. broad, convex to nearly plane, glabrous, somewhat hygrophanous, reddish brown to vinaceous brown, the margin becoming paler and finally fading over all to nearly whitish, margin slightly striate when moist, sometimes wavy and irregular and becoming upturned in age. FLESH thin, pallid, taste mild. LAMELLAE free to adnexed, crowded, narrow, whitish or tinged reddish. STIPE 2–4 in. or sometimes more in length, ⅛–¼ in. thick, equal, hollow, glabrous above, whitish tomentose below, densely cespitose and often bound together with the white mycelium, reddish brown or vinaceous brown, often darker than the pileus. SPORES smooth, white, nonamyloid, narrowly oblong to ellipsoid, with a prominent apiculus, 5–7 $\times$ 2–3 μ.

Densely cespitose on the ground or rotten wood. Aug.–Oct.

It is reported to be edible although Smith (1949) describes the taste as bitter. The dense clusters with reddish brown stipes more or less bound together toward the base are a characteristic feature of this fungus. When dried the stipes retain the reddish brown color and are darker than the pilei.

Collybia familia (Pk.) Sacc. and *C. abundans* (Pk.) Sacc. are two other densely cespitose species which have amyloid spores and are differently colored, whitish or buff to pale brown, more gray than *C. acervata*, and with whitish to brownish stipes. In *C. abundans* the pileus is depressed in the center and in *C. familia* it is not. Both of these species are edible.

COLLYBIA CONFLUENS (Pers. ex Fr.) Kummer — Edible

Figure 223, page 135

PILEUS ¾–2 in. broad, convex to nearly plane, reddish brown to buff-brown when moist, becoming grayish to pinkish buff or whitish when dry, glabrous to pruinose, sometimes minutely subsquamulose on the disk, obtuse to subumbonate, even or striatulate on the margin when moist. FLESH thin, white, rather tough, odor and taste mild. LAMELLAE free, narrow, crowded, whitish. STIPE 2–4 in. long, 1/16–3/16 in. thick, equal or nearly so, often compressed, tough, cartilaginous, reddish brown under a dense white pubescence, hollow. SPORES hyaline, narrow-fusoid, 6–9 × 2.5–4 μ.

Cespitose or gregarious, usually among fallen leaves on the ground. July–Oct. Common.

This species tends to revive when moistened, and in this way forms a connecting link between *Collybia* and *Marasmius*. It occurs rather commonly and is sometimes abundant. The densely whitish pubescent stipe is a good distinguishing character. *C. hariolorum* Fr. is somewhat similar but is reddish on the disk, the stipe is usually shorter and it does not revive so well, tending to become soft when moistened.

COLLYBIA DRYOPHILA (Bull. ex Fr.) Kummer — Edible

Figures 225, 226, page 135

PILEUS 1–2 in. broad, rather thin and pliant, convex becoming expanded with margin finally elevated and irregularly wavy, disk remaining slightly obtuse or becoming depressed, surface smooth, glabrous, moist when fresh, color variable, ranging from deep reddish tan to yellowish fawn, fading with age. FLESH thin, whitish, odor and taste not distinctive. LAMELLAE adnate to adnexed, narrow, crowded, white or pallid. STIPE 1¼–2½ in. long, about ⅛ in. thick, cartilaginous, equal or tapering upward, often compressed, central or slightly excentric, hollow within, surface smooth, glabrous, more or less concolorous with pileus, often tufted with white mycelium at the base. SPORES smooth, white, ellipsoid, 5–7 × 3–3.5 μ.

In groups or small clusters, on the ground in woods, common. June–Sept.

There are several species such as *C. butyracea* (Bull. ex Fr.) Kummer, *C. lentinoides* Peck, and *C. aquosa* (Fr.) Kummer that are closely related to *C.*

dryophila and sometimes difficult to distinguish. However, these are all edible species as well. *C. dryophila* may be found throughout the growing season and is sometimes abundant.

COLLYBIA MACULATA (A. & S. ex Fr.) Kummer — Edible

Figure 224, page 135

PILEUS 2–6 in. broad, rather firm, convex becoming expanded, disk tending to remain obtuse, margin at first inrolled, then decurved, becoming somewhat wavy when fully expanded, surface smooth, glabrous, white, spotted with rusty brown stains, in age the entire plant becoming rusty-stained. FLESH white, compact. LAMELLAE adnexed to almost free, narrow, crowded, white. STIPE white, 2½–6½ in. long, up to ½ in. thick, equal or slightly ventricose, the tapering base extending down some distance into the soil, cartilaginous, becoming hollow, fibrous-striate to slightly grooved. SPORES broadly oval to subglobose, smooth, yellowish, 5–7 × 4–5 μ.

In groups or in clusters of 2 or 3, on the ground in woods. June–Sept.

This is a fairly large white species which can be recognized by the rusty stains that develop as it matures. The spores are not pure white but have a yellowish tinge. It is reported to be edible but not of particularly good flavor.

COLLYBIA PLATYPHYLLA (Fr.) Kummer — Edible

Figure 227, page 135; Figure 416, page 297

PILEUS averaging 2½–5 in. broad, occasionally up to 8 in. broad, convex to subcampanulate at first, becoming expanded, disk often remaining obtusely umbonate or sometimes becoming slightly depressed, margin at first incurved, becoming expanded or recurved, in age splitting raggedly, color drab grayish brown to brownish drab, streaked with darker innate fibrils, paling toward the margin to grayish white, surface dry or moist, smooth or slightly scurfy. FLESH thin, white, taste unpleasant, odor mild to unpleasant. LAMELLAE white or whitish, very broad, subdistant, adnexed, becoming sinuate at the stipe. STIPE stout, 3–5 in. long or sometimes longer, ⅜–¾ in. thick, equal or tapering upward slightly, white or whitish, fibrous-striate, cuticle slightly cartilaginous, center stuffed, becoming hollow. SPORES smooth, white, broadly oval, 7–10 × 5–7 μ, immature spores smaller.

Single or in groups of several, on decaying stumps and logs or in rich soil. June–Oct.

C. platyphylla is edible but is often infested with insect larvae and the flavor is said to be strong. The gray color and very broad lamellae are distinguishing characters.

Figures 234-243

234. *Marasmius rotula.*
235. *M. rotula.*
236. *M. scorodonius.*
237. *Schizophyllum commune.*
238. *Panus rudis.*
239. *P. rudis.*
240. *Trogia crispa.*
241. *Pluteus admirabilis.*
242. *P. atromarginatus.*
243. *P. atromarginatus.*

234
235
236
237
238
239
240
241
242
243

Figure 244. *Hygrophorus eburneus.*

Figures 245-254

245. *Pluteus cervinus.*
246. *P. cervinus.*
247. *Volvariella bombycina.*
248. *V. bombycina.*
249. *Entoloma rhodopolium.*
250. *E. salmoneum.*
251. *Leptonia asprella.*
252. *L. formosa.*
253. *Clitopilus abortivus.*
254. *C. abortivus.*

245
246
247
248
249
250
251
252
253
254

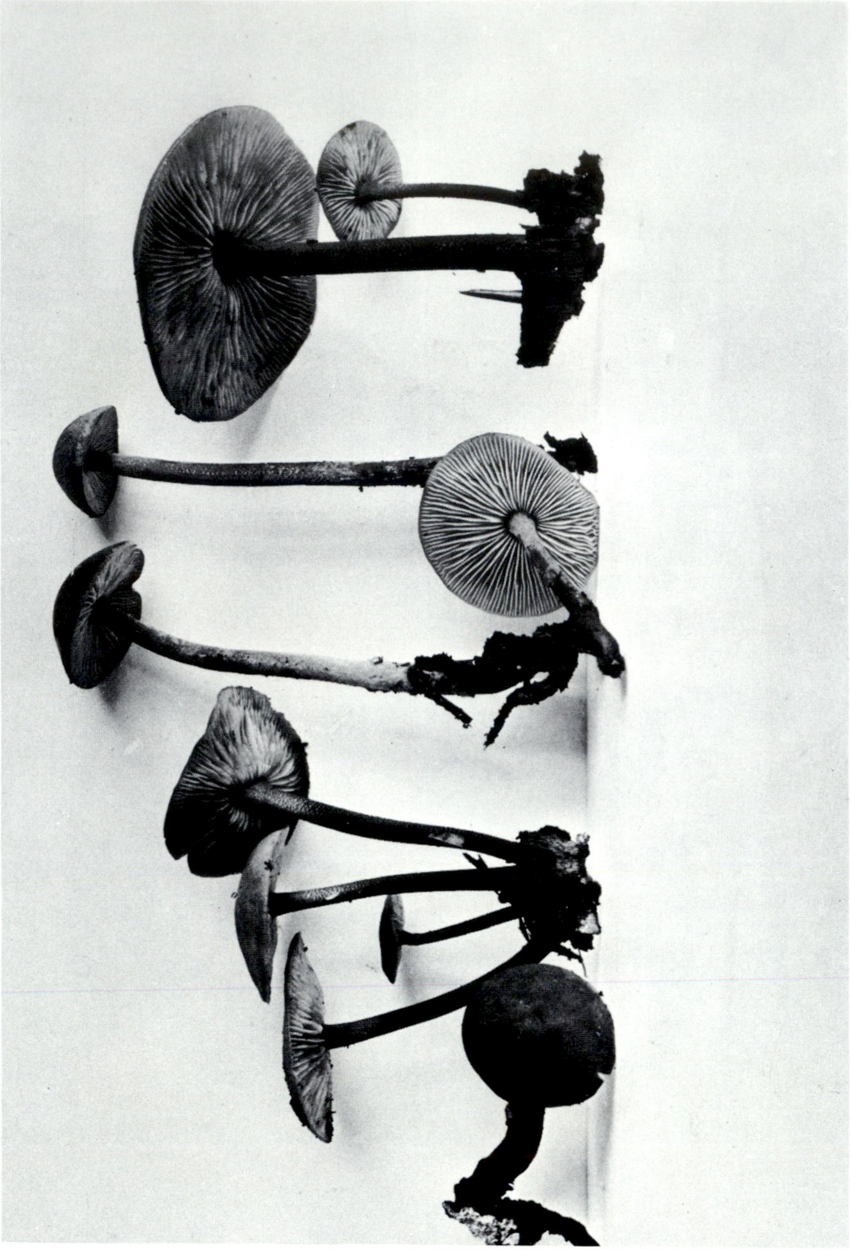

Figure 255. *Xeromphalina tenuipes.*

COLLYBIA RADICATA (Fr.) Quél. Edible

Figure 292, page 194

PILEUS 1¼–4 in. broad, convex, becoming expanded, often slightly umbonate, surface viscid when wet, glabrous, smooth or wrinkled around the umbo, color varying from grayish brown to grayish fawn or pallid. FLESH thin except on disk, whitish, odor slight or lacking, taste mild. LAMELLAE adnexed, often with fine decurrent lines on the stipe, broad, subdistant, pure white, several tiers of shorter lamellulae not reaching the stipe. STIPE tall, 2–8 in. above ground, with a long tapering root-like underground extension, ¼–½ in. thick at the ground, tapering upward slightly, cartilaginous, stuffed, becoming hollow, surface varying from even to twisted-striate or sulcate, glabrous to densely pruinose, white at the apex, elsewhere brownish to mouse colored. SPORES smooth, white, broadly oval, obliquely apiculate, 14–17 × 9–11 μ.

On the ground in woods, solitary or in small groups, common. June–Sept.

The 'rooting' stipe and the pure white, subdistant lamellae are the field marks of this species. *C. longipes* (Fr.) Kummer also has a long rooting stipe but is smaller with a dry, velvety pileus and brownish tomentose stipe. *C. radicata* exhibits a great range of variation in size and color and several varieties have been described. It is a very common species and sometimes may be found when other mushrooms are scarce.

COLLYBIA TUBEROSA (Bull. ex Fr.) Kummer

Figure 228, page 135

PILEUS tiny, ⅛–⅜ in. broad, convex to expanded, sometimes slightly umbonate, white to creamy or tinged tan, surface glabrous, unpolished or almost chalky in appearance. FLESH thin, white. LAMELLAE adnate, or with a slight, decurrent line, distant or subdistant, rather narrow, whitish, many short ones not reaching the stipe. STIPE ⅜–¾ in. long, slender, filiform, surface minutely scurfy or powdery especially toward the base, glabrous at the apex, white or tinged reddish brown, hollow, attached at the base to a small reddish brown or blackish sclerotium. SCLEROTIUM 1/16–¼ in. in diameter, varying in shape from elongated to pip-shaped. SPORES smooth, white, ellipsoid, 4–5.5 × 2–3 μ.

In groups on decayed mushrooms or occasionally on the ground. July–Oct.

This tiny little species is of no interest as food, but it is fairly common and will be encountered by any collector. It can be recognized by the small brownish sclerotia from which the fruit bodies arise. It is usually found on much-decayed mushrooms.

C. cirrhata (Schum. ex Fr.) Quél. is a similar small species but does not have sclerotia; *C. cookei* (Bres.) Arnold has sclerotia that are yellowish and more rounded than those of *C. tuberosa*. *C. albipilata* Peck is another small species growing on pine cones.

COLLYBIA VELUTIPES (Curt. ex Fr.) Kummer — Edible

Figures 229, 230, page 135; Figure 417, page 298

PILEUS ¾–2 in. broad, convex-expanded, often slightly excentric and irregular, surface glabrous, viscid, pellicle separable, color yellowish to reddish yellow or reddish brown, usually darkest on disk. FLESH moderately thick, white or tinged yellow or reddish, taste mild. LAMELLAE sinuate-adnexed, rather broad, subdistant, creamy to yellowish, edges minutely fimbriate. STIPE ¾–2½ in. long, ⅛–¼ in. thick, tough, subequal or tapering slightly to base or apex, stuffed to hollow, surface densely velvety-tomentose, bright cinnamon, usually yellowish toward the apex and dark brown to blackish toward the base. SPORES smooth, white, long-elliptic, obliquely apiculate, 7–9 × 3–4 μ.

In small clusters or singly on decaying logs and stumps, and on bark of living trees, chiefly in late autumn but found also in early spring or in summer.

C. velutipes seems to be hardy in cold weather. Fruiting bodies are sometimes found during mild spells in January and February. The reddish brown to yellowish, viscid pilei and dark velvety stipes are distinctive characters for this species.

Kauffman says the pellicle should be removed from the pileus before cooking.

MARASMIUS

Marasmius is a large genus of white-spored mushrooms, mostly small in size, and characterized by the ability to shrivel up during dry periods and revive again when moistened. This character is not very clear-cut and some species may readily be mistaken for *Collybia* or *Mycena*. There is no veil and the stipe is of a different texture from the pileus.

Many of the species are small and membranous and with the exception of *M. oreades* are of little interest as food. Some are, however, attractive little mushrooms and a few of the commoner species are included.

Key

1. Stipe glabrous 2
1. Stipe not glabrous; pileus buff or tan colored; growing in rings in grassy places .... *M. oreades*

2. Odor of garlic .... *M. scorodonius*
2. Odor not of garlic 3

3. Pileus ochraceous red, sulcate .... *M. siccus*
3. Pileus whitish with dark umbilicus; lamellae attached to a free collar .... *M. rotula*

MARASMIUS OREADES Fr. Edible

Figures 231, 232, page 135; Figure 418, page 298

Fairy Ring

PILEUS ¾–2 in. broad, rather fleshy for this genus, pliant, at first bell-shaped with slightly incurved margin, expanding to convex with or without a broad obtuse umbo, finally broadly expanded with margin elevated and disk plane or slightly umbonate, varying from dull reddish to light brown or tan, fading to yellowish buff when dry, smooth to somewhat uneven or lumpy, glabrous, margin more or less striate when moist. FLESH thin at the margin, thicker on the disk, pallid and watery when moist, whitish when dry, odor faint, taste not distinctive. LAMELLAE almost free, subdistant, somewhat interveined, rather broad, rounded behind, very thick next to the pileus, pallid whitish buff. STIPE tough, 1–2½ in. long, about ⅛ in. thick, equal or tapering downward, sometimes compressed at the apex, concolorous with the pileus or paler, smooth to minutely scurfy, stuffed to hollow. SPORES smooth, white, somewhat irregular in shape, mostly slightly subfusiform, prominently apiculate, 7–9 × 4–5.5 μ.

On the ground in lawns and grassy places, often in rings or arcs, common. May–Oct.

This is a good edible species and can be dried for winter use. However, the collector should be careful to avoid getting fruiting bodies of the poisonous *Clitocybe dealbata* (Figures 210, 211, p. 116) mixed in with his collection. The two species are likely to be found growing together, and are sufficiently similar in size and coloring to be accidentally included in the same collection. They can be distinguished readily by the lamellae, which are broad and subdistant in the *Marasmius*, and narrow and close to crowded in the *Clitocybe*. The *Clitocybe* is much whiter in color as a rule but it might be mistaken for faded specimens of the *Marasmius* if color alone were relied upon.

M. oreades is commonly known as the fairy ring mushroom from its habit of growing in circles on lawns or grassy places. The circles tend to increase in size from year to year and the grass at the periphery of the circle is usually a richer, darker green than the surrounding grass.

MARASMIUS ROTULA (Fr.) Kummer

Figures 234, 235, page 153

PILEUS ⅛–½ in. broad, thin, tough and pliant, hemispheric to convex, unexpanded, umbilicate, radiately grooved from disk to margin in an umbrella-like fashion, dry, unpolished, white or whitish, darker in the depression of the disk. FLESH whitish, membranous, odor and taste not distinctive. LAMELLAE distant or subdistant, broad, attached not to the stipe but to a free collar at the apex of the stipe, concolorous with the pileus. STIPE ½–2 in. long, filiform, tough, shining blackish, pallid at the apex. SPORES smooth, white, pip-shaped, 6–9 × 3–4 μ.

In dense groups on decaying wood and debris, common in eastern Canada. June–Oct.

This attractive little species often occurs in great abundance in the woods. The whitish pileus with dark umbilicus and the attachment of the lamellae to a free collar are the distinctive field characters.

MARASMIUS SCORODONIUS (Fr.) Kummer

Figure 236, page 153

PILEUS ¼–¾ in. broad, pliant, convex, expanding to plane, unpolished, smooth or slightly wrinkled, at first tinged reddish or tan, finally whitish. FLESH thin, odor of garlic when crushed. LAMELLAE adnate, not broad, moderately close, pallid. STIPE 1–2 in. long, about 1/16 in. thick, subequal or tapering downward, smooth, dark reddish brown to blackish at the base, pallid at the apex. SPORES smooth, white, ellipsoid, tapering to the apiculate end, 6–8 × 3–4 μ.

In groups on twigs and debris, fairly common. June–Sept.

The striking feature of this little species is its strong odor of garlic and it is sometimes used as a seasoning for roasts and gravies.

MARASMIUS SICCUS (Schw.) Fr.

Figure 114, page 51

PILEUS ½–1 in. broad, at first conical, becoming campanulate, often depressed in the center, dry, glabrous, pinkish red to ochraceous red to rusty red, striate-sulcate to the disk. FLESH membranous, mild. LAMELLAE free to adnexed, distant, white or tinged like the pileus. STIPE about 1½–3 in. long, very slender and thread-like, horny, glabrous, blackish brown, paler at apex, tubular, mycelioid at base. SPORES elongated, narrowed toward one end, white, smooth, 13–18 × 3–4.5 μ.

Gregarious in the woods on leaves, twigs and forest debris. July–Sept. Common.

Although of small size this little mushroom will attract notice because of its bright color and abundance. The color, deeply furrowed pilei and distant lamellae are good field characters.

LENTINUS

Lentinus includes a rather small group of white-spored species growing on wood and characterized by their hard, tough consistency and the serrate-torn edge of the lamellae. The stipe may be present or excentric, or lacking.

LENTINUS LEPIDEUS Fr. — Edible

Figures 303, 304, page 196

PILEUS 2–6 in. broad, tough to leathery, at first hemispherical or convex with margin incurved, expanding to plane or nearly so, dry, whitish to dingy yellowish or drab brownish buff, decorated, especially on the disk, with brown appressed spot-like or triangular scales arranged more or less concentrically, the entire surface usually at length becoming areolate or subscaly. FLESH white, firm to tough, becoming hard when old and dry, very thick on the disk. LAMELLAE variable in attachment, usually sinuate at the stipe and decurrent by narrow lines, broad, close to subdistant, white to dingy yellowish, edges saw-toothed. STIPE stout, 1½–4 in. long, ¼–1 in. thick, tapering toward the base, often more or less excentric and curved, especially if growing from the sides of stumps, etc., dry, solid, very hard and tough, white, browning in age, surface varying from nearly smooth to conspicuously scaly, sometimes with brown patches similar to those on the pileus. VEIL membranous, covering the lamellae in young stages, leaving an annular ridge on the apex of the stipe or often disappearing entirely in age. SPORES smooth, white, long-ellipsoid, many slightly irregular in shape, 9–12 $\times$ 4–5.5 μ.

Solitary or in clusters of 2 or 3, on timbers, railway ties, fence posts, stumps, etc., usually on conifer wood, common. May–Sept.

Although this species is reported to be edible when young, it soon becomes so tough and woody that it is of no value. It is often found growing in dry situations such as on railway ties or old timbers and it sometimes causes serious rotting of the wood.

Several other species are fairly common. *L. tigrinus* Fr. is a large, scaly, deeply umbilicate species that is sometimes much deformed. *L. cochleatus* Fr. usually grows in clusters with the stipes confluent and deeply furrowed. *L. vulpinus* Fr. grows in shelving clusters without stipes, and is remarkable for the very rough, coarsely hairy surface of the pileus. *L. haematopus* Berk. is a small species with a short blood-red stipe.

PANUS

The species of *Panus* occur on wood and have white spores, and the stipes are excentric or lateral or sometimes lacking. They differ from *Pleurotus* in being of tough consistency, more or less reviving when moistened, but some species might readily be mistaken for a *Pleurotus*. The consistency is similar to *Lentinus* but in *Panus* the edges of the lamellae are entire and even, whereas in *Le*[illegible] they are serrate-torn. *Panus* species are too tough to be of much value as food although some have been used to flavor soups and gravies.

Key

1. Stipe lateral or lacking 2
1. Stipe excentric 3

2. Pileus cupulate; lamellae covered by a veil when young .... *P. operculatus*
2. Lamellae not covered by a veil; taste astringent .... *P. stipticus*

3. Pileus densely hairy .... *P. rudis*
3. Pileus glabrous or delicately tomentose to slightly scaly .... *P. torulosus*

PANUS OPERCULATUS B. & C. — Not edible

Figure 175, page 110

PILEUS ¼–¾ in. broad, leathery, more or less pendent to convex, narrowed behind or above to a short stipelike base, brown or grayish brown, at first minutely flocculose-tomentose, becoming glabrous, margin strongly incurved, even. FLESH whitish, thin. LAMELLAE radiating from an excentric point, close, narrow, pale brown. VEIL a membranous tissue, covering the lamellae and splitting, leaving fragments on the margin. STIPE lateral or dorsal, very short or almost wanting, about 1/16–⅛ in. in diameter, concolorous with pileus. SPORES white, allantoid, smooth 4–5 × 1–1.5 μ.

In groups on twigs of deciduous trees. Sept.–Nov.

Panus salicinus Peck is very similar in appearance but lacks the veil. These two species have been discussed by Overholts (1938). They are not closely related to other species placed in *Panus*, but would be better placed in the genus *Tectella*, of which *T. patellaris* (Fr.) Murr. is the type. Some authors regard *Panus operculatus* as a synonym of *Tectella patellaris*, but Overholts questioned this since Fries did not describe or illustrate a veil for *T. patellaris*.

The species is of no importance from the standpoint of edibility, but it is an interesting and puzzling form that is not very common.

PANUS RUDIS Fr. — Edible; tough

Figures 238, 239, page 153

PILEUS ¾–2¾ in. broad, depressed toward the stipe, varying in shape from vase-like with central stipe to ear-shaped with almost lateral stipe, leather color to tawny or reddish, surface with a rather coarse, velvet-like, hairy coating, sometimes slightly tufted, margin at first inrolled, sometimes irregularly lobed. FLESH tough, somewhat corky when dry, taste slightly bitter, odor not distinctive. LAMELLAE decurrent on stipe, crowded, narrow, paler than the pileus. STIPE short and stout, sometimes almost lacking, equal or tapering downward, slightly excentric to almost lateral (never truly lateral), tomentose, concolorous with or paler than the pileus. SPORES smooth, white, ellipsoid, 5–6 × 2–3 μ.

Clustered on logs and stumps, common. June–Sept.

This is a very common species occurring on old stumps or logs. Its tough consistency and velvety, hairy pilei are distinctive. It is said to be edible but it is not recommended because of its tough consistency.

PANUS STIPTICUS Fr. Poisonous

Figures 278, 279, 280, 281, page 176

PILEUS ½–1¼ in. broad, tough and leathery, reviving when moistened, variable in shape, mostly kidney-shaped to shell-shaped, convex, with a slight depression toward the lateral stipe, pale cinnamon-buff, fading, surface very minutely scurfy, margin entire or irregularly lobed. FLESH leathery, taste very astringent, disagreeable. LAMELLAE cinnamon color, moderately broad, close to crowded, interveined, becoming tough when dry, ending in an even line at the stipe. STIPE a lateral continuation of the pileus, somewhat flattened, short and stubby, scarcely visible from above but distinct from below, paler than the pileus, solid, tough, surface somewhat mealy. SPORES minute, smooth, white, oblong, 4–5 × 2 μ.

In crowded, overlapping, shelving clusters, on stumps and logs, common. July–Oct.

Panus stipticus has a very unpleasant taste and is reported to be poisonous. In addition it is remarkable for its phosphorescent properties. When fresh and moist the fruiting bodies and mycelium glow in the dark, sometimes producing a ghostly effect in the woods.

PANUS TORULOSUS Fr. Edible; tough

Figures 315, 316, page 214

PILEUS 1½–4 in. broad, occasionally larger, tough and leathery, plane to depressed or infundibuliform, pale tan to brownish, tinged violet or reddish when moist, glabrous, or delicately tomentose when young, sometimes slightly scaly near the center, margin inrolled, even. LAMELLAE decurrent, close, narrow, some forked, pale tan, often tinged violet when moist. STIPE ¼–1 in. long, ¼–¾ in. thick, short and stout, excentric to lateral or sometimes nearly central, with a fine violaceous to gray tomentum, solid, tough. SPORES white, elliptical, 5.5–7 × 3–3.5 μ.

Clustered or occasionally solitary, on old stumps and logs. June–Aug.

Various authors differ as to whether or not *P. torulosus* and *P. conchatus* Fr. are distinct. Lange (1935–40) attempts to separate them on the basis of the habit of growth. He would place the forms with regular pilei and more or less central stipes in *P. torulosus*, and the clustered forms with lateral stipes in *P. conchatus*. It will probably require cultural studies to determine whether or not these forms are different species.

This fungus might be mistaken for a *Clitocybe* if collected when young and moist. It is said to be edible, but is too tough to be of much interest from that standpoint.

SCHIZOPHYLLUM

The name *Schizophyllum* means 'split leaf' and refers to the characteristic lamellae, which appear to be split along the edge and to fold back during dry weather.

Only one species is known in Canada but it is rather common. It suggests a *Pleurotus* in habit but is very different in texture. It is tough and leathery and inedible, although according to Singer the islanders in the Dutch East Indies and Madagascar chew the fruit bodies. The species of *Schizophyllum* may be of some importance as wood destroyers.

SCHIZOPHYLLUM COMMUNE Fr. — Not edible

Figure 237, page 153

PILEUS ¼–1 in. broad, thin, leathery-tough, densely matted-tomentose, pale grayish buff to grayish white, whitish when dry, sessile, fan-shaped to shell-shaped when laterally attached, vase-shaped expanding to saucer-shaped when attachment is central or excentric, margin incurved and lobed. LAMELLAE radiating from the point of attachment, uncrowded, thick and tough, grayish white or with a faint pinkish cast, densely hairy under a lens, appearing double on the edges because of a groove running the length of the edge of each lamella. SPORES smooth, creamy to deep yellowish, cylindric, about 3.5–5.5 × 1–1.5 μ.

On dead wood, often with bases confluent when growing in crowded clusters, common throughout the growing season.

This little species is of no interest as food because of its small size and tough, leathery consistency, but it frequently attracts attention because of its abundance. It is interesting because of the adaptations to withstand dry conditions that it exhibits. If the fungus is examined when moist it will be seen that each of the lamellae is split along the edge, and as the fruit body dries out, the edges curl back and roll outward, protecting the sides of the lamellae from loss of moisture, while at the same time the margin of the pileus also rolls inward and in this condition the fungus can survive considerable periods of drying. When re-moistened the margin of the pileus spreads out, the lamellae unroll and the fruit body commences to shed spores again.

TROGIA

Only a single species of this genus occurs in Canada. It is a small, white-spored, rather tough, wood-inhabiting fungus, that tends to fold up in dry weather and to revive again when moistened. It has no stipe. The most characteristic feature is the lamellae which are thick, blunt on the edge, and very much crisped and irregular.

TROGIA CRISPA Fr. — Not edible

Figure 240, page 153; Figure 419, page 299

PILEUS ¼–¾ in. broad, sessile, sometimes resupinate, shell-shaped, or shelf-like, sometimes overlapping, incurved when dry, reviving and spreading out when moist, reddish yellow to tan colored, covered with whitish hairs when young. LAMELLAE whitish to bluish gray, vein-like, very narrow, irregular, crisped, edges blunt. SPORES white, cylindric to allantoid, 3–4 × 1–1.5 μ.

Crowded to scattered on dead branches of frondose trees, fairly common. May–Nov.

This tiny mushroom is of no interest as food and is not very conspicuous when it is dry, but when it is moist the large clusters and beautifully crisped lamellae are sure to attract attention.

PLUTEUS

Pluteus includes species with a pink spore deposit, lamellae free from the stipe, and lacking both annulus and volva. The stipe is central and easily separable from the pileus. *Pluteus* species usually occur on old logs and stumps or on some form of decayed wood. Except for the common *P. cervinus* most of them are too small to be of much value as food. They are typically soft in consistency and decay rapidly.

Key

1. Pileus and stipe yellow and glabrous .... *P. admirabilis*
1. Pileus brown 2

2. Lamellae black on the edge .... *P. atromarginatus*
2. Lamellae not differently colored on edge .... *P. cervinus*

PLUTEUS ADMIRABILIS (Pk.) Pk.

Figure 241, page 153

PILEUS ½–1½ in. broad, fleshy-pliant, campanulate when very young, becoming expanded-campanulate to expanded-convex, at first subumbonate, at length slightly depressed on the disk, moist, hygrophanous, glabrous, unpolished, somewhat wrinkled especially on the disk, deep yellow, at times tinged olive-yellow, dusky yellow around wrinkles of the disk, margin striate when moist. FLESH very thin, white or whitish, dry and pithy, odor and taste not distinctive. LAMELLAE free, close, broad, broadest next to the stipe, soft and fragile, at first very pale yellowish, becoming sordid pinkish. STIPE 1–2 in. long, about ⅛ in. thick, equal, hollow within, fragile, splitting readily, glabrous, moist, clear yellow, paler than the pileus. SPORES smooth, subglobose, 5.5–6.5 × 5–6 μ, pinkish in deposits. CYSTIDIA slightly ventricose, with an elongated neck, obtuse at the apex, not horned.

In groups or scattered, on badly decayed logs and stumps, fairly common. June–Sept.

This is one of the more common of the smaller species and is an attractive little mushroom. One of the most distinctive field characters is the yellow stipe. *P. flavofuligineus* Atk. is another small yellow species with a pinkish stipe, and *P. leoninus* (Fr.) Kummer has a white stipe and the pileus is not wrinkled. These species are all too small to be of value as food.

PLUTEUS ATROMARGINATUS (Sing.) Kühner — Edible

Figures 242, 243, page 153

PILEUS 1–2½ in. broad, convex to expanded-convex, sometimes broadly umbonate or subumbonate, dark brown, moist, somewhat uneven, more or less fibrillose-streaked, the disk squarrose-scaly with minute upright tufts of fibrils, margin not striate. FLESH moderately thick on the disk, very thin on the margin, white, odor and taste mild. LAMELLAE free and not quite reaching the stipe, close to crowded, moderately broad, whitish, then dingy flesh color with smoky brown edges, many shorter lamellae of various lengths interspersed. STIPE 1–3 in. long, ⅛–¼ in. thick, subequal or slightly enlarged at the apex or base, whitish or tinged the color of the pileus, solid within, glabrous at the apex, fibrillose with smoky brown fibrils below. SPORES smooth, broadly ovoid, 5.5–7 × 4.5–5.5 μ, dingy flesh color in spore print. CYSTIDIA fusoid-ventricose, horned at the apex.

Solitary or scattered on decaying logs and stumps, not common. July–Oct.

The dark edges of the lamellae which provide a good field character for this species, are caused by the dark contents of the cystidia. This species is usually smaller and darker colored than *P. cervinus* and has been called *P. umbrosus* (Pers. ex Fr.) Kummer, but the latter has cystidia of a different type.

PLUTEUS CERVINUS (Schaeff. ex Secr.) Kummer — Edible

Figures 245, 246, page 155

PILEUS 1½–4 in. broad, fleshy, soft, at first convex to broadly campanulate, expanding to almost plane, the broad umbo either persisting or disappearing, smooth and glabrous to somewhat fibrillose, moist to dry, drab, varying from dull dark brown to pale dingy fawn (white in var. *albus*), darkest on the disk, paler toward the margin, sometimes streaked with darker innate fibrils, fading with age, margin even. FLESH very thin on the margin, thicker toward the disk, white, odor and taste mild to somewhat disagreeable. LAMELLAE close, free, broad, rounded next to the stipe, soft, in youth whitish, then flesh-pink to flesh-tan. STIPE 2–6 in. long, 3⁄16–3⁄8 in. thick, slightly enlarged downward, solid, smooth, whitish or tinged dingy yellow or brown, often bearing scattered appressed fibrils, apex at first pruinose. SPORES flesh-pink, smooth, broadly ellipsoid to ovoid, bluntly rounded or somewhat flattened on the ends, $5–8 \times 4–6\ \mu$, variable in size and shape. CYSTIDIA abundant, fusoid with long necks, about $60 \times 14\ \mu$, bearing 2–4 short horns at the apex.

Solitary or in groups of several, on decaying stumps and logs or associated with buried wood, common, especially in hardwoods. May–Oct.

This is the commonest species of the genus and is well known as a good edible mushroom. It varies considerably in size and color. *P. salicinus* Fr. is another brown species but usually smaller, with bluish or greenish tints toward the base of the stipe, and an unpleasant taste. *P. tomentosulus* Pk. may be as large as *P. cervinus* but is white and the pileus is floccose-tomentose. It can be distinguished microscopically by the cystidia which are not horned. *P. aurantiorugosus* (Trog) Sacc. is another fairly large but rare species. It is brilliantly colored, orange to reddish orange and is also known under the names *P. coccineus* Mass., *P. aurantiacus* Murr. and *P. caloceps* Atk.

Singer (1956) discusses a large species, *P. magnus* McClatchie, described from California but which he has also found in the eastern United States. This species would likely be mistaken for a pale *P. cervinus* unless examined microscopically. It differs in the cystidia which, instead of being horned as in *P. cervinus*, are fusoid with a long conical point, sometimes with a few small spines along the side. So far as is known there are no Canadian records of this species but it is one that might be found in this country.

VOLVARIELLA

Volvariella includes species with a pink spore deposit, lamellae free from the stipe, and a volva, but no annulus. Thus it is comparable to *Amanitopsis* in the white-spored group. The stipe is readily separable from the pileus and the flesh is usually rather soft. These species may be found either on wood or on the ground. They were formerly placed in *Volvaria* but this name cannot be maintained under the International Code of Nomenclature.

It is a relatively small genus and the species are not very frequently collected. In the literature there are conflicting reports as to their edibility. It seems to be established that *V. bombycina* is edible, but other species of the genus are best avoided.

VOLVARIELLA BOMBYCINA (Pers. ex Fr.) Sing. — Edible

Figures 247, 248, page 155

PILEUS 2–8 in. broad, at first ovate, becoming subcampanulate or convex-expanded, white and silky when young, becoming dingy or yellowish-stained, especially on the disk, at length finely fibrillose-scaly all over, cuticle separable to the disk, margin floccose, exceeding the lamellae and tending to split slightly at the edge. FLESH pure white, thick on the disk, thinning toward the margin, at the extreme margin almost lacking, odor and taste not distinctive. LAMELLAE free, not reaching the stipe, very crowded, very broad, whitish with a faint pink cast, becoming dingy flesh color, then brownish pink, edges somewhat uneven. STIPE 3–8 in. long, 3/8–3/4 in. thick, sometimes thicker at the base, tapering upward, often curved, white, silky-shining, smooth and glabrous, solid, no annulus. VOLVA large, thick, membranous, sac-like, clinging loosely around the base of the stipe. SPORES deep flesh color to brownish pink, smooth, ovoid, 6–8 × 5–5.5 μ.

Singly or several together on trunks of living trees or on dead wood, uncommon. July–Sept.

This is a very striking, although not common species. It sometimes reaches considerable size and the silky-tomentose pileus and large membranous volva are distinctive characters.

VOLVARIELLA SPECIOSA (Fr.) Sing. — Not edible

Figures 327, 328, 329, page 216

PILEUS 1½–4 in. broad, at first globose to ovate, becoming expanded to plane, slightly umbonate, white to grayish, viscid, glabrous, margin even. FLESH thin, soft, odor disagreeable. LAMELLAE free, crowded, broad, narrowed toward ends, rosy flesh color. STIPE 2–6 in. long, ½–¾ in. thick, equal or slightly enlarged at base, at first minutely hairy, becoming glabrous, tomentose at base, white, solid. VOLVA large, white, somewhat lobed. SPORES pink, ellipsoid, smooth, 14–20 × 9–12 μ.

Solitary or in groups on manure or rich soil, sometimes in grass or field. June–July.

There is a difference of opinion among authors as to whether or not this species is poisonous. It might be confused with *Lepiota naucina* or *Agaricus campestris* but both these species have an annulus and no volva. *V. gloiocephala* is very similar but the margin is striate and the spores slightly smaller. Some authors consider it to be a variety. This species has been reported as poisonous and could very easily be confused with *V. speciosa*. These species should not be eaten.

ENTOLOMA

Entoloma includes the pink-spored species roughly comparable to *Tricholoma* of the white-spored group. The stipe is fibrous to fleshy, sometimes splitting longitudinally very easily and there is no volva or annulus. The lamellae are sinuate-adnate to adnexed, sometimes seceding. The spores are more or less angular (Figure 33), varying from elliptical to spherical in general outline and sometimes almost square.

There is no very clear-cut distinction between the genera *Entoloma*, *Leptonia*, *Nolanea*, *Eccilia*, and the section of *Clitopilus* including species with angular spores. Many authors believe that they should be combined into one genus but this raises some difficult nomenclatural problems. Some authors have placed these species in the genus *Rhodophyllus* but this name is illegitimate because it was published after some of the other generic names mentioned above, and it is also of questionable validity because of its similarity to the name of the algal genus *Rhodophyllis*. Of the names listed above, *Entoloma* is the earliest, but there is also the older generic name *Acurtis* to be taken into account. This name was based on the so-called 'abortive' fruit bodies of *Clitopilus abortivus* and for a long time was disregarded because it was considered to be based on an abnormality. However, it has recently been shown that these fruit bodies produce normal basidia and spores and there is good reason to consider them to be a normal structure in the life cycle of the fungus. If this is so, then *Acurtis* will be the correct name for this group of species, but so far this name has not been taken up by mycologists and to use either *Acurtis* or *Entoloma* would require the creation of a good many new combinations. Thus, until either *Acurtis* is accepted or *Entoloma* officially conserved, it is thought preferable to use the other generic names rather than make new combinations in a book of this type.

None of these genera is of any importance as food. In fact some of the species of *Entoloma* are known to be poisonous and this whole group should be avoided. This genus provides a good illustration of the danger of attempting to lay down general rules regarding edibility. It has often been said that any mushroom that is pink underneath is good to eat, but *Entoloma* and its relatives provide a whole group of species with pink lamellae, and some of these species are definitely known to be poisonous, and others are suspected.

ENTOLOMA GRISEUM Pk.

PILEUS 1¼–3 in. broad, at first firm, becoming fragile, campanulate-convex to nearly plane, grayish brown, more umber when moist, slightly hygrophanous, glabrous, with a delicate separable pellicle, margin even, decurved, wavy. FLESH thin, easily splitting, odor and taste farinaceous. LAMELLAE adnexed, close to subdistant, moderately broad, at first grayish white, slowly becoming flesh colored. STIPE 1–3 in. long, 3/16–3/8 in. thick, equal or nearly so, silky-fibrillose, whitish or grayish, stuffed to hollow. SPORES pink, angular, 7–9 × 6.5–8 μ.

Solitary or in groups on the ground in the woods. July–Oct.

E. sericeum is close to this species but is smaller, and has a dark brown pileus with an umbo.

ENTOLOMA RHODOPOLIUM (Fr.) Kummer

Figure 249, page 155

PILEUS 1–3 in. broad, firm, campanulate, becoming expanded to nearly plane, hygrophanous, umber brown or smoky brown when moist, fading to pale brownish gray, silky-shining when dry, glabrous, not viscid but the surface slightly slippery, margin even, wavy. FLESH whitish, splitting easily, taste mild. LAMELLAE adnate, becoming emarginate, subdistant, broad, whitish at first, becoming deep rose. STIPE 1½–4 in. long, ¼–½ in. thick, equal or tapering up or down, sometimes curved, white, glabrous, somewhat floccose at apex, stuffed then hollow, easily splitting longitudinally. SPORES rosy pink, angular, 8–10.5 × 7–9 μ.

Solitary, in groups, or in clusters of two or three, on the ground in mixed or deciduous woods. July–Oct.

The pileus of this species is almost cartilaginous in texture and this character together with the white stipe and rosy spores and lamellae form its distinguishing characters. In *E. griseum* the stipe is more grayish and the spore color is not so bright.

ENTOLOMA SALMONEUM Peck

Figure 250, page 155

PILEUS ½–1½ (2) in. broad, fragile, conical to campanulate, with a slight umbo or papilla, rosy salmon to orange-salmon, becoming more brownish in age, glabrous at first, the older ones appearing slightly fibrillose-tomentose, margin straight, even, becoming upturned in age. FLESH very thin. LAMELLAE adnexed, subdistant, broad, narrowed at ends, yellowish salmon to pinkish salmon. STIPE 2–3 in. long, 1⁄16–3⁄16 in. thick, equal, glabrous, pruinose at the apex, concolorous, hollow. SPORES pink, 4-angled, nearly square, 11–13 μ measured diagonally.

In groups, usually among mosses in damp woods. July–Sept.

This is a beautiful and delicate species. It can easily be confused with *Hygrophorus amoenus* (Lasch) Quél. which is similar in coloring and stature but has smooth, white spores. *E. cuspidatum* Peck is somewhat similar in stature but is yellow and has a prominent papilla in the center of the pileus.

LEPTONIA

This genus includes a group of rather small, pink-spored species closely related to *Entoloma*. The genus is distinguished from *Entoloma* principally on the character of the stipe, which is cartilaginous in texture rather than fleshy-fibrous, but the distinction here is not very clear-cut. The spores are angular, as in *Entoloma*. The margin of the pileus is incurved when young so that at maturity it is usually expanded rather than conical or campanulate. The pileus is usually umbilicate or centrally depressed. The lamellae are adnate to adnexed often seceding. The genus is distinguished from *Nolanea* principally by the margin being inrolled when young, and from *Eccilia* by the attachment of the lamellae.

The species are mostly small and not well known. They are of no value as food but some of them are attractively colored. Usually they grow on the ground, occasionally on rotten wood or among sphagnum.

LEPTONIA ASPRELLA (Fr.) Kummer

Figure 251, page 155

PILEUS ½–1¼ in. broad, fleshy, convex, becoming expanded, umbilicate, gray-brown to umber, hygrophanous, somewhat scaly in the umbilicus, then fibrillose to glabrous, silky-shining when dry, margin striate, often splitting. FLESH whitish, thin, fragile, odor and taste mild. LAMELLAE adnate to adnexed or with a decurrent line, whitish or grayish white, then becoming pink, subdistant, 1/16–⅛ in. broad. STIPE 1½–3 in. long, 1/16–⅛ in. thick, cylindric to somewhat compressed, sometimes twisted, glabrous, smoky brown with a bluish cast, white mycelioid at base, hollow. SPORES pink, angular, more or less elongated, 9–13 × 6–8 μ.

Singly or in groups on the ground in woods. June–Sept.

The bluish stipe and the brown pileus which is striate and hygrophanous are the most distinctive characters of this species. *L. serrulata* (Fr.) Quél. is distinguished by the black serrulate margin of the lamellae. *L. placida* (Fr.) Quél. is a dark bluish species usually occurring on rotten wood, with a scaly pileus and dark squamules on the stipe. *L. lampropoda* (Fr.) Quél. occurs on the ground and has a bluish black pileus and smooth bluish stipe.

LEPTONIA FORMOSA (Fr.) Quél.

Figure 252, page 155

PILEUS ⅜–1½ in. broad, fleshy, convex-umbilicate, becoming plane. waxy-yellowish to grayish yellow, covered with small blackish scales especially in umbilicus, margin striate, becoming somewhat plicate in older specimens, FLESH thin, grayish or faintly yellowish, odor and taste mild. LAMELLAE adnate, sometimes with a slight decurrent tooth, broad, subdistant, tinged yellow, nearly white at first, becoming flesh-colored. STIPE 1½–2½ in. long $\frac{1}{16}$–⅛ in. thick, equal, smooth, somewhat striate, yellow, white-mycelioid at base, hollow, central. SPORES pink, angular, 10–12 × 6–7 μ.

Singly or in groups in swampy woods. Aug.–Sept.

A rather distinctive species in which the whole plant is more or less yellowish and the pileus is scaly.

NOLANEA

Nolanea includes a group of species with angular pink spores, closely related to *Leptonia*. In the young fruiting bodies the margin of the pileus is straight on the stipe rather than inrolled and the mature pileus is usually more or less conical to campanulate. This is the principal distinction from *Leptonia* although in *Nolanea* the pileus is usually umbonate or papillate whereas in *Leptonia* it is umbilicate or depressed. The lamellae are adnate to adnexed, often seceding. The cartilaginous stipe distinguishes it from *Entoloma*.

The species are not well known and are small and of no value as food. The one described here is fairly common.

NOLANEA MAMMOSA (Fr.) Quél.

PILEUS ½–1½ in. broad, conic to campanulate, umbonate, slightly hygrophanous, umber when moist, becoming grayish brown to fuscous, innately fibrillose and shining when dry, margin decurved. FLESH thin, brownish to whitish, odor and taste of rancid meal. LAMELLAE adnate, seceding, subdistant, broad, at first pale gray, then becoming pinkish, edges often uneven. STIPE 2–3½ in. long, $\frac{1}{16}$–$\frac{3}{16}$ in. thick, equal, sometimes compressed, glabrous, pruinose at the apex, brownish gray, hollow. SPORES pink, angular, 9–11 × 6–7 μ.

In groups on the ground in woods or grassy places. July–Sept.

N. fuscogrisella Peck is somewhat similar but is usually smaller; it has smaller spores, and the lamellae are at first white rather than gray. *N. papillata* Bres. is also very close to *N. mammosa* and is separated principally on the basis of smaller size and closer lamellae.

Figures 256-265

256. *Clitopilus orcellus*.
257. *Cortinarius collinitus*.
258. *Phyllotopsis nidulans*.
259. *P. nidulans*.
260. *Cortinarius armillatus*.
261. *C. semisanguineus*.
262. *Inocybe fastigiata*.
263. *Cortinarius violaceus*.
264. *Inocybe geophylla*.
265. *Pholiota acericola*.

256
257
258
259
260
261
262
263
264
265

Figure 266. *Mycena alcalina.*

267. *Pholiota aurivella.*
268. *P. aurivella.*
269. *P. caperata.*
270. *P. caperata.*
271. *P. marginata.*
272. *P. marginata.*
273. *P. flammans.*
274. *P. spectabilis.*
275. *P. squarrosoides.*
276. *Flammula spumosa.*

267
268
269
270
271
272
273
274
275
276

Figure 277. *Mycena galericulata.*

Figures 278-281. *Panus stipticus.*

CLITOPILUS

Clitopilus, as used here, corresponds roughly in the pink-spored group to *Clitocybe* in the white-spored group. The forms included have pink spores, the lamellae are broadly adnate to decurrent, and a volva and annulus are lacking. The stems are fleshy-fibrous, more or less similar in texture to the pileus and not separating from it readily.

The species included in the genus are probably not all closely related. The type species is *Clitopilus prunulus* (Scop. ex Fr.) Kummer in which the spores are marked with longitudinal ridges (Figure 32). Some students consider that only those species with longitudinally ridged spores are true *Clitopilus* species. The species with angular spores are probably more closely related to *Entoloma*. Kauffman recognized eleven species in Michigan.

Key

1. Spores longitudinally ridged .... *C. orcellus*
1. Spores angular or nearly smooth 2

2. Pileus usually more than 2 in. broad, gray, usually accompanied by whitish abortive fruit bodies .... *C. abortivus*
2. Pileus usually less than 2 in. broad 3

3. Spores strongly angular, taste and odor farinaceous; pileus not marked with concentric lines .... *C. albogriseus*
3. Spores smooth or very slightly angular, taste bitter; pileus with more or less concentric brownish lines .... *C. noveboracensis*

CLITOPILUS ABORTIVUS Berk. & Curt. — Edible

Figures 253, 254, page 155; Figure 420, page 300

PILEUS 2–4 in. broad, at first convex, becoming plane to depressed, gray to grayish brown, dry, at first delicately silky, becoming glabrous, margin at first inrolled, becoming wavy to lobed. FLESH white, rather fragile, odor and taste farinaceous. LAMELLAE decurrent, close, rather narrow, at first grayish then becoming pink. STIPE 1–3 in. long, ¼–½ in. thick, nearly equal, minutely floccose, grayish, paler than the pileus to whitish, solid, fibrous. SPORES pink, elongated, angular 8–10 $\times$ 5–7 μ.

In groups or sometimes in clusters, around stumps, or on rotten or buried wood. Sept.–Oct.

This fungus is remarkable in that some of the fruiting bodies frequently do not develop normally but form malformed whitish structures that at first sight would be taken for puffballs. These abortive fruiting bodies may be globoid or depressed or very irregular in shape. The abortive fruit bodies may be found alone or associated with normal fruit bodies. Both the abortive and normal forms are said to be edible.

CLITOPILUS ALBOGRISEUS Peck — Edible

PILEUS ½–1 in. broad, convex at first, becoming plane and depressed to umbilicate, pale gray, glabrous to slightly tomentose at the center, margin inrolled, odor and taste farinaceous. LAMELLAE adnate to decurrent, close, rather broad, at first grayish, becoming pinkish. STIPE ½–1 in. long, 1/16–3/16 in. thick, equal, glabrous, pale gray, whitish at base, solid. SPORES pink, elongated, angular, 9.5–12 × 6–8 μ.

On the ground in mixed woods, in groups or solitary. July–Sept.

This small species is fairly common. It is reported to be edible but is too small to be of any importance from that standpoint. *C. subplanus* Peck is very close but is said to lack the farinaceous odor. *C. micropus* Peck is supposed to have slightly smaller spores, shorter stipe, and a more silky pileus. Some of our specimens have rather short stipes, and pilei that are not entirely glabrous, but they have the larger spores of *C. albogriseus*.

CLITOPILUS NOVEBORACENSIS Peck — Doubtful

Figure 340, page 234

PILEUS ¾–2 in. broad, at first convex, becoming plane or slightly depressed, ashy gray to whitish, somewhat zoned toward the margin, concentrically rivulose, glabrous, margin inrolled. FLESH thin, white, taste bitter, odor farinaceous. LAMELLAE decurrent, crowded, narrow, ashy gray to pinkish flesh colored. STIPE 1–2 in. long, 1/16–¼ in. thick, nearly equal, pruinose to minutely tomentose, white-tomentose at base, colored like the pileus or paler, stuffed, becoming hollow. SPORES pink, ovoid, very slightly angular, nearly smooth, 4–6 × 3.5–4.5 μ.

Usually in groups on the ground in woods. July–Oct.

This species might be mistaken for *Clitocybe cyathiformis* (Bull. ex Fr.) Kummer, which is similar in coloring and stature. The pink spores and concentrically rivulose pileus distinguish the *Clitopilus*, which appears to be the more common of the two.

CLITOPILUS ORCELLUS (Bull. ex Fr.) Kummer — Edible

Figure 256, page 173

PILEUS 1–2 in. broad, sometimes larger, convex at first, becoming plane to depressed, grayish white to buff or faintly yellowish, slightly viscid, silky, margin inrolled, undulate or lobed. FLESH white, rather thick, odor and taste farinaceous. LAMELLAE decurrent, close, narrow, white, becoming pinkish. STIPE 1–2 in. long, ⅛–½ in. thick, nearly equal or sometimes slightly swollen in the middle, somewhat floccose-fibrillose, white, solid, sometimes excentric. SPORES pinkish, fusiform, longitudinally striate or ridged, 9–11 × 5–6 μ.

Solitary or in groups on the ground in open woods. July–Oct.

This fungus may not be distinct from *C. prunulus* (Scop. ex Fr.) Kummer. The latter is not viscid, and has subdistant lamellae, but the viscidity of *C. orcellus* is not very pronounced. Both species have longitudinally ridged spores.

PHYLLOTOPSIS

The genus *Phyllotopsis* is based on the fungus that has been commonly known as *Claudopus nidulans* (Pers. ex Fr.) Karst. However, this species is obviously not closely related to the type species of *Claudopus*, *C. byssisedus* (Pers. ex Fr.) Gill., and should not be placed in the same genus. It seems to be more closely related to *Pleurotus*, especially *P. serotinus* (Fr.) Kummer, but the spores are colored. It has also been placed in *Panus* because of the tough consistency but the spore color excludes it here also. Sometimes the spore color has been interpreted as ochre and it has been placed in *Crepidotus* but the color seems to be more pink than ochre.

The tomentose pileus, lack of stipe, tough consistency, colored lamellae, and small, cylindric-allantoid pinkish spores are the chief characters of the genus.

One other species, *P. subnidulans* (Overholts) Singer, has been described and is said to differ in having globose spores.

PHYLLOTOPSIS NIDULANS (Pers. ex Fr.) Singer — Not edible

Figures 258, 259, page 173; Figure 421, page 300

PILEUS ½–3 in. broad, rather tough, sessile, attached laterally or sometimes narrowed behind to a stem-like base, nearly circular to reniform or fan-shaped, sometimes laterally confluent forming a shelf up to 6 in. long, convex, bright yellow, fading to buff, densely tomentose-hairy, margin strongly inrolled, odor pungent, disagreeable when fresh or sometimes lacking. LAMELLAE adnate, close, rather narrow, bright orange-yellow. STIPE lacking, but there may be a tomentose base next to the lamellae. SPORES flesh pink in mass, allantoid, smooth, $6–8 \times 3–4\ \mu$.

In groups or clusters on decaying logs of both deciduous and coniferous trees. June–Oct.

The peculiar, pungent odor of this species is a good means of recognizing it when it is present, but sometimes it appears to be very faint or lacking.

CORTINARIUS

Cortinarius is a very large and difficult genus containing several hundred species. The spores are dark brown or rusty brown. The principal distinguishing character of the genus is the cortina or veil which covers the lamellae in the young stages. It is composed of loose silky hyphae that suggest a cobweb. If copious, the veil may remain as a ring or annular zone on the stipe, or it may disappear quite early. The typical rough spores and rusty brown spore deposit are diagnostic when the cortina has disappeared. An outer, universal veil may also be present and may leave a sheath or several annular zones on the stipe. In one section of the genus the cortina is glutinous.

In mature fruit bodies the lamellae are usually dark brown, but in young ones they may be of different colors such as white, yellow, olivaceous, or lilac, and the color of the young lamellae is very important in the identification of species. In a great many *Cortinarii* it is necessary to have a series of stages from young buttons to mature plants before the species can be identified. Microscopic characters, especially the size and shape of the spores, are also important in distinguishing species.

Kauffman (1932) has given the most complete account of the North American species. He recognized seven sections of the genus as follows:

1. *Myxacium*, in which both the pileus and the stipe are viscid;
2. *Bulbopodium*, in which only the pileus is viscid and the stipe is furnished with a marginate bulb;
3. *Phlegmacium*, in which the pileus is viscid and if the stipe is bulbous the bulb is not marginate;
4. *Inoloma*, in which the pileus is neither viscid nor hygrophanous but is usually fibrillose or scaly, and the stipe rather stout and usually clavate-bulbous;
5. *Dermocybe*, in which the pileus is neither viscid nor hygrophanous, but is usually silky, and the stipe rather slender;
6. *Telamonia*, in which the pileus is hygrophanous and a universal veil is present;
7. *Hydrocybe*, in which the pileus is hygrophanous and a universal veil is lacking.

***Cortinarius* is a genus which, in general, should be avoided by the amateur who is collecting for the table. Some of the species are known to be deadly and some are unpleasant to the taste. In addition, the difficulty of determining the species accurately is so great that they are best left alone.**

Only a very few of the commoner and more striking species that are not likely to be confused with others are described here.

CORTINARIUS ALBOVIOLACEUS (Fr.) Kummer — Edible

Figure 341, page 235

PILEUS 1–2½ in. broad, fleshy, at first campanulate, then convex and broadly umbonate, at first pale violaceous, soon becoming silvery-white and shining, usually slightly violaceous tinged, surface dry, appressed-silky, margin decurved. FLESH tinged violet, odor and taste mild. LAMELLAE adnate to emarginate or slightly decurrent, close, rather broad, at first pale violet, finally becoming cinnamon-brown. STIPE 1½–3 in. long, ¼–½ in. thick near the apex, becoming thicker below and somewhat clavate, usually sheathed by a thin, whitish, universal veil, violaceous in the upper part and beneath the veil, spongy-stuffed. SPORES brown, elliptical, slightly rough, 6.5–10 × 4.5–6.0 μ.

In groups on the ground in mixed woods. Aug.–Oct.

This species is fairly common and can be recognized by the silvery-shining appearance with the slight violet tinge, which is especially noticeable in the stipe and young lamellae.

CORTINARIUS ARMILLATUS (Fr.) Kummer — Edible

Figure 260, page 173

PILEUS 2–5 in. broad, fleshy, convex becoming expanded to plane, tawny reddish to brick-red, moist, innately fibrillose, margin decurved, becoming flat in age. FLESH rather thin, pallid, somewhat spongy, odor slightly of radish, taste mild. LAMELLAE adnate, sometimes sinuate, broad, distant, at first pale cinnamon, then dark rusty brown. STIPE about 2½–5½ in. long, ⅜–¾ in. thick at the apex, up to 1½ in. thick below, clavate, brownish or reddish brown, with several orange-red or cinnabar-red bands from the universal veil, solid. SPORES brown, ellipsoid, rough, 10–13 × 7–8 μ.

Solitary or in groups on the ground in coniferous woods. Usually Aug.–Sept.

The distinguishing character of this species is the series of red bands on the stipe, and it is not likely to be confused with any other species. *C. haematochelis* (Bull.) Fr. has a single red band on the stipe instead of several, but some authors regard this as only a form of *C. armillatus*.

CORTINARIUS COLLINITUS Fr. — Edible

Figure 257, page 173

PILEUS 1¼–3 in. broad, fleshy, convex to plane, variable in color, usually yellowish to orange-yellow, sometimes whitish when young and sometimes with lilac tints near the margin, very viscid when moist, the margin at first incurved, finally becoming upturned. FLESH whitish to pale yellowish buff; odor and taste mild. LAMELLAE adnate with a tooth, close, fairly broad, at first pale violet or pallid, when mature becoming dull reddish brown. STIPE about 2½–4½ in. long, ¾–1¼ in. thick, equal or tapering slightly at the base, spongy-stuffed, covered with the pale violaceous or whitish, viscid, universal veil which cracks transversely leaving thick, irregular bands or patches, at first whitish then becoming stained rusty or yellowish especially toward the base, with the cortina forming a collapsed ring above, and the apex of the stipe white and silky. SPORES rusty brown, almond-shaped, rough, 11–15 × 7–8.5 μ.

Usually in groups on the ground in either coniferous or deciduous woods. Aug.–Oct.

This species was described by Kauffman under the name *Cortinarius mucifluus* Fr. but, according to the International Rules of Nomenclature, *C. collinitus* is the correct name. It is a fairly easy species to recognize because of the very viscid pileus and stipe, the pale-yellowish colors, and the whitish patches on the stipe. It shows considerable variation in the amount of lilac color present on the stipe, young lamellae, and pileus, and several varieties and forms have been recognized on this basis. The collections in the herbarium of the Plant Research Institute can be placed in two groups depending on spore size. In one group the spores are as noted above, and Smith (1944) considers this to be the typical form. In the other group the spores are smaller, 10–13 × 6–7.5 μ, and Smith would call these *C. collinitus* v. *trivialis* (Lange) Smith.

CORTINARIUS SEMISANGUINEUS (Fr.) Gillet — Probably edible

Figure 261, page 173

PILEUS ¾–2½ in. broad, fleshy, campanulate-convex, subumbonate, in age becoming expanded, tawny yellow to cinnamon-yellow, silky to delicately fibrillose-scaly, margin even, sometimes splitting in age. FLESH yellowish white, odor and taste mild. LAMELLAE adnate to slightly decurrent, close to crowded, narrow, blood-red. STIPE 1–2½ in. long, ⅛–¼ in. thick, equal, yellow, tawny-fibrillose, solid. SPORES brown, elliptical, nearly smooth, slightly rough under high magnification, 5–8 × 3.5–5.0 μ.

In groups in moist swamps or sphagnum. Aug.–Oct.

This is a typical member of the section *Dermocybe*, and is recognized by the combination of blood-red lamellae and yellowish pileus and stipe. In *C. cinnabarinus* Fr. the pileus and stipe as well as the lamellae are blood-red. *C. croceofolius* Peck has bright orange lamellae and *C. cinnamomeus* Fr. has yellow lamellae. *C. semisanguineus* appears to be the commonest species of the group in the Ottawa district at least. Members of this group are probably all edible but there does not seem to be much definite information about them available.

CORTINARIUS VIOLACEUS (L.) Kummer — Edible

Figure 263, page 173

PILEUS 2–5 in. broad, fleshy, convex, obtuse, finally becoming plane or slightly umbonate, dark violet, sometimes metallic shining, covered with small erect tufts or scales, margin fibrillose or fringed. FLESH thick, firm, violet-gray to dark violet, not turning purple when bruised, odor and taste mild. LAMELLAE adnate, becoming adnexed, broad, subdistant, dark violet. STIPE 2½–5 in. long, ⅜–1 in. thick above, wider below, clavate to bulbous, fibrillose, dark violet, violaceous within. SPORES rusty cinnamon, broadly ellipsoid, rough, 12–17 × 7–10 μ.

Singly or scattered on the ground in coniferous forests. Aug.–Oct.

This species is not very common but has been included because it is one of the most strikingly beautiful of the mushrooms. The dark violet colors and erect scales are very characteristic. It is considered to be the type species of the genus *Cortinarius*.

INOCYBE

The genus *Inocybe* is a large one but it is not of much interest to the amateur collector. Most of the species are small and can be identified only by the use of microscopic characters. Only a few species become large enough to attract the attention of the mycophagist, and these are best left alone. Some of the species are known to be poisonous and the amateur collector is well advised to avoid the entire genus.

The spores are ochre-brown in mass and may be rough or smooth or more or less angular or tuberculate (Figure 34). Many species have cystidia on the lamellae and the characters of the cystidia and spores are important in the determination of species. The pileus is usually conic to campanulate and more or less fibrillose or scaly, sometimes silky and often splitting radially. The colors are usually rather dull, mostly shades of grown, gray, or ochre, although there are a few species more distinctively colored.

INOCYBE FASTIGIATA (Schaeff. ex Fr.) Quél.

Figure 262, page 173

PILEUS ¾–2½ in. broad, at first conic-campanulate, then more or less expanded, umbonate, splitting readily on the margin, varying in color from tawny to ochraceous or dull yellowish, innately fibrillose-streaked, soon becoming conspicuously long-rimose with streaks of paler color showing as the cuticle becomes rimose. FLESH white, rather thin except at the umbo. LAMELLAE adnexed, moderately broad, close, pallid, then grayish olive, finally brownish. STIPE 1–2½ in. long, ⅛–¼ in. thick, equal, fibrillose, white or tinged the color of the pileus. SPORES ellipsoid to somewhat bean-shaped, smooth, dull ochre-brown, 9–12 × 5–6 (7) μ, cystidia lacking.

In groups on the ground in woods and on lawns beneath trees. June–Oct.

This species has been included because it is one of the more common species of *Inocybe* and will likely be found by anyone collecting mushrooms. The ochraceous colors, radiating fibrils, prominent umbo, and splitting of the pileus are distinctive features.

INOCYBE GEOPHYLLA (Sow. ex Fr.) Kummer

Figure 264, page 173

PILEUS ½–1½ in. broad, at first subconic to campanulate, with incurved margin, then expanded-campanulate to nearly plane with a small umbo persisting, white, dry, radiately fibrillose-silky, splitting readily on the margin. FLESH white, thin except on the umbo. LAMELLAE adnate to adnexed, moderately broad, close, whitish to grayish, finally pale clay-colored. STIPE ⅝–1½ in. long, about ⅛ in. thick, equal, solid, silky-fibrillose, concolorous with the pileus. SPORES smooth, pale brown, ellipsoid, slightly inequilateral, 7–9.5 × 4.5–5.5 μ. CYSTIDIA ventricose, fusoid, about 40–55 × 10–20 μ.

In groups on the ground in woods, occasionally on lawns beneath trees, fairly common. July–Oct.

This is an interesting little species that is entirely white but has brown spores. The lamellae become pale clay-colored in age. It is a common species and can be recognized easily.

Inocybe lilacina (Boud.) Kauffm. is very similar except for its color, and some authors regard it as merely a variety of *I. geophylla*. Its pileus and stipe are lilac colored at first but soon fade. It is frequently found associated with *I. geophylla*.

PHOLIOTA

Pholiota is an important genus for the mycophagist because it includes a number of good edible species that are fairly large and often occur in large clusters providing abundant material for food. However, as in other genera, some care must be taken to identify the species correctly because at least one is known to be poisonous. Some of the species are important as wood destroyers.

The genus includes those species that have rusty brown or ochre-brown spores, lamellae attached to the stipe, and a membranous annulus but no volva. They may be found either on wood or on the ground. Some of the species are markedly scaly but others are smooth.

The North American species have been pretty well known since the excellent monograph by Overholts (1927). He recognized 56 species and a few have been added since then. However, recent authors tend to divide *Pholiota* into several genera, with *P. squarrosa* (Pers. ex Fr.) Kummer as the type of *Pholiota* in the narrower sense.

Many of the characters used in distinguishing these genera are based on microscopic structures not readily determined by the amateur collector, hence for the purposes of this book it is considered preferable to use the genus in the wider, more traditional sense.

Key

1. Occurring on the ground 2
1. Occurring on wood or sawdust 3

2. Pileus whitish to cream, glabrous; spores smooth with truncate apex .... *P. vermiflua*
2. Pileus ochraceous buff to cinnamon-buff, unevenly wrinkled, at first with fine whitish fibrils or with a hoary bloom; spores large, rough-walled .... *P. caperata*

3. Pileus glabrous at all stages 4
3. Pileus scaly or becoming so 5

4. Pileus cinnamon-brown, fading; spores rough, ellipsoid .... *P. marginata*
4. Pileus yellowish buff or paler; spores smooth with a truncate apex (Figure 31, page 7) .... *P. acericola*

5. Pileus viscid 6
5. Pileus not viscid 7

6. Edge of lamellae white-crenulate; spores 11-14 μ long .... *P. albocrenulata*
6. Edge of lamellae even, not white; spores 7-9 μ long .... *P. aurivella*

7. Stipe fibrillose; taste bitter; spores rough, 7-9 μ long .... *P. spectabilis*
7. Stipe squarrose-scaly; taste not bitter; spores smooth, less than 6 μ long 8

8. Pileus bright golden yellow to orange; scales soft and floccose .... *P. flammans*
8. Pileus dull colored, cinnamon-buff to tawny; scales stiff .... *P. squarrosoides*

PHOLIOTA ACERICOLA Peck

Figure 265, page 173

PILEUS 1–3 in. broad, fleshy, at first convex with an inrolled margin, becoming expanded, at times broadly subumbonate, nonstriate, smooth, glabrous, slightly hygrophanous, pale watery brownish to yellow-buff when moist, unpolished and paler when dry. FLESH white, rather thin except on the disk. LAMELLAE adnate, sinuate to slightly subdecurrent on the stipe, moderately broad, close, pallid, sometimes with a purplish tint, then dull brown from the spores. STIPE 2–4 in. long, ¼–½ in. thick, equal or enlarged at the base, stuffed within, glabrous to fibrillose-striate, pallid, usually with strings of whitish mycelium at the base. ANNULUS persistent, ample, membranous, white or pallid, tending to be remote from the apex. SPORES dull brown, smooth, ovoid with a truncate apex, 8.5–10.5 × 5–6.5 μ. CYSTIDIA flask-shaped to ventricose, occasionally divided at the tip to form two or three blunt projections.

Single or in groups on decaying stumps and logs, occasionally on the ground, rather common. June–Sept.

This species is most likely to be confused with *P. praecox*. The latter usually grows on the ground in open grassy places and *P. acericola* usually is on rotten wood but is sometimes found on the ground in woods. The white mycelial strands at the base of the stipe of *P. acericola* are a helpful distinguishing character. The slightly purplish cast of the lamellae is another field character to watch for.

The edible qualities of this species are not known but it might be confused with the poisonous *P. autumnalis*. If a microscope is available they can easily be distinguished by the spores. *P. acericola* is usually a taller, more slender plant, paler and more yellowish in color.

PHOLIOTA ALBOCRENULATA Peck

Figure 351, page 236

PILEUS 1–4 in. broad, fleshy, at first convex or campanulate, becoming expanded, nonstriate, margin often appendiculate with veil fragments, very viscid, ochre-brown to tawny or chestnut, decorated with darker appressed scales which on drying become faded and subsquarrose. FLESH moderately thick, whitish, odor not distinctive, taste unpleasant. LAMELLAE sinuate-adnate to subdecurrent, broad, close to subdistant, at first grayish, becoming rusty brown, white-crenulate on the edge. STIPE 2–5 in. long, ⅛–½ in. thick, equal or slightly enlarged at the base, often curved, pallid or tinged yellow or brown below, white-floccose at the apex, sparsely to densely scaly up to the annulus with squarrose or fibrillose scales. ANNULUS slight, ragged, disappearing, often clinging in fragments to the pileus margin and failing to form a ring on the stipe. SPORES smooth, brown, inequilateral, subfusiform, 11–14 × 6–7 μ.

Solitary or in groups of several, on stumps and logs, and on trunks of living trees, especially maple. June–Sept.

The white-crenulate edge of the lamellae is the most striking single character of this species. The fruiting bodies are darker colored than those of *P. aurivella* and usually are not in clusters. The scales on the pileus are rather easily rubbed off or washed off. The large spores will also distinguish it readily from similar species. Its edible qualities are not known.

PHOLIOTA AURIVELLA (Batsch ex Fr.) Kummer — Edible

Figures 267, 268, page 175

PILEUS 1½–4 in. broad or sometimes larger, fleshy-pliant, hemispheric to convex with inrolled margin at first, becoming expanded, smooth, viscid when moist, yellow to tawny, darkest on the disk, concentrically spotted with darker squamules which in age become appressed and sometimes wash off in wet weather, margin even, somewhat appendiculate. LAMELLAE adnate or sinuate at the stipe, broad, close, pallid yellowish, then rusty brown. STIPE stout, central or excentric, 1½–4 in. long, ¼–½ in. thick, subequal, solid or stuffed, viscid in wet weather, more or less concolorous with the pileus, somewhat squamulose-scaly below the annulus, often curved. ANNULUS slight, yellowish, disappearing. SPORES smooth, ellipsoid, rusty brown, 7–9 × 4–5 μ.

Solitary or in clusters on stumps and logs of deciduous trees, fairly common. Aug.–Oct.

This species is well known in North American literature under the name of *Pholiota adiposa* (Fr.) Kummer, but it now appears that the true *P. adiposa* is a European species with smaller spores than those of our fungus. *P. aurivella* is fairly common and can be recognized by the rather bright colored, scaly, viscid pileus. The viscid layer should be peeled before cooking.

Collectors on the west coast are likely to confuse this species with *P. squarroso-adiposa* Lange. This latter species occurs in large clusters, often on alder, and has pale yellow, viscid caps with darker scales, and a dry, very scaly stipe. It has smaller spores than *P. aurivella* and can be distinguished from *P. adiposa* by the dry stipe.

PHOLIOTA CAPERATA (Fr.) Kummer — Edible

Figures 269, 270, page 175; Figure 422, page 301

PILEUS 2–4 in. broad, fleshy, convex then expanded, sometimes subumbonate, glabrous or at first with fine whitish fibrils and thin patches of whitish bloom, usually more or less uneven to wrinkled, cinnamon-buff to ochraceous buff, paler on the margin or evenly colored. FLESH thick, white. LAMELLAE adnate to adnexed, broad, close, often transversely marked with light and dark bands, at first pallid, then brown from the spores. STIPE 2½–5 in. long, ¼–½ in. thick, equal, solid, pallid, glabrous to subfibrillose, at times somewhat scurfy with minute white floccules at the apex. ANNULUS large, membranous, remote from the apex. VOLVA usually not evident, sometimes leaving a few traces at the base of the stipe. SPORES rusty brown, rough, broadly ovoid, inequilateral in one view, tapering to one end, 11–16 × 7–10 μ, or varying larger.

Solitary to scattered, on the ground in woods, fairly common. July–Oct.

This species is fairly common and is an easy one to recognize. The color, the uneven surface and somewhat hoary appearance of the pileus, and the large membranous annulus are quite distinctive. The spores are more like those of *Cortinarius* than *Pholiota* and the presence of a rudimentary volva is also a character distinguishing it from other species of *Pholiota*. Most modern authors separate this species from *Pholiota* and put it in the genus *Rozites*.

PHOLIOTA FLAMMANS (Batsch ex Fr.) Kummer

Figure 273, page 175

PILEUS 1–2 in. broad or occasionally larger, campanulate or convex with inrolled margin at first, becoming expanded, bright golden yellow or tinged orange, covered when young with a dense coating of dry, lemon-yellow floccules, even on the margin, appendiculate. FLESH rather thin, yellow. LAMELLAE adnate or with a slight decurrent tooth, close, not broad, yellow at first, then somewhat rusty from the spores. STIPE 1–3 in. long, ⅛–⅜ in. thick, subequal, stuffed or hollow, yellow, densely coated up to the annular zone with lemon-yellow floccules or squarrose scales. ANNULUS yellow, ragged, disappearing. SPORES minute, ellipsoid, smooth, 3–6 × 2–3 μ.

Solitary or in clusters of several, on stumps and logs, rare. Aug.–Sept.

This species is rare but is included because of its striking appearance. The bright colors and scaly pileus and stipe make it attractive. The very small spores are also a distinctive character. There is no information regarding its edibility.

P. kauffmaniana Smith is a species occurring on the Pacific Coast that is similar in appearance to *P. flammans* but has viscid pilei.

PHOLIOTA MARGINATA (Batsch ex Fr.) Quél. Dangerous

Figures 271, 272, page 175

PILEUS ¾–3 in. broad, fleshy, convex to plane or slightly depressed, sometimes slightly umbilicate, dark cinnamon-brown, hygrophanous, fading to yellowish buff or ochraceous orange, glabrous, margin even or slightly striatulate, extending somewhat beyond the lamellae. FLESH thin, concolorous with the pileus, odor and taste none (Overholts says farinaceous). LAMELLAE adnate to slightly decurrent, subdistant, moderately broad, ⅛–¼ in., at first yellowish brown, becoming darker and concolorous with the pileus. STIPE ¾–2 in. long, ⅛–⅜ in. thick, equal or slightly enlarged near apex, fibrillose, pruinose above the annulus, hollow, concolorous with pileus or paler. ANNULUS somewhat fibrillose and sometimes disappearing. SPORES brown, ellipsoid, at first smooth finally slightly rough, 7–10 × 4.5–6 μ.

On rotten wood. May–Oct.

There are several small species with smooth pilei that resemble *P. marginata*. This group of species probably belongs in *Galerina* rather than *Pholiota*, and while *P. marginata* itself may be edible, the group as a whole is dangerous because it contains species known to be poisonous. *P. autumnalis* Peck is like *P. marginata* but is viscid and this species is definitely known to be poisonous. *P. marginella* Peck (*Kuehneromyces vernalis* (Pk.) Sing. & Sm.) is similar to *P. marginata* in gross appearance but has smaller, smooth spores and lacks cystidia on the lamellae. *P. unicolor* is said to differ in having a more persistent annulus and a thinner stipe, and in drying a brighter color. There is a species occurring on lawns on the Pacific Coast that belongs in this complex and is also definitely known to be poisonous. This species has been described as *Galerina venenata* Smith. It is said to be up to 1¼ in. broad, glabrous, moist, hygrophanous, cinnamon-brown, fading to dingy yellowish white, with a farinaceous odor and taste but the taste slowly becomes bitter and disagreeable. The stipe is brownish and has a thin apical annulus. Its occurrence on lawns makes this a very dangerous species. All of these small species with smooth pilei should be avoided as food.

PHOLIOTA SPECTABILIS (Fr.) Gill. Not edible

Figure 274, page 175

PILEUS 1½–4 in. broad, convex becoming expanded to nearly plane, rather bright colored, buff-yellow to tawny orange, dry, glabrous at first, becoming fibrillose to fibrillose-scaly, margin even, sometimes wavy, incurved at first. FLESH yellowish, thick, taste bitter. LAMELLAE adnate to adnexed with a decurrent tooth, crowded, narrow to moderately broad, yellow becoming rusty reddish. STIPE 1½–6 in. long, ¼–1 in. thick, nearly equal to ventricose or enlarged at base, concolorous or darker below the annulus, more yellow above, fibrillose below the annulus, pruinose to floccose above, solid, sometimes very hard. ANNULUS superior, membranous, persistent, yellowish. SPORES rusty brown, ellipsoid, rough-walled, 7–9 × 4.5–6 μ.

Singly to cespitose on stumps or trunks or sometimes from buried wood. June–Oct.

Because of the bitter taste this species is not recommended as an edible fungus although it is not known to be poisonous. However, it is likely to attract attention because of its size and bright colors. It is similar in color to *Phaeolepiota aurea* but lacks the granulose covering of the pileus of the latter.

PHOLIOTA SQUARROSOIDES (Pk.) Saccardo — Edible

Figure 275, page 175

PILEUS 1–3 in. broad, at first subglobose with inrolled margin, becoming expanded-convex or broadly subumbonate, even on the margin, often appendiculate, densely squarrose-scaly with dry, coarse, tawny scales, between the scales whitish to cinnamon-buff and viscid. FLESH whitish, moderately thick on the disk, odor and taste not distinctive. LAMELLAE sinuate-adnate, moderately broad, close, pallid, then brown from the spores. STIPE stout, 1½–4 in. long or longer, about ¼ in. thick, equal, stuffed, pallid, brownish toward the base, scaly up to the annulus with recurved tawny squamules, white at the apex. ANNULUS pallid, fibrillose-torn, often disappearing. SPORES smooth, brown, ellipsoid 4–6 × 3–4 μ.

In dense clusters on deciduous wood. Aug.–Sept.

P. squarrosoides is fairly common and frequently occurs in large clusters providing plenty of material for a meal. It is not likely to be confused with any poisonous species.

Pholiota squarrosa, a similar species, is a yellower fungus, with a dry pileus and slightly larger spores. It is also edible but may have an unpleasant taste when old.

PHOLIOTA VERMIFLUA Peck — Edible

Figure 362, page 254

PILEUS ¾–2½ in. broad, occasionally larger, firm, fleshy, subhemispherical, becoming expanded, even on the margin, at times appendiculate with veil fragments, creamy whitish or tinged yellowish, smooth, glabrous, moist to subviscid, almost shining when dry, becoming areolate-cracked on the disk. FLESH white, moderately thick on the disk, thin toward the margin, odor mild, taste mild to slightly unpleasant. LAMELLAE adnexed to sinuate-adnate or with a slight decurrent tooth, seceding, rather broad, close, pallid at first, then pale grayish brown, finally dark brown with edges white, alternate lamellae short. STIPE 1¼–4 in. long, usually swollen at the apex up to ¼–½ in. thick, tapering downward, sometimes subequal, solid or with a narrow tubule, at the apex whitish and minutely scurfy, below the annulus glabrous and concolorous with the pileus. ANNULUS disappearing or persistent, small, membranous, white, staining brown from spore deposit. SPORES smooth, ovoid, thick-walled, slightly inequilateral, truncate at one end, dark brown, 10–14 × 6–8 μ. CYSTIDIA pear-shaped or subglobose, scattered or scarce.

In groups in lawns, cultivated fields, and grassy places in open woods. May–Aug.

This species often appears early in the season in May or June on lawns or in gardens. It is most likely to be confused with *P. praecox* (Pers.) Fr. which can be separated with certainty by the spores, which are mostly less than 10 μ long. Both species are edible.

PHAEOLEPIOTA

This genus has been separated from *Pholiota* on the basis of the mealy-granulose covering of the pileus. It is best characterized as a brown-spored *Cystoderma* and only the one species is known.

PHAEOLEPIOTA AUREA (Mattuschka ex Fr.) Maire ex Konr. & Maubl. — Suspect

Figure 373, page 256

PILEUS 2–6 in. broad, convex to plane, slightly umbonate, dry, with a granulose, powdery covering that is easily rubbed off, ochraceous yellow to golden yellow, or ochraceous tawny, margin incurved at first, somewhat appendiculate. FLESH thick, yellowish, odor none, taste mild. LAMELLAE adnexed, rounded behind, close, broad, light buff to ochraceous buff or cinnamon. STIPE 1½–5 in. long, ½–¾ in. thick, enlarged at the base, concolorous or lighter than the pileus, granular-scurfy below the annulus, glabrous above, stuffed, sometimes becoming hollow. ANNULUS large, pendulous, membranous, dark buff below, lemon color above, disappearing in old plants. SPORES pale ochraceous buff, ovoid to elongate-ovoid, smooth or sometimes slightly rough in age, 9–12 × 4–6 μ.

Singly or gregarious on the ground. Sept.

This is a rare but very striking fungus. It appears to be western in its distribution. It has the appearance of a large *Cystoderma* with brown spores.

FLAMMULA

The genus *Flammula* includes species with ochre spores, fleshy to fibrous, central stipes, lamellae usually rather bright colored, and an annulus usually lacking although sometimes a trace of one may be found. Most of the species occur on wood although there are a few exceptions.

They are distinguished from *Pholiota* by the lack of an annulus, from *Hebeloma* by the brighter lamellae and spores and from *Naucoria* by the fibrous stipe.

The species are not very well known and are difficult to identify. Many of them have a bitter or unpleasant taste and the genus is not one to attract amateur collectors. They are not recommended for eating although there is no evidence that they are poisonous. Only the one species, *F. spumosa* (Fr.) Kummer, which is one of the commonest of the genus, is described here.

FLAMMULA SPUMOSA (Fr.) Kummer

Figure 276, page 175

PILEUS ¾–2 in. broad, convex then expanded, smooth, glabrous, viscid when moist, pale sulphur-yellow except on the disk which is fulvous-tinged, margin even. FLESH thin, tinged yellowish. LAMELLAE adnate to slightly sinuate or decurrent by a tooth, moderately broad, close, pallid yellowish, then ochre-brown. STIPE 1–2½ in. long, ⅛–¼ in. thick, equal, stuffed or becoming hollow, fibrillose, yellowish above, stained brownish toward the base. VEIL fibrillose, yellowish, disappearing. SPORES smooth, ellipsoid, ochre-brown (6) 7–8 (9) × 4–4.5 (5) μ. CYSTIDIA numerous, fusoid-ventricose.

In groups on decaying wood or on the ground in woods, fairly common. July–Oct.

This is one of the commonest species of the genus and is principally characterized by the sulphur-yellow to greenish yellow margin of the pileus and the fulvous to tawny disk.

HEBELOMA

Hebeloma includes a group of rather poorly known brown-spored species. They have viscid pilei, adnexed or emarginate lamellae, and a volva and annulus are lacking but some species have a fibrillose veil. The spores are dull colored in mass, alutaceous or dull brown, never bright rust-colored as in *Inocybe* or *Cortinarius*, and the colors of the pileus are usually somewhat dull also.

The species are not well known and are difficult to identify. Some of the species are known to be poisonous and none are recommended as food. It is a genus to be avoided by the amateur. Only one species is described here.

HEBELOMA SINAPIZANS (Fr.) Gillet — Not edible

Figure 384, page 282

PILEUS 2½–4½ in. broad, fleshy, firm, convex, obtuse, margin even, inrolled at first, becoming expanded, glabrous, viscid, pinkish buff to cinnamon. FLESH thick, compact, odor and taste of radish or mustard. LAMELLAE adnexed, broad, close, pallid then pale brownish from the spores. STIPE white, 2½–4½ in. long, ½–1 in. thick, subequal, stuffed then hollow, white-floccose-scaly, especially toward the apex. SPORES large, rough, almond-shaped, inequilateral, brown, 11–13 (13.5) × 7–8 (8.5) μ.

Scattered or in groups on the ground in woods, uncommon. Sept.–Oct.

This is one of the largest of the hebelomas and has been included as a representative of the genus. The floccose-scaly stipe and broad lamellae together with the large size are the distinguishing characters. Definite information as to its edibility is lacking but it is not recommended since other members of the genus are known to be poisonous.

CONOCYBE

Conocybe includes a group of small, fragile, *Mycena*-like mushrooms with rusty brown spores. They were formerly included in the genus *Galera* but this name cannot be used under the International Code of Nomenclature because the name had been given earlier to a genus of flowering plants. The old genus *Galera* has been divided into two genera based on the structure of the cuticle. In *Conocybe* the cuticle is cellular and in *Galerina* it is filamentous. The species are unimportant as food because they are so small and fragile and they are difficult to identify. One species is included here because it is commonly found on lawns and is fairly easily recognized.

CONOCYBE CRISPA (Longyear) Singer

Figure 385, page 283

PILEUS ½–1¼ in. broad, conic to campanulate, sometimes slightly umbonate, striate to rugulose, glabrous, atomate, whitish buff, more brownish on disk when moist. FLESH thin, membranous. LAMELLAE adnexed, close to subdistant, narrow, crisped and interveined, ferruginous brown. STIPE 1¼–3 in. long, about 1/16 in. thick, white, or tinged ochraceous, equal, slightly bulbous at base, hollow. SPORES ellipsoid to ovoid, rather variable, smooth, rusty brown, 11–16 (18) $\times$ 8–12 μ.

Gregarious on lawns and grassy places. June–July (Sept.).

The distinctive character of this species is the crisped lamellae. A similar species in which the lamellae are not crisped is also common on lawns. It has been generally known as *Galera tenera* Fr., but according to Smith (1949) the true *G. tenera* is rare and the common species that has been called *G. tenera* is, in reality, *Conocybe lactea* (Lange) Métrod. Although this species is, perhaps, more common than *G. crispa* it was thought desirable to choose the latter as a representative of this group because it could be identified with more certainty. All the species of this group are too small and fragile to be of any value as food.

Figures 282-291

282. *Naucoria semiorbicularis.*
283. *Paxillus involutus.*
284. *P. atrotomentosus.*
285. *P. atrotomentosus.*
286. *Agaricus campestris.*
287. *A. haemorrhoidarius.*
288. *A. diminutivus.*
289. *A. diminutivus.*
290. *A. silvicola.*
291. *A. silvicola* and *Amanita virosa.*

282
283
284
285
286
287
288
289
290
291

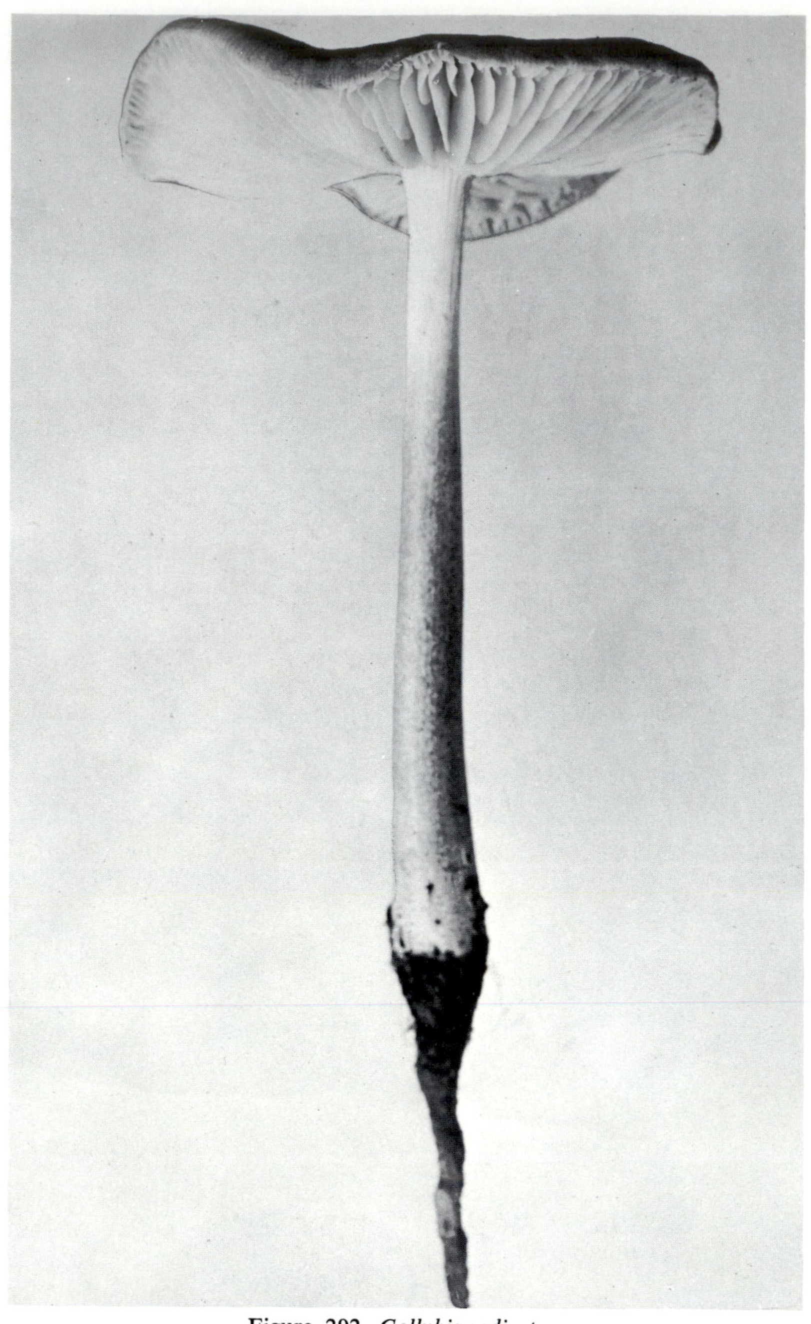

Figure 292. *Collybia radicata*.

Figures 293-302

293. *Crepidotus fulvotomentosus*.
294. *Stropharia coronilla*.
295. *S. hornemannii*.
296. *S. hornemannii*.
297. *S. semiglobata*.
298. *Naematoloma capnoides*.
299. *Psathyrella candolleana*.
300. *P. candolleana*.
301. *Coprinus comatus*.
302. *C. quadrifidus*.

293
294
295
296
297
298
299
300
301
302

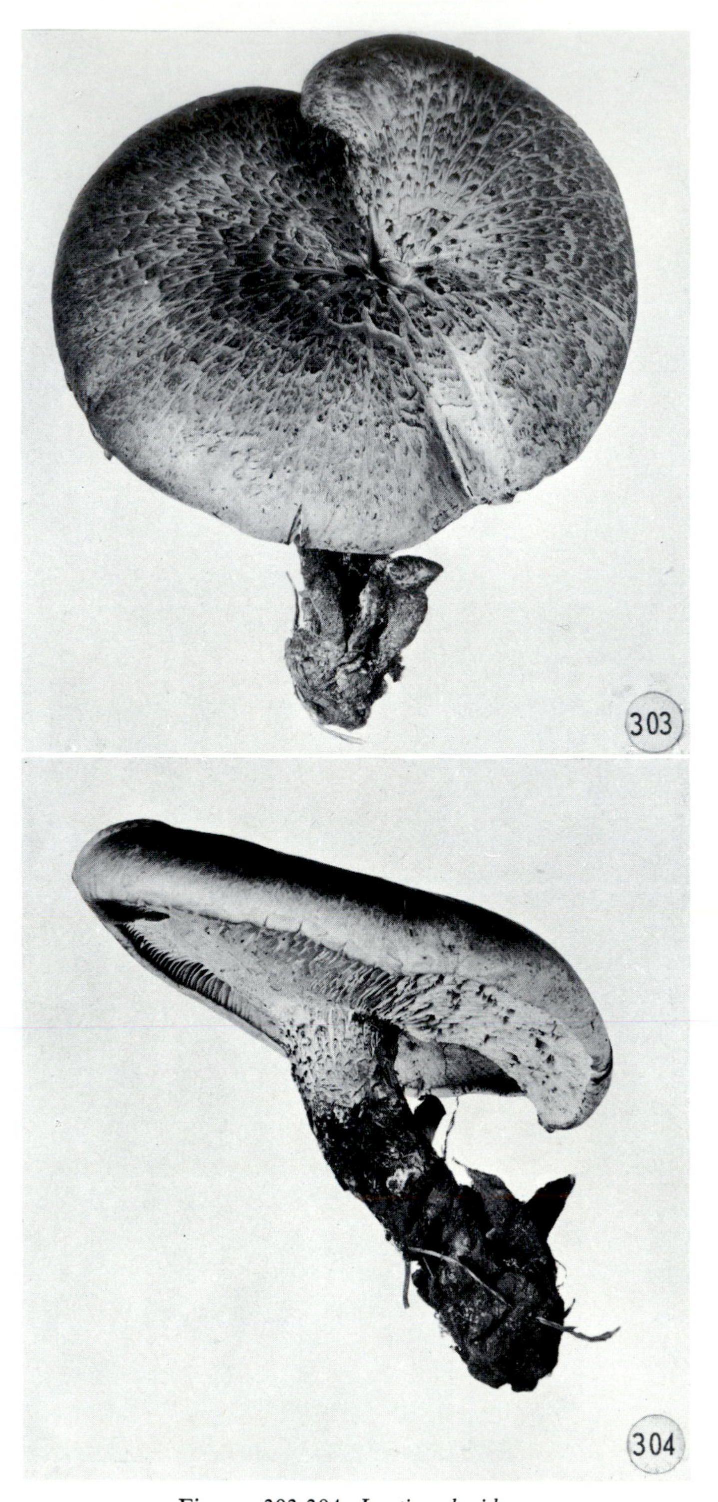

Figures 303-304. *Lentinus lepideus.*

NAUCORIA

Naucoria includes a group of small, brown-spored species that are not well known. The spores are ochre-brown to rusty brown. The stipe is subcartilaginous and there is no annulus. They grow either on the ground or on wood, occasionally on dung. Because of their small size they are of no value as edible mushrooms and not much is known about their edibility. Only one rather common species is described here.

NAUCORIA SEMIORBICULARIS (Bull.) Fr. Edible

Figure 282, page 193

PILEUS ¼–1 in. broad, hemispherical to convex, smooth, glabrous, viscid when moist, dull yellowish varying to tan or tinged reddish brown, unicolorous or darker on the disk, margin even. FLESH thin, pallid. LAMELLAE adnate, broad, close, pallid, then dull brown from the spores, edges pallid. STIPE 1–2½ in. long, up to ⅛ in. thick, subequal or slightly thickened at the base, stuffed, glabrous to minutely fibrillose, concolorous or paler than the pileus. SPORES smooth, thick-walled, ovoid, inequilateral in one view, dull brown (10.5) 11–13.5 (15) $\times$ 7.5–8 (9) μ.

In groups on the ground in grassy places, common. May–Sept.

This species is common on lawns and grassy places, appearing after rains throughout most of the season. The small size, yellowish tan color, brown spores and lack of an annulus are its distinguishing characters.

TUBARIA

Tubaria is a small genus of poorly known species. They are small brown-spored forms with decurrent lamellae and fragile, cartilaginous stipes. The species are mostly rather rare, and are of no interest to those collecting mushrooms for food.

One species is included here because it is one of the earliest mushrooms to appear in the spring and for that reason may attract attention.

TUBARIA FURFURACEA (Pers. ex Fr.) Gill.

PILEUS ⅜–1 in. broad, occasionally larger, fleshy, convex to plane or slightly depressed, cinnamon-brown, hygrophanous, fading to buff or pinkish buff, long-striate on the margin when moist, sometimes with whitish patches from the veil, appressed-fibrillose to glabrous on the disk. FLESH thin, watery brownish, no odor, taste mild. LAMELLAE close to subdistant, broad, adnate to slightly decurrent, concolorous with pileus. STIPE about ¾–1½ in. long, and about ⅛ in. thick, equal or slightly enlarged at the base, fibrillose, concolorous with pileus or slightly paler, hollow. SPORES ellipsoid, flattened slightly on one side, pale ochraceous in mass, smooth, 7–9 × 4–5.5 μ.

Growing singly or in groups on sticks and debris in the woods. May–Oct.

This is a rather common little brown mushroom with broad lamellae and a fibrillose pileus, often found early in the spring. *T. pellucida* (Bull. ex Fr.) Gill. is similar but has smaller spores, 5.5–7 × 4–5 μ.

CREPIDOTUS

Crepidotus corresponds to *Pleurotus* of the white-spored group and includes those species with ochre-brown to rusty spore deposit and in which the stipe is excentric, lateral, or lacking. The lamellae may be whitish in young fruiting bodies but become brown as the spores mature. Most of the species occur on decayed wood and they are mostly rather small and of no importance as food.

CREPIDOTUS FULVOTOMENTOSUS Peck — Edible

Figure 293, page 195

PILEUS ½–2½ in. broad, convex becoming expanded, sessile, laterally attached, often semicircular or kidney-shaped in outline, margin incurved at first, striatulate when moist, surface coated when young with a dense tawny tomentum, which, as the pileus expands, becomes separated into tawny, fibrillose scales, exposing the paler ground color beneath. FLESH thin, pliant, pallid or tinged yellowish. LAMELLAE radiating from the lateral point of attachment, moderately broad, close, pallid then dull ochre-brown, edges white. STIPE lacking, the pileus attached to the substrate at a lateral point. SPORES ovoid, slightly inequilateral in one profile, dull ochre-brown, 7.5–9 (10) × 5–6 μ.

In groups on decaying hardwood, common. May–Oct.

This is one of the larger species of the genus and can be recognized by the tawny scales on the pileus. *C. calolepis* (Fr.) Karst. also has a brown tomentum but has smaller spores. *C. dorsalis* Peck is reddish yellow and has globose spores. *C. versutus* Peck has a white tomentum. *C. mollis* Peck is glabrous and somewhat gelatinous, and *C. haerens* Peck is viscid. *C. malachius* B. & C. has globose spores and broad lamellae, and *C. applanatus* (Fr.) Kummer has globose spores and narrow lamellae. These five species are all white or whitish. *C. cinnabarinus* Peck is bright scarlet and more common in the West.

The genus is not important as far as food is concerned but several of the species are fairly common and will be encountered by the collector.

PAXILLUS

Paxillus includes species with ochre-yellow spores, the stipe more or less excentric, and the lamellae easily separable from the flesh of the pileus. The lamellae are usually more or less decurrent and anastomose on the stipe, sometimes becoming poroid. In this respect they show relationship with the Boletaceae and some authors consider that they should be placed in this family. It is a relatively small genus and we have two fairly common species, neither of which is recommended as food.

PAXILLUS ATROTOMENTOSUS (Batsch ex Fr.) Fr. Not recommended

Figures 284, 285, page 193

PILEUS 2–5 in. broad, occasionally larger, tough, convex at first, becoming plane to depressed, rusty brown to blackish brown, appressed-tomentose, dry, margin inrolled. FLESH white, firm, rather thick, odor and taste mild. LAMELLAE adnate-decurrent, easily separable from the pileus, close, rather narrow, forked and anastomosing on the stipe, yellowish tawny. STIPE excentric to lateral, 1–3 in. or more long, ½–1¼ in. thick, stout, straight or curved, enlarged toward base, rooting, covered with a dense, blackish brown, velvety tomentum, solid. SPORES yellow, oval, smooth, 5–7 × 3–4 μ.

Singly or clustered on old logs or stumps or from buried wood. July–Sept.

The blackish tomentose stipe is a very distinctive character of this species. It is reported to be edible but tough and of poor flavor.

PAXILLUS INVOLUTUS (Batsch ex Fr.) Fr. Not recommended

Figure 283, page 193

PILEUS 2–5 in. broad, at first convex, becoming expanded, then depressed, yellowish brown to reddish brown or olive-brown, with darker spots, downy-tomentose, becoming smooth, margin persistently inrolled and often somewhat ridged. FLESH pale yellowish, becoming brownish when bruised, thick. LAMELLAE decurrent, easily separable from the pileus, crowded, broad, forked and anastomosing on the stipe, olive-yellow, becoming brown when bruised. STIPE 1½–3 in. long, ½–1¼ in. thick, equal or tapering slightly downward, sometimes enlarged at the base, glabrous, colored like the pileus or paler, often streaked or spotted with darker brown, solid, central or excentric. SPORES yellowish brown, elliptical, smooth, 7–9 × 4–5.5 μ.

Solitary or in groups on the ground in woods or at the base of stumps. July–Oct.

This is an unattractive plant because of the rather dingy colors and the brown stains that develop. Some authors maintain that it is edible but as there are reports that it has caused poisoning in Europe, it is not recommended. It is fairly common and easily recognized by the inrolled margin and the close, decurrent, yellowish lamellae that separate readily from the pileus.

AGARICUS

Agaricus is one of the most important genera for those who are interested in mushrooms as food. It includes both the common cultivated mushroom and the meadow mushroom, which is probably the wild mushroom most frequently eaten, at least in English-speaking countries.

The genus is characterized by the purple-brown spores, presence of an annulus, and free lamellae. The stipe is a different texture from the pileus and separates readily from it. The genus is relatively clear-cut and it is comparatively easy to recognize an *Agaricus* but many of the characters used to distinguish species seem to intergrade and some of the species are difficult to identify.

Agaricus is usually regarded as one of the safest genera to use as food, and critical identification of the species is not very important from this standpoint. However, *A. xanthodermus* Genev. has been reported to cause illness in some people, and Smith reported that a form of *A. arvensis* Fr. found growing in a swamp in Michigan caused illness. *A. placomyces* Pk. and *A. hondensis* Murr. have also been reported to cause illness on occasion. Hence, with *Agaricus* species as with any other mushroom, unfamiliar species should be tried cautiously at first.

The name *Psalliota* has frequently been used for this genus but under the International Code of Nomenclature *Agaricus* is the correct name.

According to Smith (1949) there are about 70 species of *Agaricus* in North America. There have been two recent studies of the European species by Möller (1950, 1952) and Pilat (1951) but they do not include all the North American species.

Key

1. Pileus small, less than 1½ in. broad .... *A. diminutivus*
1. Pileus mostly 2 in. or more broad 2

2. Flesh quickly turning red when broken .... *A. haemorrhoidarius*
2. Flesh not reddening when broken 3

3. Growing in fields, pastures, open places or lawns 4
3. Growing in the woods 5

4. Annulus double, lamellae narrow .... *A. edulis*
4. Annulus single .... *A. campestris*

5. Pileus scaly with brown to black fibrils .... *A. placomyces*
5. Pileus smooth, white, disk turning yellowish when bruised .... *A. silvicola*

AGARICUS CAMPESTRIS Fr. — Edible

Figure 286, page 193

Meadow Mushroom

PILEUS 1½–3 in. broad, fleshy, firm, convex, somewhat flattened, becoming nearly plane, white, sometimes tinged brownish when old, at first silky, becoming delicately fibrillose-scaly or glabrous, the margin extending beyond the lamellae and usually fringed with veil remnants. FLESH white, thick, firm, not changing color when bruised, odor and taste pleasant. LAMELLAE free, crowded, rather narrow, at first pink, becoming gradually purple-brown and finally black. STIPE 1–2 in. long, 3/8–5/8 in. thick, equal or sometimes narrowed below, white, silky above the annulus, becoming brownish below, slightly fibrillose to glabrous, stuffed. ANNULUS thin, single, sometimes evanescent or remaining attached to the margin of the pileus. SPORES dark chocolate-brown, ellipsoid, smooth, 5.5–7.5 × 3.5–4.5 μ; basidia four-spored.

It grows singly or in groups in grassy places, lawns, pastures, fields, etc. It is usually found in the fall but sometimes occurs in the spring.

The meadow mushroom is probably the best known of all the wild species and for some people it is the only true mushroom, all other species being regarded as unsafe to eat. About the only danger in connection with this mushroom is that the button stage may be collected carelessly and a young *Amanita virosa* gathered by mistake. In most cases the habitat is sufficiently different to prevent such errors, since the *Agaricus* is usually found in open fields and the *Amanita* in the woods. Nevertheless the danger is sufficiently great that button stages should not be used unless the collector is certain that he can distinguish between *Agaricus* and *Amanita* at this stage and that no universal veil is present.

The cultivated mushroom was for a long time considered to be a form or variety of the meadow mushroom but it is now recognized as a distinct species, *Agaricus hortensis* Cke. The spores, basidia, and cystidia are different in the two species, and *A. campestris* grows in grassland and will not grow under the same conditions as *A. hortensis*. However, the two species are so similar in appearance that an acquaintance with the cultivated species will enable one to recognize *A. campestris* in the field.

A. arvensis Fr., the field mushroom or horse mushroom, is a larger, more robust species with a flatter pileus. The spores are larger also. Larger specimens of *A. campestris* might be confused with it but this is unimportant since both are edible, although Smith reported that specimens of what he considered to be a variety of this species growing in a swamp and to which he gave the name *A. arvensis* var. *palustris*, caused illness.

AGARICUS DIMINUTIVUS Peck — Edible

Figures 288, 289, page 193

PILEUS 3/4–1 1/2 in. broad, fragile, convex becoming plane, whitish or grayish, more brownish on disk, silky-fibrillose, the fibrils more or less reddish or pinkish to reddish brown. FLESH thin, whitish, odor and taste mild. LAMELLAE free, close to crowded, moderately broad, at first whitish, finally dark purplish brown. STIPE 1–2 in. long, 1/16–3/16 in. broad, equal or slightly bulbous at base, whitish, glabrous or somewhat fibrillose, stuffed, then hollow. ANNULUS delicate, whitish, persistent. SPORES purple-brown, ellipsoid, smooth, about 5–6 × 3–3.5 μ.

Usually single, occasionally in groups, among grass or moss. One collection in our herbarium is on wood. Aug.–Sept.

This is an attractive but rather delicate species, usually too small to be of any interest as food but said to be edible. Several other small species have been recognized but they are not well known. *A. auricolor* Krieger is yellow with yellow floccose patches on the stipe below the annulus, *A. micromegethus* Peck is merely white fibrillose below the annulus, and the stipe of *A. comptuliformis* Murr. is glabrous below the annulus.

AGARICUS EDULIS (Vitt.) Moeller & J. Schaeffer — Edible

Figure 343, page 235

PILEUS 1½–4 in. broad or sometimes larger, firm, fleshy, at first broadly convex to hemispherical, somewhat depressed on the disk, expanding and becoming plane, white or whitish to slightly yellowish when old, glabrous to slightly silky, the margin at first incurved and exceeding the lamellae. FLESH thick, firm, white, odor and taste mild. LAMELLAE free, crowded, narrow, at first pink, then purple-brown to blackish brown. STIPE 1–1½ (2) in. long, ½–1 in. thick, short, stout, equal, glabrous or somewhat scurfy above the annulus, white, solid. ANNULUS double, usually about midway up the stipe. SPORES purplish brown to chocolate-brown, broadly ellipsoid to subglobose, smooth, 5–6 × 4–5 μ.

Singly or in groups, usually in cities along pavements or on lawns or sometimes in barren areas where the soil is packed hard. June–Oct.

This species has been better known in North America as *A. rodmani* Peck but *A. edulis* seems to be the correct name for it. The rather squatty stature, double annulus, and firm flesh are the principal characters of this species. It is a fine edible mushroom and usually appears in towns and cities where it is probably collected by mistake for *Agaricus campestris*.

AGARICUS HAEMORRHOIDARIUS Fr. — Edible

Figure 287, page 193

PILEUS 2–4 in. broad, fleshy, at first subglobose to ovoid, expanding and becoming campanulate-convex or finally plane, fibrillose-scaly, vinaceous brown to grayish brown. FLESH white, quickly turning to blood-red when cut or bruised, odor and taste mild. LAMELLAE free, crowded, moderately broad, at first whitish, then pinkish, finally purple-brown. STIPE 2–5 in. long, ¼–½ in. thick, equal or bulbous at the base, somewhat fibrillose to glabrous, whitish becoming brownish, stuffed, then hollow. ANNULUS large, conspicuous, white, persistent. SPORES purple-brown, ellipsoid, smooth, 5–7 × 3–4 μ.

In groups or scattered, sometimes in small clusters, on the ground in mixed woods. July–Oct.

The outstanding character of this species is the almost instantaneous change to blood-red in the color of the flesh when fresh specimens are broken or bruised. Other species also exhibit this color change but it is usually slower.

AGARICUS PLACOMYCES Peck — Probably edible for most people

Figure 344, page 235

PILEUS 2–5 in. broad, rather fragile, at first broadly ovate, becoming convex and finally plane, whitish beneath the blackish brown, fibrillose scales, the disk blackish brown from the unbroken fibrillose covering. FLESH white or slightly yellowish under the cuticle, sometimes becoming pinkish, thin, odor slight to somewhat disagreeable, taste mild. LAMELLAE free, crowded, rather narrow to moderately broad, at first white to grayish, becoming pink and then purple-brown. STIPE 2½–5 in. long, ¼–½ in. thick, tapering upward, more or less bulbous at the base, glabrous, whitish, sometimes staining yellow, stuffed becoming hollow. ANNULUS large, conspicuous, double, whitish above, the lower layer cracking into brownish patches. SPORES chocolate-brown, ellipsoid, smooth, 5–6 × 3.5–4 μ.

Solitary or in groups or sometimes in clusters in mixed woods. June–Sept.

The scaly pileus, large annulus, and tapering, somewhat bulbous stipe are the chief distinguishing characters of this species. There have been occasional reports of it causing illness so that it should be tried with caution at first.

AGARICUS SILVICOLA (Vitt.) Sacc. — Edible for most people

Figures 290, 291, page 193

PILEUS 2½–6 in. broad, moderately firm, at first convex, expanding and becoming plane, white or creamy white, staining yellow on the disk when bruised, somewhat silky-fibrillose. FLESH moderately thick, brittle, white, becoming yellow when bruised, odor and taste mild. LAMELLAE free, crowded, narrow to moderately broad, at first whitish, then pink and finally blackish brown. STIPE 3–6 in. long, ¼–¾ in. thick, equal or tapering upward slightly, with an abrupt bulb at the base, or bulb lacking and base flattened, somewhat silky to glabrous, creamy white, staining yellow when bruised, stuffed then hollow. ANNULUS large, double, smooth above with the lower layer cracking and forming yellowish patches which may disappear. SPORES purplish brown to chocolate-brown, ellipsoid, smooth, 5–6.5 × 3–4.5 μ.

Solitary or in groups, occasionally in clusters of two or three, usually growing in the woods. July–Sept.

This species is edible although it should be tried with caution as occasional cases of illness in individuals have been reported. The remarks relative to the dangers of using button stages in *A. campestris* apply with much greater force to *A. silvicola* because it grows in much the same habitat as *Amanita virosa*.

A. silvicola is rather variable in size from a slender plant to a very robust form easily confused with *A. arvensis*. Usually the slender forms have an abrupt, flattened bulb at the base of the stipe but this character may vary also. The annulus is large and conspicuous. *Agaricus abruptibulbus* Peck is considered to be a synonym.

STROPHARIA

Stropharia includes the species that have purple-brown spores, an annulus, and the lamellae attached to the stipe. Usually the pileus is viscid. It differs from *Naematoloma* in possessing an annulus and might be confused with that genus if the annulus has disappeared.

It is not a large genus. About 35 species are known from North America but only a few are common, and the species are sometimes difficult to identify. Some are suspected of causing poisoning and the genus should be avoided by amateurs.

Key

1. Pileus bright green, fading to yellow .... *S. aeruginosa*
1. Pileus never green 2

2. Growing on dung; pileus yellow, hemispherical .... *S. semiglobata*
2. Not on dung 3

3. Pileus mostly more than 2 in. broad, brownish or smoky purplish; stipe squarrose-scaly. .... *S. hornemannii*
3. Pileus less than 2 in. broad, yellowish; stipe smooth or slightly fibrillose .... *S. coronilla*

STROPHARIA AERUGINOSA (Curt. ex Fr.) Quél. — Reported poisonous

Figures 386, 387, page 283

PILEUS ¾–2 in. broad, fleshy, campanulate-convex, becoming plane, slightly umbonate, viscid, at first bright green from the thick gluten, fading slowly to yellowish, sometimes with white scales near the margin, becoming glabrous. FLESH whitish to bluish, soft. LAMELLAE adnate, close, broad, at first whitish, then grayish, finally chocolate-brown, slightly purplish, the edges white and minutely flocculose. STIPE 1½–3 in. long, ⅛–⅜ in. thick, equal, viscid, scaly to fibrillose below the annulus, bluish green, hollow. ANNULUS evanescent. SPORES dark brown, slightly purplish, ellipsoid, smooth, $7–9.5 \times 4–5\ \mu$.

It occurs in the woods or sometimes in gardens. Sept.–Oct.

This is a striking and beautiful species when it is young and fresh but the bright green color fades with age. It is said to be common in Europe but has only been collected occasionally in the Ottawa district.

STROPHARIA CORONILLA (Bull. ex Fr.) Quél. — Suspected

Figure 294, page 195

PILEUS ¾–2 in. broad, convex to nearly plane, pale yellow to whitish or buff to pale ochre-yellow, glabrous, moist to slightly sticky. FLESH white, soft, fairly thick, odor slightly unpleasant. LAMELLAE adnate, rounded behind, lilac-flesh color becoming purplish black, close, moderately broad, edge white-fimbriate. STIPE short, ¾–1½ in. long, ⅛–½ in. thick, equal, white, dry, flocculose above the annulus, fibrillose below, becoming smooth, stuffed to hollow. ANNULUS membranous, distant from apex, persistent, striate on upper side. SPORES purple-brown, ellipsoid, smooth, 7–9 × 4.5–5 μ.

Gregarious to scattered on lawns or grassy places. Aug.–Oct.

This small species is not common but because of its occurrence on lawns and the fact that it is suspected of being poisonous it has been included. It might be mistaken for an *Agaricus* but the lamellae are not free from the stipe.

STROPHARIA HORNEMANNII (Fr.) Lund. & Nannf. — Not recommended

Figures 295, 296, page 195

PILEUS 1–5 in. broad, fleshy, firm, convex to plane, sometimes slightly umbonate, viscid, glabrous or with some white floccose scales at the margin at first, brownish or smoky reddish brown, to purplish brown, becoming olive-brown near the margin and more yellowish brown on the disk, margin inrolled when young, then decurved and sometimes elevated. FLESH whitish to watery buff or yellowish, thick, thin on margin, odor slight, taste somewhat disagreeable. LAMELLAE adnate with a decurrent tooth, close, broad, pale grayish then becoming dull purple-brown, some shorter. STIPE 2–5 in. long, ¼–¾ in. thick, equal or nearly so, whitish to yellowish, stuffed becoming hollow, covered with white, fibrillose or floccose scales below the annulus, glabrous and silky above. ANNULUS at first erect, then pendant, white to brownish. SPORES ellipsoid, smooth, purple-brown in mass, 11–13 × 5.5–7 μ.

It usually is solitary or in groups in mixed woods. Sept. and Oct.

This species can be recognized by the large size, scaly stipe, and dull-brown color, often with a smoky purplish tinge. It has been known under the name *Stropharia depilata* (Pers. ex Fr.) Quél. It is one of the largest and most conspicuous species of the genus. Although there does not seem to be any definite information regarding its edibility, it is not recommended because some of the *Stropharia* species are under suspicion as causing poisoning.

STROPHARIA SEMIGLOBATA (Fr.) Quél. — Not recommended

Figure 297, page 195

PILEUS ½–1½ in. broad, fleshy, hemispherical, finally convex to nearly plane, bright light yellow, fading to dull yellow, sometimes with an olive tinge, glabrous, very viscid, margin even. FLESH thick on disk to thin on margin, pale watery yellowish, odor and taste mild. LAMELLAE adnate, close to subdistant, broad, at first olive-gray, becoming purplish brown. STIPE 2–4½ in. long, ⅛–3/16 in. thick, equal or slightly enlarged at the base, stuffed becoming hollow, viscid below the annulus, slightly fibrillose above, whitish to pale yellowish.

ANNULUS delicate, whitish, often evanescent. SPORES purple-brown, ellipsoid, smooth, 15–20 × 8.5–11 μ.

It occurs singly or in groups on the dung of cattle and horses. June–Sept.

The veil in this species is very delicate and frequently tears in such a way that no annulus is formed, or the annulus may be very evanescent. The most striking characters are the yellow color, viscid pileus and stipe, hemispherical pileus, and the habitat on dung. Kauffman attempted to separate the forms that become convex to plane as *S. stercoraria* Fr. but they seem to be too close to be recognized as distinct species. It has been reported edible but, in general, *Stropharia* species should be avoided.

NAEMATOLOMA

The best-known species of this genus were formerly placed in the genus *Hypholoma* but Smith (1951) has pointed out that the old genus *Hypholoma* was composed of diverse elements and under the International Code of Nomenclature *Naematoloma* should be the correct name for the group of species typified by *N. sublateritium* (Fr.) Karst.

Naematoloma thus includes a rather small group of species with purple-brown to dull cinnamon-brown spores. The pileus is usually rather bright colored and may or may not be viscid. The lamellae vary from adnexed to adnate or subdecurrent and the color of the young lamellae is sometimes important in distinguishing species. The stipe may be thick and fleshy to fibrous-tough, or slender and cartilaginous.

Naematoloma is distinguished from *Stropharia* by the presence of an annulus in the latter. However, the distinction is not clear-cut because a veil is sometimes present in *Naematoloma* but usually remains attached to the margin of the pileus; in some of the dung-inhabiting species of *Stropharia* the annulus is evanescent.

Naematoloma is also very close to *Psilocybe* and can only be separated with certainty from this genus by microscopic characters. A characteristic type of cystidium called a gloeocystidium is present in the lamellae of *Naematoloma* but absent in *Psilocybe* (see Addendum).

Only two of the larger species of *Naematoloma* are described here and both are considered to be edible. Many of the other species are quite small and of no value as food, and *N. fasciculare* (Fr.) Karst., which has been reported to be poisonous, is not likely to be eaten because of its very bitter taste.

NAEMATOLOMA CAPNOIDES (Fr.) Karst. — Edible

Figure 298, page 195

PILEUS ½–2 in. broad, firm, convex, expanding to plane, sometimes slightly umbonate, brightly colored, orange-reddish or yellowish brown on the disk, paler and more yellowish on the margin, glabrous or at first slightly fibrillose, the margin inrolled at first and appendiculate with fibrils of the veil. FLESH whitish, fairly thick, firm, taste mild. LAMELLAE adnate-seceding, close, rather narrow to moderately broad, at first whitish to grayish, becoming purple-brown. STIPE 2–3 in. long, sometimes longer, ⅛–⅜ in. thick, equal or slightly enlarged at the base, slightly fibrillose up to the faint annular zone, yellowish above, rusty brown below, hollow. SPORES purple-brown, ellipsoid, smooth 6–7.5 × 3.5–4.5 μ.

In clusters on wood of conifers. August to November and occasionally in May.

This species is usually a little smaller and more orange to tawny than *N. sublateritium*. It might be confused with *N. fasciculare* (Fr.) Karst. but in the latter the lamellae are at first pale yellow and become greenish to olive-green. *N. fasciculare* also has an intensely bitter taste and has been reported to be poisonous.

NAEMATOLOMA SUBLATERITIUM (Fr.) Karst. — Edible

Figure 342, page 235; Figure 423, page 301

Brick-top

PILEUS 1–3 in. broad, firm, fleshy, convex-expanded, sometimes with a slight obtuse umbo, brick-red on the disk, paler on the margin to whitish, glabrous on disk to more or less whitish to yellowish-fibrillose on the decurved margin. FLESH thick, firm, whitish or becoming yellowish in age or when bruised, no odor, taste mild to slightly bitter. LAMELLAE adnate, close to crowded, narrow, at first whitish or in some collections yellow, then becoming gray or olive-gray and finally purple-brown. STIPE 2–4 in. long, ¼–½ in. thick, equal, whitish above to reddish brown below, the veil leaving a fibrillose annular zone and the surface more or less fibrillose below this, solid. SPORES purple-brown, ellipsoid, smooth, 6–7.5 × 3–4 μ.

It grows in dense clusters or occasionally in groups on hardwood logs, stumps, or roots, common. Aug.–Nov.

This species is commonly known as the brick-top mushroom because of the brick-red color of the pileus. It is common and rather variable. One form with bright yellow lamellae in the young mushrooms has been called *Hypholoma perplexum* (Pk.) Sacc. but it is now generally regarded as merely a form of the brick-top. Partly to nearly completely sterile fruiting bodies may be found.

In European literature it has been reported poisonous but North American mycophagists seem to be unanimous in claiming it to be edible.

PSATHYRELLA

Psathyrella is now used to include a large group of species formerly distributed throughout other genera such as *Hypholoma*, *Psilocybe*, *Psathyra*, and *Stropharia*. The spore color is typically purple-brown but may vary from pinkish to brick-red, dark brown, or blackish. They are mostly small, fragile species that can be identified only by microscopic characters and, in general, they are of no value as food.

Of the other dark-spored genera, *Coprinus* differs from *Psathyrella* by its deliquescing lamellae, *Panaeolus* by its mottled lamellae, and *Pseudocoprinus* by its plicate-striate pileus and the presence of paraphyses of the type found in *Coprinus*. The remaining genera, *Agaricus*, *Stropharia*, *Naematoloma*, and *Psilocybe* are all distinguished from *Psathyrella* by the structure of the cuticle which in them is composed of filamentous hyphae but in *Psathyrella* is composed of pear-shaped to vesiculose cells arranged in a palisade layer.

PSATHYRELLA CANDOLLEANA (Fr.) Smith — Edible

Figures 299, 300, page 195

PILEUS 1–3 in. broad, sometimes larger, fragile, at first oval, then conic to convex, finally more or less umbonate with the margin upturned, buff or honey colored, fading to whitish or creamy, hygrophanous, at first with some whitish flocci, then glabrous and atomate, margin thin, often splitting, sometimes a dirty violet color, often appendiculate with white fragments of the veil. FLESH thin, white, fragile, odor and taste mild. LAMELLAE adnate, crowded, narrow, at first whitish to grayish, then purplish, finally purple-brown. STIPE 2–4 in. long, ⅛–¼ in. thick, equal, smooth, somewhat mealy at the apex, white, hollow, rigid and easily splitting lengthwise. ANNULUS membranous, sometimes remaining attached to the margin of the pileus, usually evanescent. SPORES purple-brown, ellipsoid, smooth, 7–8.5 × 4–5 μ.

Common in lawns, fields and occasionally in woods. June–Sept.

This is a common species on lawns and grassy places, sometimes occurring in considerable abundance after rains. It may be found throughout the growing season. The pilei are rather thin and fragile but they are of good flavor and anyone interested in mushrooms as food should become acquainted with this species.

P. candolleana was formerly in the genus *Hypholoma* and *H. appendiculatum* Fr. and *H. incertum* Peck are synonyms. The buff color, appendiculate margin, rather cartilaginous, hollow stipe, and the series of color changes of the lamellae as the spores mature are its distinguishing characters. Sometimes the pileus glistens as if small particles of mica were scattered on the surface.

PSATHYRELLA HYDROPHILA (Fr.) Smith — Suspected

Figure 388, page 284

Pileus ¾–2½ in. broad, fragile, campanulate-convex, becoming nearly plane with a slight umbo, watery cinnamon-brown to chestnut-brown, fading to ochraceous buff, hygrophanous, glabrous, or with silky white fibrils especially at the margin, margin somewhat striate when moist. FLESH thin, fragile, brownish, odor and taste mild. LAMELLAE adnate-seceding, crowded, narrow, at first grayish brown, then purplish brown to dark brown. STIPE 1–2½ in. long, ⅛–¼ in. thick, equal, white, glabrous or somewhat fibrillose, somewhat pruinose at the apex, hollow, splitting easily. SPORES purple-brown, ellipsoid, smooth, 4–6 × 2.5–3.5 μ.

Cespitose to densely gregarious on very rotten wood. July–Sept.

This is a fairly common species and usually occurs in large clusters on rotten wood. It is typical of a number of small, fragile, reddish brown species placed in this genus. They are difficult to identify accurately and must be studied microscopically. Kauffman listed this species as suspected and, with the exception of *P. candolleana* above, this whole group is best avoided as food.

PSATHYRELLA VELUTINA (Fr.) Sing. — Edible

Figure 389, page 285

PILEUS 1¼–3 in. broad, convex to convex-campanulate, then plane and obtusely umbonate, tawny brown or yellowish brown, darker in center, hygrophanous, fading to buff-brown, appressed-fibrillose, becoming fibrillose-scaly, margin not striate but often fringed or appendiculate from the veil, and splitting. FLESH watery brownish, thick, odor and taste earthy. LAMELLAE adnate, close to crowded, broad, at first yellowish, then dark purple-brown, the edges white-flocculose and beaded with drops of moisture in wet weather. STIPE 1–3 in. long, ⅛–⅜ in. thick, equal, fibrillose or somewhat floccose-scaly up to the annulus, whitish above, brownish below, hollow. ANNULUS evanescent, fibrillose, whitish at first, becoming blackish from the spores. SPORES dark purple-brown, ovoid-ellipsoid, rough, 9–12 × 7–8 μ.

In clusters, scattered, or solitary, along roadsides or in the woods. July–Sept.

The tendency for the lamellae to become beaded with drops of moisture is one of the noteworthy characters of this species which has also been known as *Hypholoma lachrymabundum* (Fr.) Quél. The rough spores are also a distinguishing character.

COPRINUS

Species of the genus *Coprinus* are commonly known as 'inky caps' because the lamellae and often the flesh of the pileus dissolve into an inky fluid at maturity. This characteristic and the black color of the spore deposit are the principal distinguishing features of the genus.

According to Smith (1949) there are about 75 species of *Coprinus* known in North America. Many of these are small delicate fungi that are little known and difficult to identify. The four species described here are the best known and are fairly common and frequently used for food. The stipes are tough and cartilaginous and should be discarded.

When *Coprinus* species are gathered for food, it is important to pick young specimens and use them immediately because of this characteristic of the lamellae and flesh dissolving at maturity. Specimens kept for any length of time will likely be found to be a revolting inky mess. The appearance of these fungi in various stages of decomposition is, at first sight, likely to create a feeling of disgust and revulsion. To the imaginative they may suggest scenes of horror and it is undoubtedly a *Coprinus* species that inspired the following lines by the poet Shelley:

> "Their moss rotted off them flake by flake
> Till the thick stalk stuck like a murderer's stake
> Where rags of loose flesh yet tremble on high,
> Infecting the winds that wander by."

However, when the process is understood it is found to be a remarkable and fascinating adaptation for spore dissemination. In most mushrooms, the lamellae are more or less wedge-shaped, the broad edge of the wedge being attached to the pileus. The spores mature evenly over the entire surface of the lamellae from where they fall down and are carried away by air currents. In *Coprinus*, however, the lamellae are not wedge-shaped, but are parallel-sided and are frequently very crowded. Consequently, if spores were matured and discharged in the usual way, they would be shot onto the surface of the neighboring lamella and their passage into the air would be interfered with. In *Coprinus* the spores do not mature simultaneously over the surface of the lamellae but in a relatively narrow zone beginning first at the outer edge of the pileus and progressing gradually back toward the stipe. As the spores mature, a process of autodigestion sets in by which the lamellae and flesh are transformed to fluid and the edge of the pileus curls back, spreading the lamellae apart (Figure 392, p. 287), thus enabling the mature spores to be discharged into the air. The spores are disseminated by air currents as in other mushrooms, and not by the drops of fluid, although if the fluid is examined under the microscope it will be found to contain many spores that have been accidentally trapped there.

From time to time reports have appeared in the literature suggesting that mild poisoning may result from eating *Coprinus* in conjunction with alcohol consumption. Recent experiments conducted by Child (1952) have given no support to this contention. It seems probable that such reports have been based upon misidentifications of *Panaeolus sphinctrinus* or perhaps other *Panaeolus* species eaten by mistake for a *Coprinus*. However, other mycologists claim that there are well-authenticated cases of poisoning by *Coprinus* where there has been no possibility of misidentification.

COPRINUS ATRAMENTARIUS (Bull. ex Fr.) Fr. Edible

Figures 390, 391, page 286

Inky Cap

PILEUS 1–3 in. broad, occasionally larger, at first ovoid, then expanding to conic or campanulate, gray to brownish on disk, often lobed and folded, somewhat silky-fibrous, smooth or sometimes squamulose, becoming tattered on the margin in age. FLESH thin. LAMELLAE free, crowded, broad, at first white, becoming black and then dissolving into an inky fluid. STIPE 2–8 in. long, ¼–¾ in. thick, equal or narrowed at the base, somewhat fibrillose below the annulus, white and silky above, hollow. ANNULUS usually toward the base of the stipe, very evanescent. SPORES black, elliptical, smooth, 8–11 × 5–6.5 μ.

Usually in clusters on the ground or in sawdust, apparently associated with buried wood. July–Sept.

This species is common and often appears in dense clusters. It is considered a very desirable edible species for the young pilei are firm and meaty. Usually the gray pilei are smooth but, especially under dry conditions, may become more or less scaly from the splitting of the cuticle. *Coprinus insignis* Peck is somewhat similar in appearance, but has rough spores.

COPRINUS COMATUS (Müll. ex Fr.) S. F. Gray Edible

Figure 301, page 195; Figures 392, 393, page 287

Shaggy Mane

PILEUS cylindrical or barrel-shaped, 2–6 in. long and 1–2 in. thick, gradually expanding and becoming somewhat conical to bell-shaped, at first covered with a brownish or ochraceous brown cuticle which becomes torn into shaggy scales except on the disk, exposing the white to pinkish flesh, margin becoming split and recurved. FLESH thin, soft and fragile. LAMELLAE nearly free, very crowded, broad, at first white, becoming pinkish and then black, gradually dissolving into an inky fluid starting at the margin of the pileus. STIPE 2–6 in. long, ¼–¾ in. thick, equal or tapering upward, slightly bulbous at the base, smooth, hollow. ANNULUS movable, usually basal. SPORES black, elliptical, smooth, 13–18 × 7–8 μ.

Common along roadsides, in city dumps, fields, or on lawns. It may be found at any time during the growing season, but is more common in the fall.

The shaggy mane is one of the most easily recognized mushrooms and probably one of those most frequently used for food. It might be confused with *Coprinus ovatus* Fr. or *C. sterquilinus* Fr. The former is more ovate than cylindrical in shape and has smaller spores, whereas the latter is usually smaller, becomes more expanded than *C. comatus*, and has larger spores. Since both of these species are also edible a misidentification is of no consequence.

COPRINUS MICACEUS (Bull. ex Fr.) Fr. — Edible

Figure 305, page 213

Glistening Inky Cap

PILEUS ½–2 in. broad, at first ovate to elliptical, becoming conic to campanulate, ochraceous tan to ochraceous brown, sometimes fading to whitish, usually darker on the disk, at first covered with minute, glistening particles which may either persist or disappear in older specimens, strongly striate to sulcate, the striae of unequal lengths, smooth on the disk, more or less lobed and uneven on the margin. LAMELLAE adnate-seceding, crowded, moderately broad, at first white, then purplish to black and dissolving into an inky fluid. STIPE 1–3 in. long, ⅛–¼ in. thick, equal, silky, white, hollow. SPORES dark brown to black, ellipsoid to ovoid, 7–9 × 4–5 μ.

Usually in dense clusters on the ground or around old stumps. Very common and may be found throughout the growing season.

This species is rather small and delicate, but usually appears in considerable abundance. Many householders regard it as a nuisance because they consider the masses of fruiting bodies appearing on the lawn to be unsightly. It is associated with buried wood and may continue to appear for several years in places where a tree has been removed and old roots remain in the soil. Successive crops of fruit bodies may develop throughout the growing season following wet periods.

COPRINUS QUADRIFIDUS Peck — Edible

Figure 302, Page 195

PILEUS 1–3 in. broad, at first oval, becoming campanulate to somewhat expanded, gray to grayish brown, at first covered with a tomentose-floccose veil which breaks up into flakes or scales and may disappear, margin long-striate, often wavy, becoming rolled back. LAMELLAE free, crowded, broad, at first whitish, then dark purple-brown to black. STIPE 1¼–4 in. long, ⅛–⅜ in. thick, equal or tapering upward, white, somewhat floccose, with an evanescent basal annulus. SPORES black, smooth, ellipsoid, 7.5–10 × 4–5 μ.

Gregarious or cespitose on rotten wood. June–Aug.

This is a good edible species often appearing early in the season and occurring on rotten wood. It differs from *C. atramentarius* in having floccose veil patches on the pileus and in its occurrence on rotten wood. The fruit bodies arise from well-developed, root-like strings of mycelium termed rhizomorphs.

Figures 305-314

305. *Coprinus micaceus.*
306. *Gomphidius glutinosus.*
307. *Panaeolus semiovatus.*
308. *P. sphinctrinus.*
309. *Boletinellus merulioides.*
310. *Boletinus spectabilis.*
311. *B. cavipes.*
312. *B. cavipes.*
313. *B. pictus.*
314. *B. pictus.*

305
306
307
308
309
310
311
312
313
314

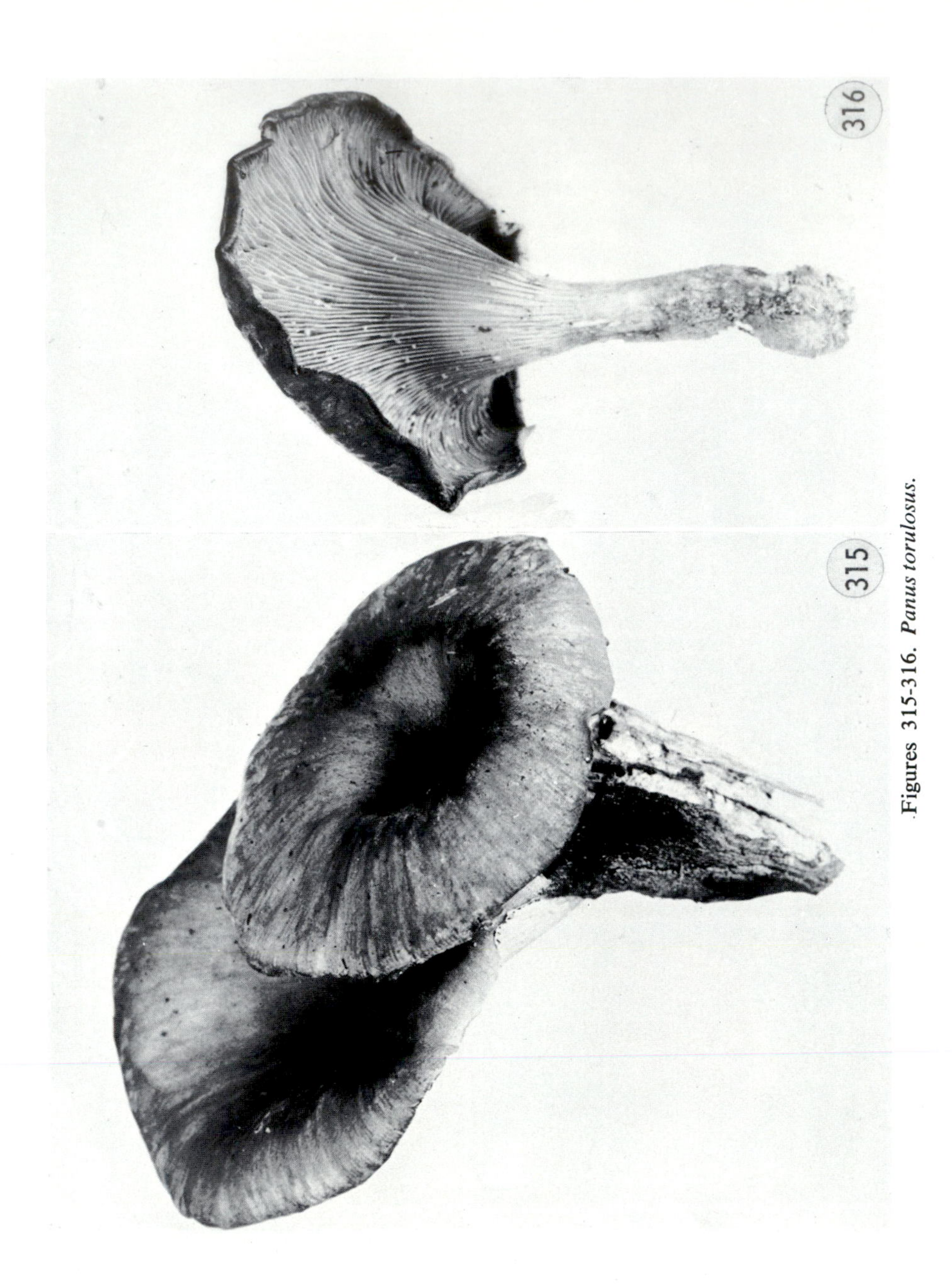

Figures 315-316. *Panus torulosus.*

Figures 317-326

317. *Gyroporus cyanescens.*
318. *G. castaneus.*
319. *Boletus edulis.*
320. *B. edulis.*
321. *B. subvelutipes.*
322. *B. subvelutipes.*
323. *Leccinum aurantiacum.*
324. *L. subglabripes.*
325. *L. chromapes.*
326. *L. chromapes.*

317
318
319
320
321
322
323
324
325
326

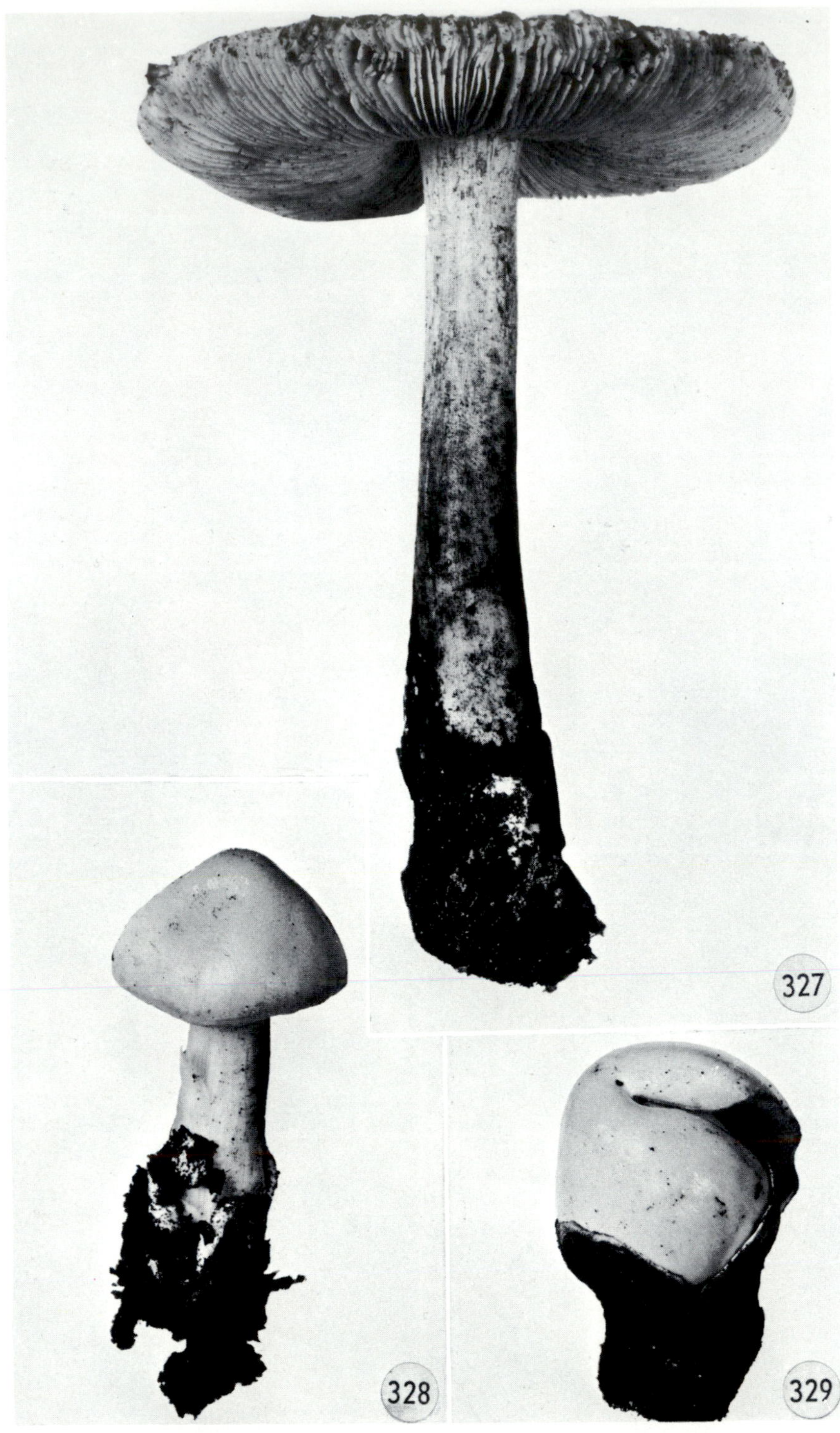

Figures 327-329. *Volvariella speciosa*. 327, mature fruiting body; 328, young fruiting body; 329, immature fruiting body emerging from volva.

PANAEOLUS

Panaeolus includes a small group of black-spored species. The lamellae have a characteristic mottled or dotted appearance as a result of the spores maturing unevenly. The lamellae do not deliquesce as in *Coprinus*. A veil is present in some species but is usually evanescent.

Species of *Panaeolus* should not be eaten. Some are known to be poisonous and to cause a form of intoxication. It is considered probable that reports of poisoning by *Coprinus* species when taken with alcohol are sometimes based on the use of *Panaeolus* species that have been mistaken for *Coprini*.

PANAEOLUS FOENISECII (Fr.) Kühner — Edible, not recommended

Figure 394, page 288

PILEUS ½–1 in. broad, convex or campanulate-convex, sometimes plane, fleshy, hygrophanous, variable in color, dark grayish brown or smoky brown to reddish brown, fading to dingy tan or buff, glabrous, the surface sometimes cracking into patches or scales, margin even, sometimes striate when moist. FLESH thin, watery brown, fading, odor and taste mild. LAMELLAE adnate, then seceding, close to subdistant, broad, variegated chocolate-brown to purplish black. STIPE 1½–3 in. long, 1⁄16–3⁄16 in. thick, equal, glabrous, pruinose at the apex, pale brownish, hollow. SPORES somewhat almond-shaped, dark purplish brown, rough, 12–20 × 8–10 μ.

In groups or scattered on lawns and grassy places. June–Sept.

This is one of the commonest of the small mushrooms occurring on lawns. It may be puzzling at first because of the great changes in color that occur as it dries out, but one soon becomes acquainted with its range of variation. Microscopically the large rough spores are very distinctive. It is reported to be edible, but all species of *Panaeolus* are best avoided.

PANAEOLUS SEMIOVATUS (Fr.) Lundell — Not recommended

Figure 307, page 213

PILEUS ¾–2 in. broad, occasionally larger, conic to ovoid to campanulate, not expanded, whitish to pale clay-colored, viscid, smooth, or becoming cracked in age. FLESH fairly thick, soft, fragile, whitish, odor and taste not distinctive. LAMELLAE adnate, separating from the stipe, ventricose, close, broad, grayish or brown, mottled black. STIPE 3–8 in. long, ¼–½ in. thick, equal or enlarged at the base, white to pallid buff, smooth or somewhat striate, hollow. ANNULUS whitish becoming black from the spores, may be striate, membranous, attached near the middle of the stipe or slightly above. SPORES black, ellipsoid, smooth, 16–20 × 8–11 μ.

Solitary or in small groups on horse dung, throughout the growing season.

Because of the membranous annulus some authors have placed this fungus in *Stropharia* and it has also been placed in a distinct genus *Anellaria* where it has been known as *Anellaria separata* (Fr.) Karst. However, it seems closely related to other *Panaeolus* species and many of them also have a partial veil which usually remains attached to the margin of the pileus rather than forming an annulus. Thus in this instance the presence of an annulus does not seem to be a character worthy of generic rank.

Panaeolus species are generally regarded as dangerous and this species is not recommended.

PANAEOLUS SPHINCTRINUS (Fr.) Quél. — Poisonous

Figure 308, page 213

PILEUS ¾–2 in. broad, fragile, at first bluntly conic or nearly ovoid, becoming conic-campanulate, brownish gray or olivaceous gray, glabrous, moist or somewhat slippery when wet, sometimes more or less areolate when dry, the margin slightly incurved and appendiculate from fragments of the veil. FLESH thin, colored like the surface, odor and taste not distinctive. LAMELLAE ascending-adnate, seceding, subdistant, broad, at first grayish, becoming mottled blackish, edges white-flocculose, shorter lamellae present. STIPE 2½–5 in. long, $\frac{1}{16}$–⅛ in. thick, equal, reddish brown with a grayish-pruinose covering, hollow, striate at the apex. SPORES black, more or less lemon-shaped, smooth, $13–19 \times 9–12\ \mu$.

It grows singly or in groups, usually on cow or horse dung in pastures. May–Sept.

This species is fairly common and is known to be poisonous. It produces symptoms of intoxication. It might be gathered by mistake for *Coprinus atramentarius* and should be carefully distinguished from that species. It was called *P. campanulatus* Fr. by Kauffman.

P. retirugis Fr. (Figure 424, page 302) is similar to *P. sphinctrinus* but has a more wrinkled or reticulate pileus. It is also believed to be poisonous.

PSEUDOCOPRINUS

Pseudocoprinus includes a small group of thin, fragile species with plicate-striate pilei and black spores. They are similar to delicate *Coprinus* species but the lamellae do not dissolve. Because of their small size and delicate texture, they are of no value as food, but one species is included here because it sometimes appears in such abundance as to attract attention.

PSEUDOCOPRINUS DISSEMINATUS (Pers. ex Fr.) Kühner Edible

Figure 395, page 288

PILEUS ¼–½ in. broad, membranous, obtusely conic to oval or campanulate, umbonate, whitish to grayish or gray-brown with the umbo buff, scurfy, becoming glabrous, margin strongly plicate-sulcate to the umbo. FLESH thin, membranous, fragile, odor none, taste mild. LAMELLAE adnate, subdistant, broad, ventricose, at first white, becoming gray to finally black. STIPE ¾–1½ in. long, very slender, scarcely 1/16 in. in diameter, at first minutely hairy, becoming glabrous, white, hollow. SPORES black or purple-black, ellipsoid, smooth, 7–10 $\times$ 4–5 μ.

It is common throughout the summer and fall, usually arising in clusters of numerous fruit bodies on old stumps or associated with buried wood, often appearing in abundance on lawns.

It might be taken for a *Coprinus* but the lamellae do not deliquesce. It can be recognized by the small size, furrowed pileus and buff umbo, and by its occurrence in extensive clusters. It is edible but so thin and fragile as to be of little value.

GOMPHIDIUS

Gomphidius includes a group of species characterized by their black or blackish spores and decurrent, waxy, usually subdistant to distant lamellae. They are usually more or less slimy-viscid, sometimes with a viscid veil that may leave traces of an annulus and cause the lower part of the stipe to be viscid also. Because of the waxy lamellae they were formerly considered to be close to *Hygrophorus* but they are now considered to be more closely related to the boletes.

At least in eastern Canada, they are generally rather rare and not much is known about their edibility. They are not attractive as food because of the slimy coating and rather watery flesh but they will certainly draw the attention of the collector by their striking and unusual appearance.

Singer (1949) made a study of *Gomphidius* in North America and recognized thirteen species. Only one is described here.

GOMPHIDIUS GLUTINOSUS (Schaeff. ex Fr.) Fr.

Figure 306, page 213

PILEUS 2–4 in. broad, convex to plane or slightly depressed, not umbonate or rarely subumbonate, glabrous, viscid to glutinous, livid purplish brown. FLESH white, unchanging when bruised, sometimes sordid or pinkish in age, taste mild to slightly acid. LAMELLAE at first whitish, then smoky gray to blackish, decurrent, forked, subdistant to distant, waxy in consistency. STIPE 1½–3½ in. long, ⅜–¾ in. thick, nearly equal or tapered at the base, glabrous to slightly fibrillose, white to pale brownish, yellow at the base, sheathed by a viscid veil that leaves an annular line near the apex. SPORES smoky gray, cylindric-fusoid, smooth, (15) 17–20 (22) × 5–7.5 μ.

Singly or gregarious in conifer woods, usually associated with spruce. Aug.–Sept.

The dark, decurrent lamellae, the brownish, slimy pileus, and the yellow base of the stipe are the principal field characters of this species. It is not known whether or not it is edible.

G. maculatus (Scop. ex Fr.) Fr. is another brownish to reddish brown species with a yellow base to the stipe but it lacks the slimy veil and grows in association with larch. *G. vinicolor* Peck is a reddish brown to vinaceous red species found with two-needle pines. Collectors on the west coast may find *G. tomentosus* Murr., an ochraceous or ochraceous orange species with a fibrillose to somewhat scaly pileus that is not viscid.

BOLETACEAE

The Boletaceae comprise a group of species that have the stature and shape of a mushroom and are soft and fleshy in consistency but in which the spores are produced on the sides of pores or tubes rather than on lamellae. They were formerly classified with the Polyporaceae but modern systematists are generally convinced that they are more closely related to the mushrooms than to the polypores. Such genera as *Gomphidius* and *Paxillus* in the mushrooms show definite relationships with the boletes.

The demarcation of genera within the family Boletaceae is a matter of some uncertainty at present. Traditionally three genera have been recognized in North America, *Boletus*, *Boletinus*, and *Strobilomyces*, but modern investigators have concluded that these genera are too broad and, especially in the genus *Boletus*, there are groups of related species sufficiently distinct to merit the rank of genera. It is considered that the splitting of the old genus *Boletus* into several other genera represents a distinct advance in our knowledge of the classification and relationships of this group, but since some of the characters that form the principal bases for the genera are microscopic they are not readily used in a popular work intended for those who may not have a microscope available.

In addition, the distinctions between some of the genera are not based on clear-cut characters but rather on combinations of characters that may be difficult for the amateur. It is therefore proposed to present two keys. The first of these is a more technical key to the genera of boletes in which the species described here would fall according to the system proposed by Slipp and Snell (1944) The second is a key to the species themselves, based only on macroscopic characters. It makes no attempt to indicate relationships but is intended only as a guide to the identification of the species.

Coker and Beers (1943) and Smith (1949) consider the boletes to be one of the safest groups for the amateur to try as food. To be sure, some of the species with red pore mouths are poisonous and all of these should be avoided, but they are relatively rare. Smith also advises against using species in which the flesh turns blue when cut or broken although *Gyroporus cyanescens* which shows this reaction to an extreme degree is reported to be edible.

One of the difficulties about using the boletes for food is that it is hard to find them free from insect larvae. They seem to become infested very early and they need to be collected carefully, the button stages being especially desirable. It is recommended that the tubes be removed and discarded before the pilei are cooked as they are of a different consistency and tend to become slimy.

Key to the genera

1. Spores globose, reticulate ..... *Strobilomyces*
1. Spores smooth 2

2. Spores small, oblong to short-elliptical 3
2. Spores long-elliptical or subfusiform, colored 4

3. Spores hyaline ..... *Gyroporus*
3. Spores colored ..... *Boletinellus*

4. Tubes more or less radially arranged, not easily separable from the pileus or from each other ..... *Boletinus*
4. Tubes not radially arranged, easily separable from the pileus and from each other 5

5. Pileus viscid and spores ellipsoid ..... *Suillus*
5. Pileus not viscid, or if viscid spores subfusiform 6

6. Tubes and spores flesh-colored ..... *Tylopilus*
6. Tubes and pores not flesh-colored 7

7. Stipe scabrous, rather slender, tapering upward ..... *Leccinum*
7. Stipe not scabrous 8

8. Stipe often more or less bulbous when young, may be reticulate, tubes sometimes stuffed when young with red mouths ..... *Boletus*
8. Stipe never subbulbous, tubes not stuffed nor with red mouths ..... *Xerocomus*

Key to the species

1. Tubes easily separable from the pileus and from each other 6
1. Tubes not easily separable from the pileus or from each other 2

2. Tubes arranged more or less in radial rows with veins between 3
2. Tubes not arranged in radial rows, pileus with prominent, erect scales, becoming blackish ..... *Strobilomyces floccopus*

3. Stipe hollow ..... *Boletinus cavipes*
3. Stipe solid 4

4. Stipe central 5
4. Stipe excentric to lateral; pileus brownish; tubes greenish yellow ..... *Boletinellus merulioides*

5. Pileus with red scales on a yellow background; spores ochraceous brown in mass ..... *Boletinus pictus*
5. Pileus with gray scales on a red background; spores purple-brown in mass ..... *Boletinus spectabilis*

6. Parasitic on *Scleroderma* ..... *Xerocomus parasiticus*
6. Not parasitic on *Scleroderma* 7

7. Tube mouths not differently colored from the rest of the tubes 8
7. Tube mouths red; colored tomentum at base of stipe ..... *Boletus subvelutipes*

8. Pileus viscid 9
8. Pileus not viscid 14

9. Annulus present 10
9. Annulus lacking 11

10. Stipe dotted with glandules ..... *Suillus subluteus*
10. Stipe not dotted ..... *Suillus grevillei*

11. Pileus glabrous 12
11. Pileus subtomentose, viscid when wet; tubes turning blue when broken ..... *Xerocomus badius*

12. Stipe dotted with glandules 13
12. Stipe not dotted; fruiting bodies small, reddish brown; taste peppery ..... *Suillus piperatus*

13. Pileus bright yellow, often streaked or spotted with red; stipe slender, ¼ inch or less in diameter ..... *Suillus americanus*
13. Pileus reddish brown, to grayish yellow or tawny; stipe stouter, usually more than ¼ inch in diameter ..... *Suillus granulatus*

14. Stipe soon hollow 15
14. Stipe solid 16

15. Flesh and tubes instantly turning blue when cut ..... *Gyroporus cyanescens*
15. Flesh not becoming blue when cut ..... *Gyroporus castaneus*

16. Stipe stout, more or less reticulate 17
16. Stipe slender, not reticulate 18

17. Tubes white becoming pinkish, taste bitter ..... *Tylopilus felleus*
17. Tubes soon yellow, taste pleasant ..... *Boletus edulis*

18. Tubes yellow or greenish yellow 19
18. Tubes whitish 20

19. Pileus glabrous ..... *Leccinum subglabripes*
19. Pileus subtomentose, cracking ..... *Xerocomus chrysenteron*

20. Pileus orange or red 21
20. Pileus brown to blackish ..... *Leccinum scabrum*

21. Pileus pinkish red; stipe bright yellow at base .... *Leccinum chromapes*
21. Pileus orange-red to brownish orange; stipe not yellow at base .... *Leccinum aurantiacum*

BOLETINELLUS MERULIOIDES (Schw.) Murr. — Edible

Figure 309, page 213

PILEUS 1½–5 in. broad, circular to reniform, depressed towards one side, olive-brown to yellowish brown or reddish brown, finely tomentose to glabrous, dry, margin usually indented and inrolled, finally spreading, even. FLESH soft but rather tough, yellowish to pinkish near the surface, sometimes turning bluish green when cut, taste mild, odor none or of raw potatoes. TUBES decurrent, strongly radiating, partly lamellate, short and wide, yellow with a greenish tinge, becoming more ochraceous in age. STIPE excentric or lateral, ½–2 in. long, ½–¾ in. thick, nearly equal or slightly swollen at the base, expanding into the pileus above, reticulate and yellowish above, reddish brown or olivaceous brown to blackish and short-tomentose below, solid. SPORES yellowish to brownish-ochraceous, ellipsoid, smooth, 7.5–10.5 × 5.5–7.5 μ.

On damp ground in woods or open places, usually gregarious and generally associated with ash. July–Sept.

This species is not likely to be confused with any other. It has been widely known also under the name *Boletinus porosus* (Berk.) Pk. It resembles *Paxillus involutus* in color and shape and shows relationships with the genus *Paxillus*. It is said to be edible but it is not a very attractive fungus.

BOLETINUS CAVIPES (Opat.) Kalchbr. — Edible

Figures 311, 312, page 213

PILEUS 1–4 in. broad or sometimes larger, broadly convex, subumbonate, tawny brown to yellowish brown, sometimes tinged reddish or purplish, fibrillose-squamulose. FLESH yellowish, taste farinaceous to bitter. TUBES decurrent, radiating, at first sulphur-yellow becoming dingy ochraceous in age. STIPE 1–3 in. long, ¼–½ in. thick, enlarged below up to 1½ in., usually more or less reticulate above the annulus, yellow above the annulus, concolorous with the pileus below, at first stuffed, soon hollow. ANNULUS white to ochraceous, delicate, evanescent, sometimes partly adhering to the margin of the pileus. SPORES olivaceous-ochraceous in a fresh deposit, changing to yellowish-ochraceous, one-celled, ellipsoid, smooth, (7) 8–10 × 3–4 μ.

Singly or gregarious in damp woods or swamps, associated with larch or pine. Sept. and Oct.

This species is usually a rich tawny brown but one collection was a bright golden yellow. It seemed to be the same in every other respect and was growing along with typical specimens. The hollow stipe is the chief diagnostic character.

BOLETINUS PICTUS Peck — Edible

Figures 313, 314, page 213

Painted Bolete

PILEUS 1½–4 in. broad, convex, at first dark red, fibrillose, soon becoming squamulose, the fibrils separating into reddish scales and revealing the yellowish flesh beneath, dry to moist or subviscid, margin somewhat appendiculate from the veil. FLESH yellow, slowly becoming reddish when bruised. TUBES adnate to decurrent, at first yellowish, becoming dingy ochraceous, drying brown, more or less radiately arranged. STIPE 1¼–3 in. long, ¼–½ (¾) in. thick, equal or swollen at the base, colored like the pileus, yellow at the apex, reddish-scaly below. ANNULUS whitish to grayish, fairly persistent. SPORES ochraceous brown, ellipsoid, smooth, (7.5) 8–10 (11) × 3.0–4.0 μ.

Singly or gregarious in woods or swamps, perhaps associated with pine. July–Oct.

This species is sometimes called the painted boletus and is one of the more beautiful of our fungi. It is sometimes confused with *B. spectabilis* but the latter has gray scales on a red background in contrast to the red scales on a yellowish background of *B. pictus*. The spore size will separate them with certainty.

BOLETINUS SPECTABILIS (Peck) Murr. — Edible

Figure 310, page 213

PILEUS 1½–3 in. or more broad, convex, bright red, at first covered with a red tomentum, then becoming scaly, the seales viscid, fading to grayish red, brownish, or yellowish, margin more or less appendiculate. FLESH whitish to pale yellow, becoming brighter yellow when wounded, taste and odor unpleasant. TUBES adnate to slightly decurrent, yellowish to ochraceous, drying dark brown, more or less radiately arranged. STIPE 1½–3 in. long, ¼–½ in. thick, swollen at the base or nearly equal, yellow above the annulus, red or yellowish red below, solid. ANNULUS reddish to yellowish, double, more or less persistent. SPORES purplish brown, ellipsoid, smooth, 11–14 × 4.5–6 μ.

Singly or gregarious, associated with larch in bogs. Aug.–Sept.

The radiating arrangement of the tubes is less marked in this species than in other *Boletinus* species. It is a very showy and beautiful species. The grayish scales and larger spores separate it from *B. pictus*. *B. paluster* Peck is somewhat similarly colored but is smaller and has strongly radiating, decurrent pores, and smaller spores.

BOLETUS EDULIS Bull. ex Fr. — Edible

Figures 319, 320, page 215

Edible Boletus (Cèpe, Steinpilz)

PILEUS 2½–6 in. broad, sometimes larger, convex to nearly plane, variable in color, yellowish brown or tawny brown to light buff or grayish red, margin often paler, glabrous, dry to subviscid when wet. FLESH white or yellowish, sometimes pinkish, unchanging, taste sweet and nutty. TUBES adnexed to nearly free, depressed around the stipe, at first white and stuffed, becoming greenish yellow. STIPE 2½–6 in. long, ½–1¼ in. thick, equal or swollen at the base up to 2¼ in., reticulate, sometimes for the entire length, sometimes only at the apex, whitish to yellowish or brownish, solid. SPORES olivaceous brown to ochraceous brown, ellipsoid-fusiform, smooth, 13–18 (21) × 4–6 μ.

Solitary or gregarious on the ground in woods and open places. June–Oct.

B. edulis is one of the best edible fungi but it is difficult to find it free from insect larvae in eastern Canada. In Europe it is a very well-known species and has a good many common names such as *cèpe* and *Steinpilz*. In some parts of central Europe special trains used to and may still run from the cities in the right season for people to go to the country and collect this bolete. It can be sliced and dried and retains its rich, nutty flavor when used in gravies or stews.

BOLETUS SUBVELUTIPES Peck — Dangerous

Figures 321, 322, page 215

PILEUS 1½–5 in. broad, convex, yellowish brown to reddish or dark brown, sometimes paler or olivaceous toward the margin, at first somewhat velvety-tomentose, becoming glabrous. FLESH yellow, changing to blue when wounded, taste mild. TUBES adnexed, depressed around the stipe, red at the mouths, elsewhere yellow, becoming blue when wounded. STIPE 2–4¾ in. long, ½–1¼ in. thick, equal or tapering upward, sometimes bulbous at the base, more or less furfuraceous to nearly glabrous, with a red or yellow, coarsely hairy tomentum at the base, reddish brown above, yellow at the apex, solid. SPORES yellowish, ellipsoid-fusiform, smooth, 12–17 × 4.5–6 μ.

Usually solitary or gregarious on the ground in mixed woods. July–Sept.

In several species the mouths of the tubes are red and some of these are known to be poisonous. The species in this group are rather difficult to identify and no species with red tube mouths should be eaten. This species, with the furfuraceous, nearly equal stipe and colored tomentum at the base appears to be the commonest one.

GYROPORUS CASTANEUS (Bull. ex Fr.) Quél. — Edible

Figure 318, page 215

PILEUS 1–2¾ in. broad, convex to nearly plane, chestnut-brown to reddish brown, sometimes paler to cinnamon, dry, minutely velvety-tomentose, margin becoming upturned. FLESH white, unchanging or sometimes brownish, mild. TUBES depressed around the stipe, at first white, becoming cream to yellow, and becoming brownish when bruised. STIPE 1–2 in. long, ½–¾ in. thick, tapering upward or nearly equal, concolorous, velvety-tomentose, hollow. SPORES yellow, broadly ellipsoid, smooth, (7) 8.5–11 (13) × (4) 5–6 (7) μ.

Singly or gregarious on the ground in woods or open places. July–Sept.

This species is easily recognized by the chestnut-brown, tomentose pileus and stipe, the ellipsoid spores, whitish to yellowish tubes, and hollow stipe.

GYROPORUS CYANESCENS (Bull. ex Fr.) Quél. — Edible

Figure 317, page 215

PILEUS 1½–4 in. broad, convex to nearly plane, pale yellowish to buff or tan, coarsely floccose-tomentose, dry, margin incurved. FLESH whitish, instantly turning blue when wounded and darkening to nearly black, mild. TUBES free, depressed around stipe, whitish, becoming yellowish, instantly turning blue when wounded. STIPE 2–3 in. long, ½–1 in. thick, tapering up, ventricose or irregularly swollen, tomentose, concolorous with pileus, instantly turning blue when wounded, stuffed, becoming hollow. SPORES yellow, oblong-ellipsoid, smooth, 8–10 (11) × 4.5–6 μ.

Singly or gregarious on the ground in woods and open places. July–Sept.

This species is easily recognized by the pale, tomentose pileus and the immediate change to blue of all parts of the fruit body when wounded. In spite of the unattractive appearance from this reaction it is reported to be edible and of good flavor.

LECCINUM AURANTIACUM (Bull.) S. F. Gray — Edible

Figure 323, page 215

PILEUS 1½–6 in. broad, convex, reddish orange to orange-yellow or reddish brown, dry, minutely tomentose to fibrillose-squamulose, rarely glabrous, margin appendiculate. FLESH white or whitish to pinkish, at times turning slightly blue when cut, finally becoming grayish to blackish, firm, taste mild, TUBES adnate to adnexed, becoming free, dirty white to gray. STIPE 2–6 in. long, ½–2 in. thick, tapering upward to nearly equal, more or less scabrous to squamulose, the projections at first whitish, then reddish brown, finally blackish, solid, sometimes changing to blue-green at base when cut. SPORES brown, ellipsoid-fusiform, smooth, 11–17 × 3.5–5 μ.

Solitary or gregarious on the ground, usually associated with birch or poplar. June–Oct.

The scabrous stipe, orange pileus, and appendiculate margin are the distinguishing characters. It is often very firm and hard, especially when young. It is one of our commonest boletes.

LECCINUM CHROMAPES (Frost) Sing. — Edible

Figures 325, 326, page 215

PILEUS 1½–4 (5) in. broad, convex to nearly plane, pinkish red, sometimes brownish or buff in older specimens, dry, slightly tomentose, margin thick. FLESH white or faintly pink, unchanging, mild. TUBES depressed at the stipe, nearly free, whitish to flesh colored, somewhat brownish in age. STIPE 2–4 in. long, 3/8–3/4 in. thick, equal or tapering upward, sometimes narrowed at the base, whitish, more or less washed with rose, bright chrome-yellow at the base, scabrous-dotted. SPORES pinkish brown, oblong-ellipsoid, smooth, 10–14 (16) $\times$ 3.5–5.5 μ.

Usually singly on the ground in woods. June–Sept.

This species is not common but is one of our most beautiful boletes and is easily recognized by the rosy pileus and the bright yellow base of the stipe.

LECCINUM SCABRUM (Bull. ex Fr.) S. F. Gray — Edible

Figure 330, page 233

PILEUS 2–5 in. broad, convex to plane, minutely velvety to glabrous, slightly viscid when wet, color variable, usually some shade of brown, from pallid to tawny brown, grayish brown, or blackish brown. FLESH whitish, unchanging or occasionally becoming slightly pinkish to grayish, not blackening, taste mild. TUBES depressed at the stipe and free or nearly so, whitish becoming light brownish, darkening when bruised. STIPE 3–5 in. long, 1/2–3/4 in. thick, tapering upward to nearly equal, whitish or grayish, with blackish, scabrous dots, solid, sometimes turning blue at the base when cut. SPORES brown, ellipsoid-fusiform, smooth, (14) 15–19 (21) $\times$ 5–7 μ.

Singly or gregarious on the ground in woods or open places. July–Oct.

This is probably our commonest bolete although Singer has recently shown that two species have been confused under this name. He has described the second species as *Leccinum oxydabile* (Sing.) Sing. The two are very similar in appearance and can be separated with certainty only by microscopic characters. If a form is collected in which the flesh turns red when cut it is probably *L. oxydabile*. The latter has slightly larger spores and the structure of the cuticle of the pileus is different. In *L. scabrum* the cuticle is composed of slender, filamentous hyphae but in *L. oxydabile* there are some much broader hyphae and chains of short cells.

A whitish form is sometimes found that Singer calls *L. scabrum* ssp. *niveum* (Figure 331, p. 233) but which others have regarded as a good species. It has smaller spores than the typical form and sometimes has greenish tints in the pileus. It can be distinguished from the whitish *L. albellum* (Pk.) Sing. by the structure of the cuticle which in the latter also has chains of short cells.

Duller-colored specimens of *L. aurantiacum* may also be mistaken for *L. scabrum*, but can be distinguished by their smaller spores and the appendiculate margin of their pilei.

LECCINUM SUBGLABRIPES (Peck) Sing. — Edible

Figure 324, page 215

PILEUS 1 1/4–3 1/2 in. broad, convex to plane, yellowish brown to reddish brown or chestnut, glabrous, dry, with a slightly projecting, sterile margin.

FLESH pale yellowish, unchanging, mild to slightly acid. TUBES adnate, becoming depressed around the stipe, bright yellow becoming greenish yellow to olive. STIPE 1½–3½ in. long, ¼–⅝ in. thick, equal or slightly tapering upward, narrowed at the base, yellowish with reddish stains, somewhat yellowish furfuraceous, solid. SPORES olive-brown, ellipsoid-fusiform, smooth, (11) 12–14 (16) × (3) 3.5–4.5 (6) μ.

Singly or gregarious on the ground in frondose or mixed woods. June–Sept.

This is a fairly common species and differs from the other *Leccinums* in having yellow tubes. However, the slender, furfuraceous stipe, and other characters seem to indicate that it is more closely related to *Leccinum* than to *Boletus*.

STROBILOMYCES FLOCCOPUS (Vahl ex Fr.) Karst. — Edible

Figure 348, page 235; Figure 425, page 302

PILEUS 2–5 in. broad, hemispherical to convex, dry, covered with large, thick, erect, floccose, blackish brown scales, margin thick and irregular, appendiculate with scales and veil fragments. FLESH whitish, soon reddening and becoming black when wounded, mild. TUBES at first whitish, changing color like the flesh, adnate, depressed at the stipe, mouths rather large, angular. STIPE 2–5 in. long, ¼–¾ in. thick, nearly equal or somewhat enlarged at base, colored like the pileus, flocccse-tomentose from the remains of the veil, solid. SPORES black in mass, globose, reticulate, 9–12 × 9–10 μ.

Usually occurring singly on the ground or attached to rotten wood in frondose or mixed woods. July–Sept.

This species has been known under the name *Strobilomyces strobilaceus* (Scop. ex Fr.) Berk., but *S. floccopus* is the correct name for it. It is sometimes called the cone-like boletus or pine cone fungus, perhaps because of the shaggy appearance caused by the large scale. It is a striking fungus but rather unattractive in appearance. It is said to be edible but of indifferent quality.

SUILLUS AMERICANUS (Peck) Snell — Edible

Figure 332, page 233

PILEUS 1–3 in. broad, convex to subconic or sometimes slightly umbonate, bright yellow, more or less streaked or spotted irregularly with red to reddish brown, viscid, glabrous, margin slightly tomentose-appendiculate when young. FLESH yellow, turning reddish when wounded, mild. TUBES adnate to decurrent, yellow to brownish yellow, drying ochraceous brown, glandular-dotted. STIPE 1–2½ in. long, ⅛–¼ in. thick, sometimes thicker, rather slender, equal or nearly so, thickly glandular-dotted, yellow between the brownish dots, blackish when dried, solid. ANNULUS lacking or very rarely present, then yellowish, floccose, evanescent. SPORES cinnamon-brown, ellipsoid, smooth, (8) 9–10 (11) × 3–4 (5) μ

Gregarious on the ground in woods or open places, probably associated with pine. July–Sept.

This species is easily confused with *S. subaureus* (Pk.) Snell. The latter has a thicker stipe, less heavily glandular-dotted, and a thicker pileus. The spores of *S. subaureus* are slightly smaller, mostly 7–9 μ long and only rarely reaching 10 μ, whereas spores of *S. americanus* are mostly 9–10 μ and some reach 11 μ.

It is usually described as lacking an annulus but one collection was found which agreed with *S. americanus* in every respect but had a clearly marked annulus present.

It is said to be edible but lacking flavor.

SUILLUS GRANULATUS (L. ex Fr.) Kuntze — Edible

Figure 333, page 233

PILEUS ¾–3¼ in. broad, convex to plane, usually reddish brown but variable to grayish pink, grayish yellow, tawny, or brown, viscid, glabrous. FLESH pale yellowish to whitish, mild. TUBES adnate, yellowish, the mouths glandular-dotted. STIPE ¾–2¾ in. long, ¼–½ in. thick, equal or nearly so, white to brownish, yellow near apex, glandular-dotted at apex to about half way down, solid. SPORES yellowish brown, ellipsoid, smooth, (6) 7–9 (10) × 2.5–3.5 μ.

Usually gregarious in woods or open places, probably associated with pine. June–Oct.

This is one of our commonest boletes. It is most likely to be confused with *S. brevipes* (Peck) Kuntze which is also associated with pines but usually appears late in the year and has a short stipe that lacks glandular dots. Both species are good to eat.

SUILLUS GREVILLEI (Kl.) Sing. — Edible

Figure 334, page 233

PILEUS 1½–4 in. broad, sometimes larger, convex to nearly plane, chestnut-brown to yellow on margin, or reddish yellow to golden yellow, glabrous, viscid, margin sometimes more or less appendiculate. FLESH yellow, mild. TUBES adnate to decurrent, bright golden yellow, becoming brown or purplish brown when wounded. STIPE 1–4 in. long, ¼–¾ in. thick, equal or slightly tapering upward, usually finely reticulate above the annulus, not glandular-dotted, variable below the annulus, more or less fibrillose to glabrous, reddish or reddish brown to yellow, bright yellow above the annulus, solid. ANNULUS whitish to yellowish to reddish brown, usually prominent and persistent. SPORES golden brown to ochraceous brown, smooth, ellipsoid, 7–11 × 3–4 μ.

Gregarious on the ground, associated with larch. Aug.–Oct.

Sometimes this species is very richly colored and striking in appearance.

It has also been known under the names *Boletus clintonianus* Peck and *B. elegans* Fr. The slimy cuticle should be removed before cooking.

SUILLUS PIPERATUS (Bull. ex Fr.) Kuntze — Not edible

Figure 335, page 233

PILEUS ¾–2¼ in. broad, convex, yellowish brown, cinnamon-brown or reddish brown, glabrous or subtomentose toward the margin when young, slightly viscid when moist. FLESH whitish or yellowish, sometimes tinged reddish, sometimes turning blue near the tubes when wounded, the blue quickly fading again, taste very acrid and peppery. TUBES adnate to slightly decurrent, deep reddish brown, irregular, slightly radially arranged near the stipe. STIPE 1–3 in. long, ⅛–¼ in. thick, equal or nearly so, straight or curved, paler than the pileus, bright yellow at the base, solid. SPORES rusty brown, ellipsoid, smooth, 8–11 $\times$ 3–4 μ.

On the ground in woods and open places. July–Oct.

The most distinctive character of this small brownish species is the very peppery taste. *S. rubinellus* (Peck) Sing. is another small species but brighter colored, with tubes entirely red and mild taste.

SUILLUS SUBLUTEUS (Peck) Snell — Edible

Figure 336, page 233

PILEUS 1–3 in. broad, convex to nearly plane, yellowish brown to reddish brown, sometimes with an olive tinge, viscid, glabrous or somewhat virgate-fibrillose to indistinctly squamulose. FLESH yellow or yellowish, unchanging, mild or slightly acid. TUBES adnate, yellow at first, becoming more olivaceous in age, glandular-dotted. STIPE 1½–3½ in. long, ¼–½ in. thick, equal above and below the annulus, yellow above, more pallid toward the base, solid. ANNULUS forming an apical to median, grayish band, not sheathing the stipe, viscid. SPORES brownish-ochraceous, ellipsoid, smooth, (7.5) 8–10 (11) $\times$ 2.5–3.5 μ.

Usually solitary to gregarious on the ground, associated with five-needle pines. July–Sept.

It is likely to be confused with *S. luteus* (L. ex Fr.) S. F. Gray which is generally a stouter plant with a stipe usually more than ½ in. thick, and the annulus forming a sheath rather than a band on the stipe. *S. luteus* is more likely to be found associated with two-needle pines.

TYLOPILUS FELLEUS (Bull. ex Fr.) Karst. — Not edible

Figure 337, page 233

PILEUS 1½–8½ in. broad, convex, becoming nearly plane, grayish brown, yellowish brown, to reddish brown, dry, glabrous. FLESH white, sometimes more or less pinkish when wounded, taste very bitter. TUBES adnate, depressed around the stipe, at first white, becoming rosy flesh colored, becoming brownish when bruised, stuffed when young. STIPE 1½–4¾ in. long, ½–2 in. thick, equal or tapering upward, sometimes bulbous at the base, more or less reticulate, concolorous with the pileus or nearly so. SPORES ellipsoid-fusiform, smooth, rose colored (9) 10–14 (17) × 3–4.5 μ.

Solitary or gregarious in woods and open places. June–Oct.

The rose colored tubes, reticulate stipe, and bitter taste are the distinguishing characters of this species. Occasionally the bitter taste may be lacking. It is one of the more common species and sometimes reaches a very large size. *T. plumbeoviolaceus* (Snell) Snell has been confused with it or regarded as a variety, but it seems to be a distinct species with violaceous colors and very firm consistency, and matures rather slowly.

XEROCOMUS BADIUS (Fr.) Kühner ex Gilbert — Edible

Figures 338, 339, page 233

PILEUS 2–3 in. broad, convex to nearly plane, bay-brown to chestnut-brown, sometimes tinged olivaceous, viscid, minutely tomentose. FLESH yellowish, turning blue when wounded, especially near the tubes, then the blue fading, mild. TUBES adnate or depressed around the stipe, pale greenish yellow, becoming blue when wounded. STIPE 2–3½ in. long, ¼–¾ in. thick, nearly equal, colored like the pileus, sometimes yellow at the apex, white mycelioid at base, more or less streaked with darker lines, solid. SPORES olive-brown, ellipsoid-fusiform, smooth, 10–15 × 3.5–5.0 μ.

Singly or gregarious on the ground, usually associated with pine. June–Sept.

The deep brown pileus and stipe and the greenish yellow pores that turn blue when wounded are the chief distinguishing characters of this species. It is not very common.

XEROCOMUS CHRYSENTERON (Bull. ex Fr.) Quél. — Doubtful

Figures 345, 346, page 235

PILEUS 1–2 in. broad, convex to nearly plane, olive-brown to reddish brown, dry, felty-tomentose, becoming cracked and showing reddish in the cracks. FLESH yellowish, slightly acid. TUBES adnate to depressed, greenish yellow, changing to blue when wounded, rather large. STIPE 1–1½ in. long, ⅜–⅝ in. thick, equal or nearly so, striate, reddish or yellowish, solid, rather tough and rigid. SPORES yellow-brown, ellipsoid-fusiform, smooth, 10–14 × (3) 4–5.5 μ.

Usually solitary on the ground or on rotten wood. June–Oct.

This species is common and variable. The felty pileus with cracks showing red, the large greenish pores that change to blue, and the slender, tough stipe usually showing some red coloration are the distinguishing features. *X. subtomentosus* (L. ex Fr.) Quél. might be confused with it but in this species the cracks on the pileus show yellow, the flesh and tubes do not turn blue or perhaps the mouths very slightly so, the tubes are more yellow, and the stipe is slightly reticulate at the apex and has no red in its coloration.

There are conflicting reports in the literature regarding the edibility of *X. chrysenteron*. Its rather tough texture does not recommend it and it is probably better avoided, although it is likely not actually poisonous.

XEROCOMUS PARASITICUS (Bull. ex Fr.) Quél.

Figure 347, page 235

PILEUS 1½–2 in. broad, convex, yellowish brown to grayish to olivaceous, dry, velvety-tomentose, becoming glabrous, sometimes cracking. FLESH white, becoming yellowish when cut. TUBES decurrent, somewhat lamellate near the stipe, yellow to olivaceous. STIPE 1–2 in. long, ¼–½ in. thick, equal or nearly so, yellow, with a velvety tomentose covering that becomes torn into flecks and patches leaving the stipe more or less dotted. SPORES dark olive-brown, ellipsoid-fusiform, smooth, 12–17 × 5–6.5 μ

Parasitic on species of *Scleroderma*. Aug.–Sept.

This is a rare species, remarkable for its unusual habit of parasitizing a puffball. It is not likely to be confused with anything else because of its unusual place of growth.

Figures 330-339

330. *Leccinum scabrum.*
331. *L. scabrum* ssp. *niveum.*
332. *Suillus americanus.*
333. *S. granulatus.*
334. *S. grevillei.*
335. *S. piperatus.*
336. *S. subluteus.*
337. *Tylopilus felleus.*
338. *Xerocomus badius.*
339. *X. badius.*

330
331
332
333
334
335
336
337
338
339

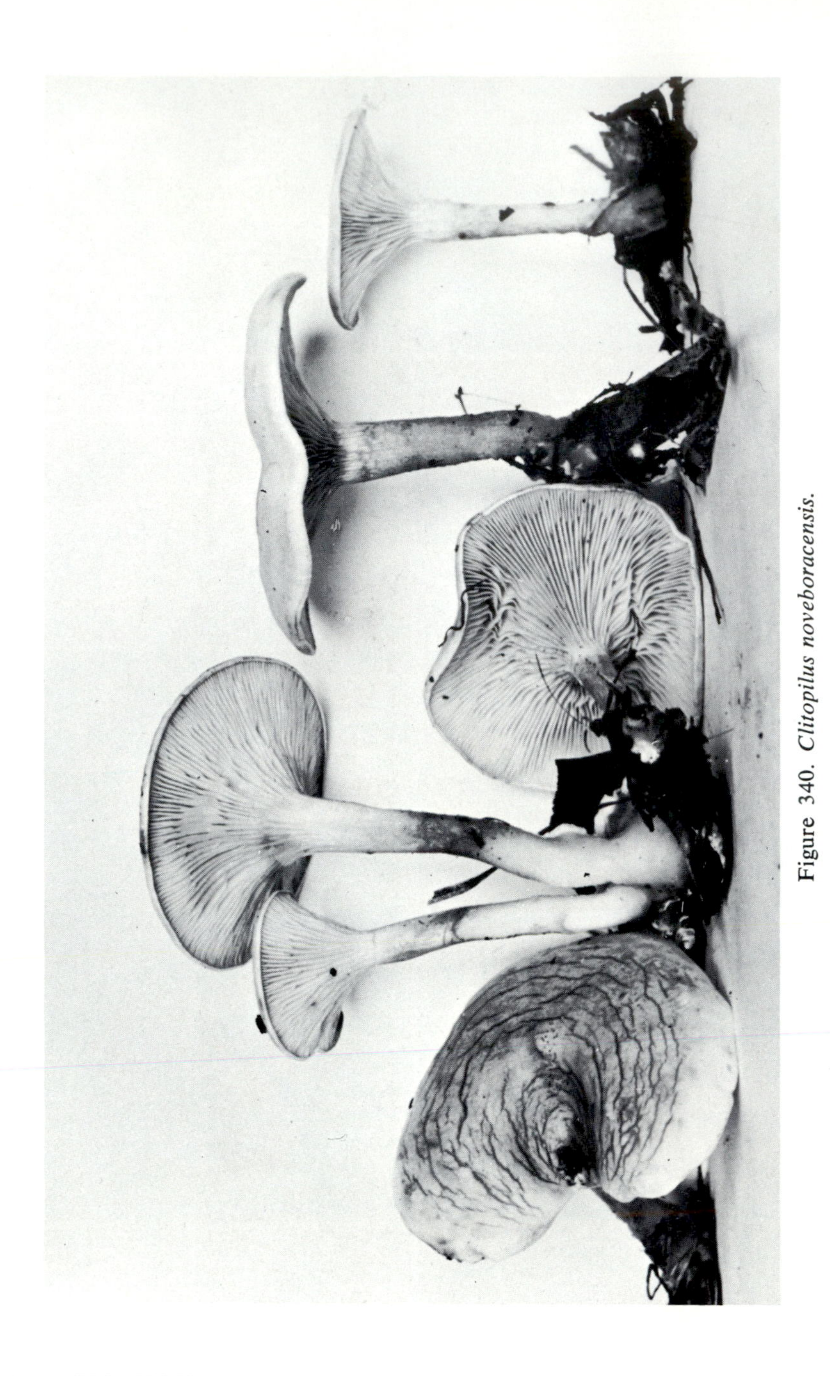

Figure 340. *Clitopilus noveboracensis.*

Figures 341-350

341. *Cortinarius alboviolaceus.*
342. *Naematoloma sublateritium.*
343. *Agaricus edulis.*
344. *A. placomyces.*
345. *Xerocomus chrysenteron.*
346. *X. chrysenteron.*
347. *X. parasiticus.*
348. *Strobilomyces floccopus.*
349. *Clavaria stricta.*
350. *Hypomyces lactifluorum.*

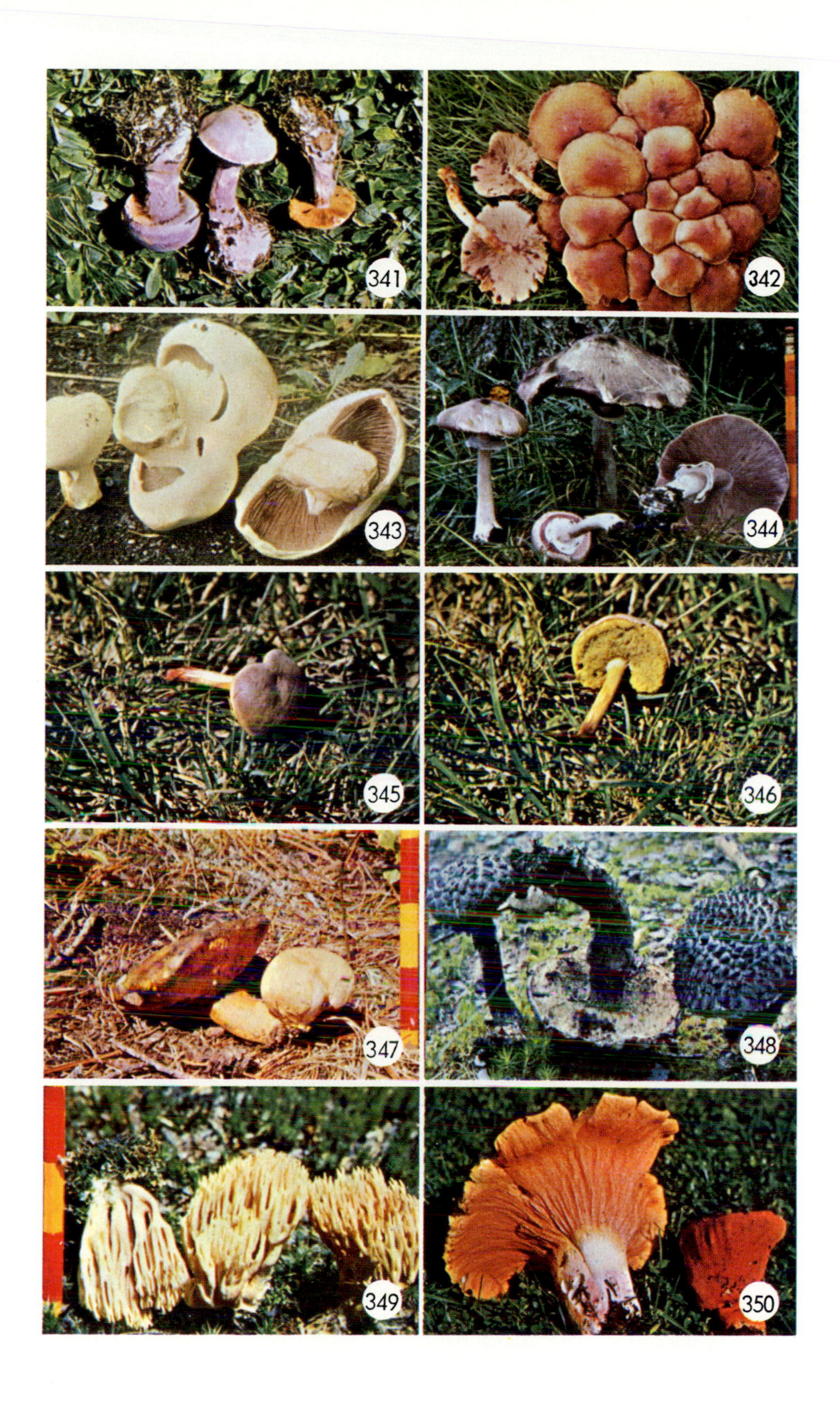
341
342
343
344
345
346
347
348
349
350

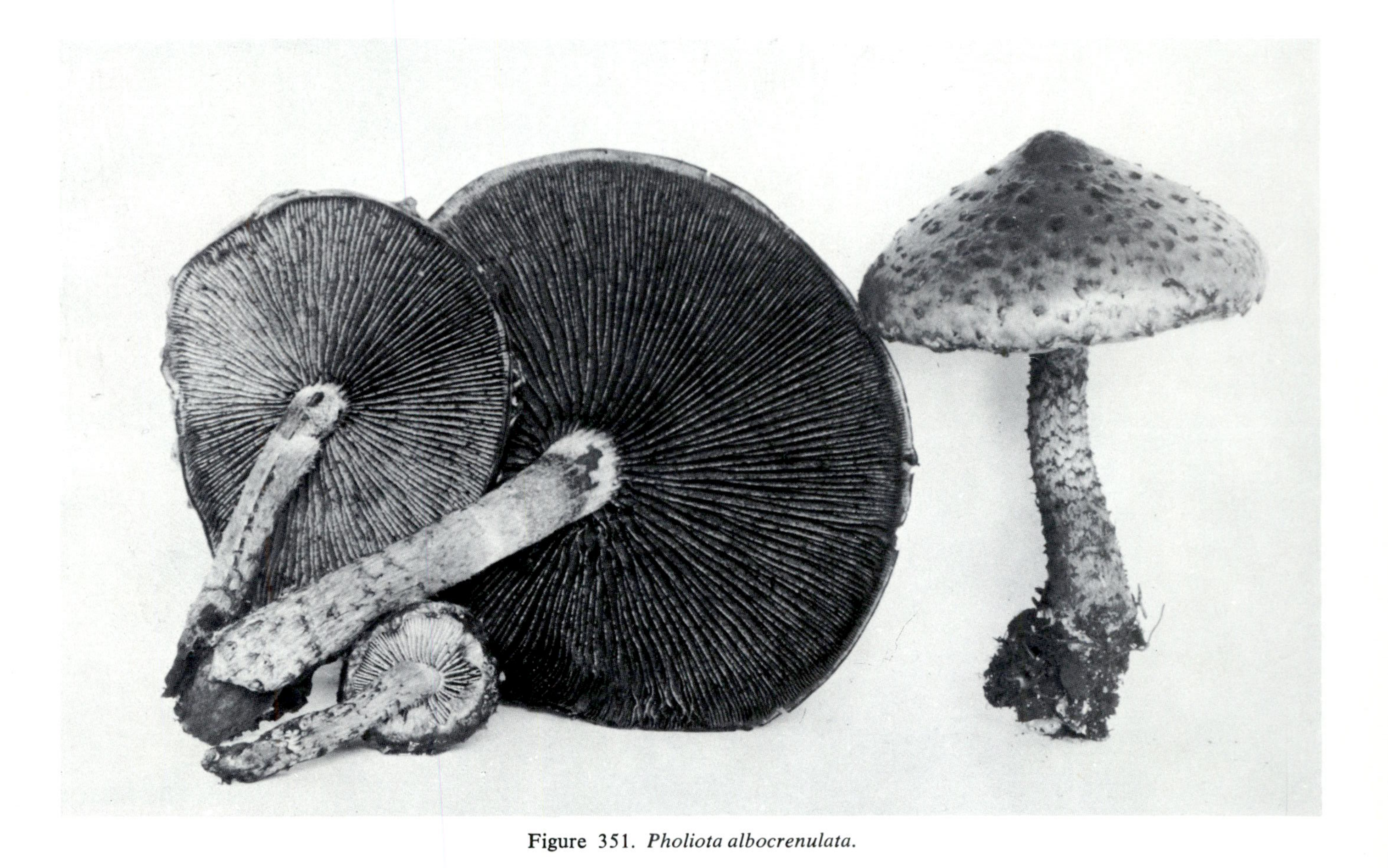

Figure 351. *Pholiota albocrenulata.*

POLYPORACEAE

The Polyporaceae include a group of fungi in which the spores are borne in the interior of tubes as in the Boletaceae but the fruiting bodies are not fleshy as in the Boletaceae but usually hard and tough, leathery, corky, cheesy, or woody in consistency. They are nearly always found on wood and are a very important group of fungi economically because of the damage they cause by rotting the wood. However, because of their tough consistency they are of little interest as food. Only a very few species are edible.

They will doubtless attract the attention of the collector because of their abundance and prominence in the fungus flora of the woods. Typically they appear as bracket-like or shelf-like fruiting bodies sometimes referred to as conks on trees or logs; some are inconspicuous, lying flat against the substrate, and some are more or less stalked and mushroom-like in shape. It might be thought that the latter type would be confused with boletes but in practice there is little difficulty in distinguishing between the boletes and polypores because the consistency is so different.

Some of the fruiting bodies of the polypores are perennial, persisting for several years and developing a new layer of tubes each season. Some of these fruiting bodies may reach considerable size. One species that occurs on the Pacific Coast, *Oxyporus nobilissimus* W. B. Cooke has been known to reach a size of 56 × 37 inches and a weight of 300 pounds.

Only seven species have been included here: one, *Ganoderma tsugae*, because its distinctive and beautiful appearance will certainly attract attention, and the others because they are fairly common and have been reported to be edible when young. The Beefsteak Fungus, *Fistulina hepatica* (Huds.) Fr., long celebrated as an excellent edible fungus, belongs in the Polyporaceae but it appears to be so rare in Canada as to be scarcely worth including. It is a fleshy tongue-shaped fungus, said to resemble a beefsteak in color, and with the pores separate from each other. It may be found in southern Ontario.

Key to the species described

1. Pileus and stipe appearing as if varnished .... *Ganoderma tsugae*
1. Pileus and stipe not varnished 2

2. Pileus yellowish, scaly; pores large .... *Polyporus squamosus*
2. Pileus not scaly 3

3. Pileus with strongly projecting, thick sterile margin, growing on birch .... *Polyporus betulinus*
3. Pileus without a projecting sterile margin 4

4. Pileus stipitate 5
4. Pileus not stipitate 6

5. Fruiting bodies occurring singly or gregarious, whitish to yellowish .... *Polyporus ovinus*
5. Fruiting bodies densely cespitose; stipes numerous, branching, arising from a fleshy mass .... *Polyporus frondosus*

6. Pileus bright sulphur-yellow to pinkish or orange .... *Polyporus sulphureus*
6. Pileus dark brown, velvety, watery-fleshy .... *Polyporus resinosus*

GANODERMA TSUGAE Murr. — Not edible

Figure 396, page 289

PILEUS 2–12 in. in diameter, more or less fan-shaped to bean-shaped, stipitate or sessile by a narrowed base, variable in color, reddish, reddish brown, mahogany colored, brownish orange to nearly black, with a white to yellow or orange margin, glabrous, with a varnished appearance. FLESH white or nearly so, except brownish near the tubes, thick, tough, corky. TUBES whitish becoming brown when bruised. STIPE when present usually lateral or excentric, colored like the pileus, 1¼–6 in. long, ½–1½ in. thick. SPORES brown, ovoid, slightly rough, 9–11 × 6–8 μ.

On and about stumps and logs of coniferous trees, especially hemlock. July–Nov., sometimes persisting through the winter.

This is a beautiful fungus. The bright colors and varnished appearance will certainly attract attention, but it is not edible. There has been some difference of opinion as to whether or not *G. lucidum* (Leyss ex Fr.) Karst. is the same fungus but it seems best to regard this form, which occurs on hardwoods, as a distinct species.

POLYPORUS BETULINUS Bull. ex Fr. — Edible when young

Figure 352, page 253

PILEUS 1¼–10 in. long, 1¼–6 in. broad, more or less elongated to circular, convex, or attached by a lateral, stem-like umbo, usually sessile, whitish to brownish, glabrous or somewhat scurfy, with a thick, sterile, inrolled margin projecting down below the tube surface. FLESH white, rather thick, rather cheesy to corky in age. TUBES white, small, usually smooth but sometimes becoming somewhat toothed. SPORES cylindric to allantoid, smooth, 3.5–5 × 1–2 μ.

On living or dead birch trees. May–Nov.

The characteristic, thick, projecting margin and the occurrence on birch only, are the chief distinguishing marks of this species. It is an easy one to recognize. It has been reported to be edible when young but is usually too tough to be of any value.

POLYPORUS FRONDOSUS (Dicks.) Fr. — Edible

Figure 397, page 289

Hen of the Woods

FRUCTIFICATION 6–12 in. or more in diameter, consisting of a fleshy mass of crowded, much-branched stipes which expand above into imbricated pilei ½–

2 in. broad, somewhat fan-shaped, whitish to gray, smooth, margin often split or lobed, taste slightly peppery. TUBES white, decurrent, very short. SPORES white, smooth, broadly ellipsoid to ovoid, 5–7 × 3.5–5 μ.

Arising from a buried sclerotium, usually around hardwood stumps. Not common. Sept.

This species is sometimes called the hen of the woods because the mass of imbricated gray pilei suggests a hen on a nest. It is said to be of good flavor. *P. umbellatus* (Pers.) Fr. has a more definite, central, stem-like mass from which the stipes arise, the pilei are more circular and centrally depressed, and the spores are cylindric. It is also said to be edible.

POLYPORUS OVINUS (Schaeff.) Fr. — Edible

Figure 353, page 253

PILEUS 1½–5 in. broad, white or whitish, becoming yellowish in age, convex to expanded or sometimes depressed, smooth or surface becoming cracked. FLESH white, becoming yellowish, rather tough, taste and odor mild and pleasant. TUBES whitish to yellowish, short, decurrent. STIPE 1–3 in. long, ¼–¾ in. thick, central to excentric, white, somewhat bulbous. SPORES white, smooth, broadly ellipsoid to subglobose, 3.5–4 × 2.5–3.5 μ.

On the ground in coniferous woods, not common. July–Oct.

This is reported to be a good edible species but is not found often. *P. confluens* (Alb. & Schw.) Fr. differs in becoming reddish when old or on drying. It is also said to be edible. *P. griseus* Peck is a whitish to smoky gray species with slightly larger, rough spores.

POLYPORUS RESINOSUS Schrad. ex Fr. — Edible when young

Figure 354, page 253

PILEUS 2¾–10 in. long, 1¼–6 in. broad, sessile or effused-reflexed, shelf-like or bracket-like, dark brown to blackish brown, at first velvety-tomentose, becoming nearly glabrous, sometimes radiately furrowed, the margin thick, exuding drops of water when young. FLESH thick, straw-colored, watery when young, becoming tough and corky when mature. TUBES whitish, becoming brown when bruised, small. STIPE lacking. SPORES allantoid, smooth, 4–7 × 1.5–2 μ.

On old logs and stumps of hardwoods or conifers. Aug.–Nov.

The form on conifers is regarded by some as a distinct species, *P. benzoinus* (Wahl.) Fr., but it is at least very close to *P. resinosus*. This species has also been reported as edible when young but soon becomes tough. The velvety brown fruiting bodies are attractive in appearance.

POLYPORUS SQUAMOSUS Micheli ex Fr. Edible when young

Figures 356, 357, page 253

PILEUS 2½–12 in. broad, or sometimes larger, nearly circular to elongated or reniform, convex to plane and centrally depressed, whitish to yellowish or brownish, dry, scaly, tough-fleshy. TUBES decurrent, large, angular, white or yellowish. STIPE lateral or excentric, ⅜–2 in. long, ⅜–1½ in. thick, sometimes nearly lacking, reticulate above, black below. SPORES elongate-cylindric, smooth, 10–15 (18) × 4–6 μ.

Singly or in clusters, usually from wounds on deciduous trees, occasionally on stumps or logs. May–July.

The large pores and scaly pileus are the chief distinguishing characters of this species. It is said to be edible when young but mostly is too tough to be of any value as food.

POLYPORUS SULPHUREUS (Bull.) Fr. Edible

Figure 355, page 253

FRUITING BODY consisting of a massive cluster of overlapping, more or less horizontal shelves, up to 12 in. or more across, variable in form, bright sulphur-yellow to yellow-orange or pink, sessile or on a stipe-like base, upper surface glabrous, uneven, lower surface bearing short yellow pores, margin at first thick and blunt, becoming narrower with age, at first soft and fleshy, becoming tougher with age, taste mild or sometimes unpleasant in age. SPORES smooth, one-celled, ovoid to subglobose 5–7 × 4–5 μ.

In clusters on dead or living trees or around stumps. Aug.–Oct.

The large, brilliantly colored fruiting bodies are very distinctive and unlikely to be confused with anything else. This species is edible and has been highly recommended when young specimens are used. Older specimens are likely to be tough and of poor flavor.

HYDNACEAE

The family Hydnaceae includes a large group of fungi in which the spores are borne over the surface of teeth or spines developing from the underside of the fruiting body. The fruiting body may vary considerably in structure from a simple layer of fungus tissue on a piece of wood to a large shelf-like or bracket-like form, to an intricately branched structure, or to a mushroom-like fruiting body with pileus and stipe. Many Hydnaceae grow on wood but some are found on the ground. Most of the species are tough and fibrous to woody and not edible but a few are quite good and none are known to be poisonous.

HYDNUM CORALLOIDES Fr. Edible

Figure 358, page 253

FRUCTIFICATION consisting of an intricate system of branches, white to buffy, soft and fleshy, up to 10 inches across, with white cylindric spines about ¼–½ in. long distributed along the under side of the branches more or less in tufts. SPORES spherical, smooth, hyaline, 5.5–7 μ.

On dead hardwood trunks or logs, frequently on beech. July–Nov.

This is a striking and beautiful species. There are several other rather similar species and it is uncertain whether or not some of these are just ecological forms. *H. caput-ursi* Fr., the bear's head fungus, has longer spines but it seems possible to build up a graded series from one type to the other through examination of many collections. *H. laciniatum* Fr. is more branched and has shorter spines and there also appears to be a difference in the spores which are smaller and ellipsoid. *H. erinaceum* Fr. is a more massive fructification with very long spines up to 1½ in. in length. All of these are edible.

In modern classifications this group of species is placed in the genus *Hericium.*

HYDNUM REPANDUM Fr. Edible

Figure 359, page 253; Figure 426, page 303

PILEUS 1½–4 in. broad, convex to plane or depressed, uneven and often irregular, whitish to buff or pinkish cinnamon, dry, minutely velvety to smooth. FLESH white, soft, fairly thick. TEETH slightly decurrent, fleshy, soft, fragile, round to slightly flattened, whitish to cream colored. STIPE ½–2½ in. long, ¼–1 in. thick, sometimes excentric, smooth, colored like the pileus, solid. SPORES white, smooth, ovoid to subglobose, 7–9 × 6.5–7.5 μ.

Singly or gregarious on the ground in frondose and mixed woods. July–Oct.

This species is now usually placed in the genus *Dentinum.* It is mushroom-like in form but is readily recognized by the teeth on the under side of the pileus, the color and the soft, fleshy consistency. Most of the other stipitate Hydnaceae are tough and fibrous or corky to woody in consistency.

HYDNUM SEPTENTRIONALE Fr. Not edible

Figure 360, page 253

FRUCTIFICATION consisting of many horizontal, overlapping sessile pilei united at the base and forming a massive cluster, single pilei 1–6 in. wide and about the same in length, sometimes much larger, at first whitish, becoming buff or yellowish, finely hairy, dry, margin slightly incurved, tough and fibrous in consistency. TEETH white to yellowish, round, pointed, about ¼–¾ in. long. STIPE lacking. SPORES white, ellipsoid, smooth, 4–6 × 2.5–3.5 μ.

On living trunks of hardwoods, especially maple and beech. Aug.–Sept.

This species is too tough to be of value as food but the massive fructifications of overlapping pilei are likely to attract attention. It will appear in successive years on the same tree. In modern classifications this fungus is placed in the genus *Steccherinum*.

CLAVARIACEAE

The Clavariaceae, sometimes called coral fungi, include forms with erect, simple or branched, fleshy or tough fructifications, the fruiting surface smooth, not differentiated into spines, pores, or lamellae. The important character in distinguishing this family from the Thelephoraceae is that there is no differentiation of the fruiting body into an upper sterile surface and a lower fertile surface. The entire surface of the fruiting body is fertile in the Clavariaceae.

The species are difficult to identify and only a few are described here. Most of the clavarias are edible but some are bitter and unpleasant to the taste or are very tough, and in Europe one species has been reported to be poisonous. This is *Clavaria formosa* Pers., which is also widely distributed in North America. It is a medium to large, much-branched form, white at the base, flesh color or pinkish above, but with the tips of the branches yellow. In age it fades to tan or ochraceous tan. This species should be avoided.

Key

1. Fruiting body simple, unbranched 2
1. Fruiting body much branched 3

2. Growing in clusters, bright yellow .... *C. fusiformis*
2. Growing singly or gregarious, ochraceous yellow to brownish, club-shaped .... *C. pistillaris*

3. Tips of branches pink or rosy .... *C. botrytis*
3. Tips of branches not pink 4

4. Taste bitter, fruiting body light tan to tawny .... *C. stricta*
4. Taste mild 5

5. Fruiting body smoky gray .... *C. cinerea*
5. Fruiting body pale yellowish .... *C. flava*

CLAVARIA BOTRYTIS Fr. — Edible

Figure 361, page 253

FRUCTIFICATION 2–4 in. high, much branched, stipitate, the stipe white, ¾–1 in. thick, slightly tapering downward, bulbous at base, the branches whitish to cream with pink to rosy tips, or sometimes lavender in age, erect, parallel or curving and with a somewhat cauliflower-like appearance, taste and odor mild. SPORES cylindric-ellipsoid to oblong-ellipsoid, longitudinally striate, 12–15 × 3.5–5.5 μ.

On the ground in woods. July–Oct.

The rosy tips of the branches provide a striking character by which this species can be recognized.

CLAVARIA CINEREA Fr. — Edible

Figure 398, page 290

FRUCTIFICATION 1–4 in. high, much branched, stipitate, the stipe smoky gray, 1/8–1/4 in. thick, nearly equal, smooth, branches smoky gray to bluish gray, surface finely powdery, erect, parallel or irregular, often wrinkled, sometimes toothed at apex, taste and odor not distinctive. SPORES white, smooth, broadly ellipsoid to ovoid, 7–10 × 5.5–7.5 μ.

Cespitose or gregarious on the ground in woods. July–Sept.

CLAVARIA FLAVA Fr. — Edible

Figure 399, page 290

FRUCTIFICATION 2–6 in. high, much branched, the main branches arising from a thick, short, whitish, stem-like base, erect, cylindric, tapering toward the apex, pale yellow, becoming brownish when bruised, whitish, sometimes brownish, sometimes toothed, taste mild, pleasant. SPORES yellow, ellipsoid, minutely rough, 7.5–10 × 3–4 μ.

On the ground in moist woods. June–Oct.

C. aurea (Schaeff.) Fr. is said to be very similar, but more robust, deeper yellow, and not turning brown when bruised.

CLAVARIA FUSIFORMIS (Sow.) Fr. — Edible

Figure 363, page 255

FRUCTIFICATION 2–4 in. high, cylindrical or compressed, not branched, pointed at the top, bright yellow, hollow, taste mild or bitter. SPORES nearly spherical 4.5–6.5 μ in diameter.

Singly or in clusters on the ground in woods. Aug.–Sept.

This is a distinctive species with bright yellow, unbranched but clustered fruiting bodies.

CLAVARIA PISTILLARIS (L.) Fr. — Edible

Figure 364, page 255

FRUCTIFICATION 4–6 in. high, 1–2 in. thick, club-shaped, unbranched, top rounded or sometimes depressed, yellowish or ochraceous to brownish, smooth or sometimes longitudinally grooved or wrinkled, soft and fleshy, cream colored within, finally hollow, taste mild. SPORES oblong-ellipsoid, smooth, 11–13.5 × 5–7 μ.

Singly or gregarious on the ground in moist woods. Aug.–Oct.

The large, rather bright colored, club-shaped fruiting bodies of this fungus are very striking and distinctive. Occasionally the tip of the club will be somewhat depressed and sterile, thus proving the exception to the rule that the clavarias are fertile all over.

CLAVARIA STRICTA Fr. — Edible, not recommended

Figure 349, page 235

FRUCTIFICATION 1½–3 in. high, much branched, the main branches arising from a thick irregular whitish base, erect, cylindric or flattened toward the base, tapering upwards, terminating in several small yellowish teeth, light tan to tawny, consistency rather tough, not brittle, taste bitter. SPORES cinnamon-buff, ellipsoid, minutely rough 7.5–9 × 3.5–4.5 μ.

Forming dense tufts on rotten wood. July–Oct.

The tough consistency and bitter taste are the distinguishing characters of this species and render it of poor quality as an edible species although it is not poisonous.

THELEPHORACEAE

The Thelephoraceae include a large group of fungi in which the spores are borne on a smooth surface and not on spines, pores, or lamellae. Most of them consist simply of a layer of fungus tissue growing on wood or bark, frequently on the under side of logs or sticks, and producing spores over the surface. Some of this group develop a definite pileus which may be more or less bracket-like, growing on wood, or upright and growing on the ground. The latter are distinguished from the Clavariaceae, which also have a smooth hymenium, by the fact that the hymenium does not cover the entire surface of the fruiting body, but there is always some differentiation into an upper sterile surface and a lower fertile surface. Only a single species is described here.

CRATERELLUS CORNUCOPIOIDES Fr. — Edible

Figure 365, page 255

Horn of Plenty

FRUCTIFICATION about 1–3 in. high, ¾–2¼ in. across the top, funnel-shaped or trumpet-shaped with a flaring margin, the margin even to wavy or lobed, sometimes becoming torn, thin, rather tough or brittle, interior or upper surface dry, rough to scaly, dark grayish brown, lower surface or hymenium ashy to blackish, smooth or somewhat wrinkled. STIPE very short or absent. SPORES ellipsoid, smooth, one-celled, 11–15 × 7–9 μ.

Gregarious on the ground in open woods. July–Oct.

Although rather unattractive in appearance because of its dark colors, this is reported to be quite good as an edible fungus and it is unlikely to be confused with anything else. It has several common names such as horn of plenty, trumpet of death, and fairy's loving cup, indicating that it is a species that attracts attention. The name trumpet of death has no reference to its edible qualities but only to its sombre appearance.

TREMELLALES

The Tremellales or jelly fungi can be recognized in the field by their more or less gelatinous or jelly-like consistency. They shrink greatly on drying and in dry weather are inconspicuous, but when moistened they swell up and are sometimes striking in appearance.

The group is separated from the other basidiomycetes on a character that is considered to be more fundamental than their jelly-like consistency, namely, the structure of the basidium. A typical basidium is one-celled with a septum at the base, but in the Tremellales the basidium itself becomes septate or deeply forked. The Tremellales are divided into three families, the Tremellaceae in which the basidium is longitudinally or obliquely septate, the Auriculariaceae in which it is transversely septate, and the Dacrymycetaceae in which it is forked and deeply divided. These characters can be determined only by microscopic examination so that for the amateur the consistency is the best field character, although there are some instances in which this character alone is misleading.

Only four species are described here, two of the Tremellaceae, one of the Auriculariaceae, and one of the Dacrymycetaceae.

PSEUDOHYDNUM GELATINOSUM (Fr.) Karst.

Figure 366, page 255

FRUITING BODY 1–2¼ in. broad, gelatinous, translucent, whitish becoming brownish, rather thick, upper surface papillose, lower surface bearing whitish, gelatinous, tooth-like spines about ⅛ in. long, short stipitate or sessile, spores white, subglobose, 5–7 μ.

On rotten wood. Aug.–Sept.

At first sight this might be taken for one of the Hydnaceae because of the teeth on the under side of the pileus, but the gelatinous consistency distinguishes it and if examined microscopically the basidia are found to be divided longitudinally. It is not of interest as food but is a pretty and unusual fungus. It is likely to be found only in wet weather as it shrinks greatly and becomes very inconspicuous when dry.

PHLOGIOTIS HELVELLOIDES (Fr.) Martin

Figure 367, page 255

FRUITING BODIES 2–4 in. high, 1½–2¼ in. in diameter, gelatinous but rather firm, drying horny, more or less funnel-shaped, usually split on one side, pinkish white to deep rose, substipitate, spores oblong. 10–12 × 4–6 μ.

On the ground under conifers or on rotten coniferous wood. Aug.–Oct.

This is not regarded as an edible species but is an attractive and striking fungus. It has been known under the name *Gyrocephalus rufus* (Jacq.) Bref.

AURICULARIA AURICULA (Hook.) Underw. — Edible

Figure 368, page 255

FRUITING BODY about ¾–3½ in. broad, sessile, somewhat cupulate or ear-shaped, irregular, smooth or wrinkled, tough-gelatinous, yellow-brown to cinnamon-brown, drying horny and nearly black, attached centrally or laterally, spores white in mass, allantoid, 12–16 × 4–6 μ.

Gregarious or cespitose on dead wood or sometimes exposed wood of living trees. July–Oct.

As the name indicates, this fungus is somewhat suggestive of a human ear. It has become involved in an ancient legend to the effect that Judas Iscariot hanged himself on an elder tree and the elder was thereupon condemned to bear this excrescence, which was known as Judas' ear and later corrupted to Jew's ear. Apparently the fungus occurs rather commonly on the elder in Europe but no doubt it did so for many centuries before the time of Judas Iscariot. It is not clear why Judas' ear should have been singled out to commemorate his evil deed but there may be some confusion here with the ear of the servant of the high priest, which was cut off by Peter at the time of the betrayal.

DACRYMYCES PALMATUS (Schw.) Bres.

Figure 369, page 255

FRUITING BODIES bright orange to orange-red, tough-gelatinous, becoming softer in age, forming irregular clusters, often wrinkled and convoluted, sometimes up to 2 in. across, usually smaller, sometimes with a whitish, stipe-like base, spores orange, cylindrical, curved, becoming seven-septate, 17–25 × 6–8 μ.

On coniferous wood. May–Nov.

The bright orange fructifications are very conspicuous when moist but on drying they shrivel to an inconspicuous, reddish orange, horny mass. This is probably one of the fungi that have been called 'witches' butter' or 'fairy butter.' It is not regarded as an edible fungus.

GASTEROMYCETES

The Gasteromycetes include the fungi commonly known as puffballs and closely related forms. In this group the spores are produced on basidia but are not forcibly discharged as in the mushrooms, boletes and other Basidiomycetes. The basidia break down and the spores are typically left as a powdery mass within the fruiting body and are dispersed through a pore or by the wearing away of the outer layers of the fruit body.

Representatives of three main groups of Gasteromycetes are discussed here. In the Lycoperdaceae, which include the puffballs proper, the spores are produced and dispersed as described above. In the Phallaceae or stinkhorns, the spores are produced in slime and elevated on the end of a stalk-like structure that emerges from an enclosing volva. They usually produce an offensive odor that is believed to attract insects which aid in the dispersal of the spores. In the Nidulariaceae or bird's-nest fungi, the spores are produced within structures called peridioles which are borne in a cup-like fructification from which they are dispersed by driving rain drops, and the spores are then freed by the decay of the peridiole wall.

The effect of rain in dispersing spores of the true puffballs may easily be demonstrated by placing a mature puffball at some distance under a burette and permitting drops of water to fall on it. As each drop strikes the puffball a puff of spores is emitted.

This is a large and extremely varied group of fungi with many qualities that attract the attention and interest of naturalists from the immense size attained by some specimens of *Calvatia gigantea* to the repulsive odors of the stinkhorns, fantastic shapes of some of the other groups, and fascinating methods of spore dispersal found among them.

The puffballs proper are generally regarded as one of the safest groups of fungi to use as food and one of the few groups in which it is possible to give a sort of rule of thumb for determining an edible species. It seems safe to say that any puffball that is white and homogeneous inside is good to eat. However, in this group as with all other fungi it is wise to proceed cautiously with any species not previously tried because of the possibility of personal sensitivity to a particular species. It is important to make certain that the fruiting body is homogeneous within since it is possible that young specimens of the deadly poisonous *Amanita virosa* might be mistaken for a puffball before the volva is ruptured. Cutting the fruit body across will reveal the outline of the young mushroom if it is an *Amanita* (Figure 91). The eggs of the phalloids can also be distinguished from the puffballs in this way and in spite of the fact that these are edible, according to some authors, they are not recommended as food. The Nidulariaceae are, of course, too small and tough to be of any value as food.

Key

1.	Fruiting body a stalked structure with a differentiated apical portion where the spores are borne; odor offensive	2
1.	Fruiting body not as above	4
2.	Apical spore-bearing portion forming a distinct pileus, separate from the stalk	3
2.	Apical spore-bearing portion a continuation of the stalk, not separate; fruiting body pinkish, odor slight	*Mutinus caninus*
3.	Plant with a lacy veil extending below the pileus	*Dictyophora duplicata*
3.	Veil membranous, inconspicuous, scarcely extending below the pileus; surface of pileus granular, greenish	*Phallus ravenelii*
4.	Fruiting body a small cup-like structure striate within, containing several tiny spore cases, suggesting a nest containing eggs	*Cyathus striatus*
4.	Fruiting body not as above	5
5.	Outer layers of fruit body splitting into several segments or rays to form a star-shaped body	*Geastrum triplex*
5.	Outer layers of fruit body not splitting into rays	6
6.	Fruiting body very large; spores dispersed by wearing away of the outer layers	*Calvatia gigantea*
6.	Fruiting body not over 3 in. in diameter; spores dispersed through an apical pore	7
7.	Inner coat rather papery; fruiting bodies easily becoming detached and blown about by the wind	*Bovista pila*
7.	Inner coat not papery; fruiting bodies remaining attached	8
8.	Outer coat of cone-shaped spines that fall off leaving distinct spots; typically growing on the ground	*Lycoperdon perlatum*
8.	Outer coat more or less persistent, rough, not of conical spines; typically growing in clusters on rotten wood	*Lycoperdon pyriforme*

MUTINUS CANINUS (Pers.) Fr. — Not edible

Figure 370, page 255

EGG white, ovoid, about ½–¾ × ½–⅝ in., attached by a rhizomorph. FRUITING BODY 2–4 in. tall, ¼–⅜ in. thick, cylindric, equal, narrowed at the apex and usually perforated, pinkish to whitish, olivaceous brown at the tip where the spores are borne, sheathed at the base by a volva. ODOR unpleasant but relatively slight. SPORES 4–5 × 1.5– 2μ.

Singly or gregarious on soil or rotten wood in open woods, sometimes in gardens. July–Sept.

The small size and pinkish colors are distinctive for this species. Usually the odor is relatively weak.

DICTYOPHORA DUPLICATA (Bosc) E. Fisch. — Not edible

Figure 400, page 291

EGG 1½–2¾ in. in diameter, subglobose to somewhat flattened or ovate, whitish, sometimes wrinkled at the base, attached by a thick white rhizomorph.

FRUITING BODY 5–8 in. high, 1–2½ in. thick at the base, tapering upward slightly, odor fetid and very disagreeable. STIPE cylindrical, spongy or honeycombed, hollow, white, sheathed at base by remains of egg forming a whitish to brownish volva. PILEUS more or less conical, attached to the apex of the stipe, perforated, reticulate, greenish black. VEIL lacy and net-like, whitish to pinkish, attached to apex of stipe beneath the pileus and projecting below it. SPORES 3.5–4 × 1.5–2.0 μ.

Singly or gregarious in the woods or in gardens, usually around dead trees or stumps. July–Oct.

The large size and lacy veil are the distinguishing characters of this species.

PHALLUS RAVENELII Berk. & Curt. — Not edible

Figure 371, page 255

EGG 1–2 in. in diameter, more or less egg-shaped to subglobose, whitish to pinkish, or tinged lilac, tough, wrinkled at the base, attached by a pinkish lilac rhizomorph. FRUITING BODY 4–6 in. high and ½–1¼ in. thick, odor fetid and very disagreeable. STIPE whitish to yellowish, somewhat spongy or honeycombed, hollow, equal or tapering upward, encircled by a white, membranous band from the veil, the base enclosed by the remains of the egg, forming a volva. PILEUS more or less conical, attached around a raised, white, perforated disk at the apex, granular, shiny, greenish to olive-gray. SPORES 3–3.5 × 1.5 μ.

Usually gregarious on sawdust or very rotten wood. June–Oct.

The eggs might be mistaken for a puffball but if they are cut open the outline of the young fruiting body can be seen surrounded by a layer of a jelly-like substance. A closely related species, *P. impudicus* Pers., has a deeply reticulated pileus.

CYATHUS STRIATUS Pers. — Not edible

Figure 372, page 255

FRUITING BODY ¼–⅝ in. tall, ¼–½ in. broad at the mouth, tapering to the base, somewhat vase-shaped, attached by a pad of brown mycelium, exterior dark cinnamon-brown, coarsely fibrillose, inner surface pallid to blackish or tinged purplish, striate, mouth at first closed by a thin fibrillose epiphragm which ruptures and disappears at maturity. PERIDIOLES more or less flattened or disk-shaped, nearly black, attached to the cup by an elastic cord. SPORES hyaline, thick-walled, 14–20 × 8–12μ.

Gregarious to cespitose on old sticks and various vegetable debris. July–Oct.

This species is distinguished by the striations on the inner surface of the cup. *C. stercoreus* (Schw.) de Toni is another common species growing on the ground and with the inside of the cup smooth. *C. olla* Pers. has a smooth cup but has smaller spores than *C. stercoreus.*

Crucibulum levis (DC.) Kambly & Lee is another species belonging in this group. It has whitish peridioles and the wall of the fructification consists of only one layer rather than three as in *Cyathus.* It is usually found on old wood also.

It has been demonstrated that these fruiting bodies of the bird's-nest fungi are adapted to dispersal of the spores by rain. The force of rain drops splashing against the inside of the cup is sufficient to drive the peridioles out for some considerable distance and the spores are freed by the gradual decay and wearing away of the wall of the peridiole. This is referred to as the splash-cup method of spore dispersal.

GEASTRUM TRIPLEX Jungh. — Edible

Figure 374, page 281

Earth Star

FRUITING BODY 1–2 in. in diameter at the widest part, more or less bulb-shaped, acute at the apex, not stalked, brown or reddish brown, the outer coat splitting at the apex into 4–6 rather uniform segments or rays, spreading back to form a star shape and often splitting into two layers of which the inner remains as a sort of cup around the spore case. SPORE CASE sessile, thin and papery, opening by a conical pore which is different in texture from the rest of the spore case, smooth but somewhat fringed, usually seated in a slight depression. SPORES brown, globose, warted, 3.5–4.5 μ.

Singly or gregarious on the ground in open woods. July–Nov.

There are several species of *Geastrum* commonly known as 'earth stars' but this is one of the larger and more common species. It is reported to be edible when young but seems to be pretty hard and tough.

CALVATIA GIGANTEA Pers. — Edible

Figures 404, 405, page 292

Giant Puffball

FRUITING BODY more or less globose, 8–20 in. or more in diameter, attached to the ground by a short, cord-like rhizomorph, surface smooth, soft-leathery, white to creamy yellow or finally brownish, white within when young, firm, fleshy, slowly becoming yellowish to olivaceous, finally rusty ochraceous and powdery. SPORES globose, minutely spiny, 3.5–4.5 μ.

Singly to gregarious in woods, pastures or fields. Aug.–Sept.

The giant puffball is one of the best-known and most widely used edible fungi. The large size, white color, and chamois-like outer skin are distinguishing features. Specimens intended for food should be cut open to make sure that they are white and homogeneous inside and not infested by worms.

Some people have the idea that these large puffballs appear very suddenly, but actually they grow and increase in size over a period of nearly two weeks before they reach maturity. Observations published by Peck (1912) indicate that their period of development is about 12–14 days and that the daily increase in circumference is about 3–4 inches.

Nearly every year reports are published concerning the finding of large puffballs with competing claims as to record size. Most of these, however, are far short of any real record. Güssow and Odell (1927) report a specimen 5 feet 1¼ inches in circumference and weighing 18¾ pounds but this is dwarfed by a report from New York State in 1877 mentioned by Ramsbottom (1953) of a specimen 5 feet 4 inches long, 4 feet 6 inches wide and 9½ inches high. It was said to have been mistaken for a sheep at a distance. The largest specimen ever collected by the author was 4 feet in circumference and weighed 11 pounds.

Buller calculated that a fruit body 16 × 12 × 10 inches would produce 7 trillion spores and Ramsbottom noted that if each of these produced one puffball of similar size and if their spores were equally successful, the resulting mass would be 800 times the weight of the earth. It is an amazing example of the reproductive potential of a living organism and it is obvious that under natural conditions the chances of a puffball spore for survival must be extremely slight indeed.

Two other species of *Calvatia* that are fairly common but do not attain the size of *C. gigantea* are *C. cyathiformis* (Bosc) Morgan and *C. craniiformis* (Schw.) Fr. Both of these have a rather thick, stout basal part that is somewhat pear-shaped to top-shaped and may sometimes be found persisting after all the spores have been dispersed. *C. cyathiformis* can be distinguished by the purplish spore mass whereas in *C. craniiformis* it is olivaceous brown to dark brown. All of these species are edible when young.

BOVISTA PILA Berk. & Curt. — Edible

Figures 401, 402, 403, page 291

FRUITING BODY 1½–2½ in. in diameter, usually globose or nearly so, attached by a small rhizomorph which breaks easily, at first white with a thin furfuraceous outer coat that soon wears off, exposing the smooth, somewhat papery, gray-brown to bronze, inner coat, which often has a somewhat metallic luster, at first white within, then becoming purplish brown and powdery. SPORES dark brown, globose, smooth, sometimes with a short pedicel 3.5–4.5 μ.

Solitary or gregarious in woods or pastures. Aug.–Oct.

The fruiting bodies frequently come loose from their attachment. They often persist through the winter and are found the following spring full of spores.

B. plumbea Pers. is a somewhat smaller species, attached to the ground by a mass of fibers rather than a rhizomorph, and with a more blue-gray inner coat. The spores are larger, more ovoid, and have very long pedicels.

LYCOPERDON PERLATUM Pers. Edible

Figure 375, page 281; Figures 406, 407, page 292

FRUITING BODY 1–2½ in. high, ½–2¼ in. thick at the widest part, typically top-shaped to pear-shaped, or irregular in shape from crowding with a tapering or nearly cylindric, stem-like base, sometimes wrinkled or folded toward the base, white at first, becoming buff or brownish, surface covered with many cone-shaped, whitish spines, some longer, some shorter, which disappear as the plant matures, leaving net-like markings on the surface, entirely white within when young. SPORES produced only in the upper part which becomes yellowish to olive-brown and powdery, basal part remaining sterile, the tissue containing small chambers. SPORES olive-brown, globose, minutely spiny, 3.5–4.5 μ in diameter.

Singly, gregarious, or cespitose on rich soil, or sometimes on rotten wood. June–Nov.

This is one of our commonest puffballs. The shape and the cone-shaped spines that leave a network of scars when they fall off are the chief distinguishing characters. It has long been known as *Lycoperdon gemmatum* Batsch but *L. perlatum* is the correct name.

LYCOPERDON PYRIFORME Pers. Edible

Figure 408, page 293

FRUITING BODY ¾–2 in. high, ¾–1½ in. thick at the widest part, typically somewhat pear-shaped to subglobose, narrowed below and attached by white rhizomorphs, pale brownish to tawny brown or rusty brown, sometimes yellowish, surface furfuraceous, scaly or with short spines, sometimes areolate, the outer coat eventually wearing away and exposing the smooth inner coat, white within at first, becoming olivaceous to olive-brown and powdery as the spores mature, basal part sterile, tissue containing small chambers. SPORES olive-brown, globose, smooth, 3–4 μ.

Usually cespitose to gregarious around old logs, stumps, sawdust piles. June–Nov.

This is a very common and widely distributed puffball. It is not very large but is usually found in considerable abundance. The old fruit bodies often persist through the winter and may be found the following spring but, of course, they are only edible when young and white within. The color, shape, and occurrence on rotten wood are distinguishing characters.

Figures 352-361

352. *Polyporus betulinus.*
353. *P. ovinus.*
354. *P. resinosus.*
355. *P. sulphureus.*
356. *P. squamosus.*
357. *P. squamosus.*
358. *Hydnum coralloides.*
359. *H. repandum.*
360. *H. septentrionale.*
361. *Clavaria botrytis.*

352
353
354
355
356
357
358
359
360
361

Figure 362. *Pholiota vermiflua.*

Figures 363-372

363. *Clavaria fusiformis*
364. *C. pistillaris.*
365. *Craterellus cornucopioides.*
366. *Pseudohydnum gelatinosum.*
367. *Phlogiotis helvelloides.*
368. *Auricularia auricula.*
369. *Dacrymyces palmatus.*
370. *Mutinus caninus.*
371. *Phallus ravenelii.*
372. *Cyathus striatus.*

363
364
365
366
367
368
369
370
371
372

Figure 373. *Phaeolepiota aurea.*

ASCOMYCETES

The Ascomycetes comprise a vast number of species of fungi that differ fundamentally from the Basidiomycetes in the manner in which the spores are formed. In the Basidiomycetes the spores are formed outside the mother cell or basidium, usually developing on little stalks that arise from it, whereas in the Ascomycetes the spores are produced within the mother cell or ascus and are not discharged until they are mature.

The great majority of the Ascomycetes are minute fungi requiring a microscope for determination of their characters but some are large enough to attract the attention of the amateur collector and a few are known to be among our best edible fungi. Although a microscope is needed to actually see the difference between an ascus and a basidium, in practice it is not difficult to recognize an ascomycete in the field.

All of the species described here, with one exception, belong in the section Discomycetes in which the asci are borne in an exposed fruiting layer or hymenium and not within a closed fruiting body. Relatively few species are described here and anyone interested in the group should consult special works dealing with them. The single pyrenomycete described, *Hypomyces lactifluorum* (Schw.) Tul., is not included in the key.

Key

1. Fruiting body with a distinct stipe and differentiated pileus 2
1. Fruiting body more or less cup-shaped, without a differentiated pileus but sometimes stipitate 8

2. Pileus pitted or honeycombed 3
2. Pileus convoluted, wrinkled, or smooth, not pitted 5

3. Base of pileus attached to stipe 4
3. Base of pileus free from the stipe .... *Verpa bohemica*

4. Pileus subglobose to ovoid; pits irregular with edges the same color or paler .... *Morchella esculenta*
4. Pileus conical; pits more or less longitudinally arranged with edges darker .... *Morchella angusticeps*

5. Pileus irregular, reddish brown, surface convoluted; growing on the ground in spring .... *Gyromitra esculenta*
5. Pileus slightly wrinkled or smooth, usually more or less saddle-shaped 6

6. Stipe deeply longitudinally fluted .... *Helvella crispa*
6. Stipe smooth or slightly furrowed at base 7

7. Pileus smoky gray to smoky yellowish or nearly black .... *Helvella elastica*
7. Pileus tan to reddish brown .... *Gyromitra infula*

8. Cups bright scarlet, whitish-hairy externally; growing on sticks in the spring .... *Sarcoscypha coccinea*
8. Cups not scarlet 9

9. Cups black, stipitate, tough .... *Urnula craterium*
9. Cup brownish, sessile to substipitate, soft, fleshy, spreading out widely; growing on wood .... *Peziza repanda*

MORCHELLA ESCULENTA Fr. Edible

Figure 376, page 281; Figure 427, page 304

Common Morel

PILEUS 2–5 in. long and ¾–1½ in. thick at the widest point, sometimes much larger, usually more or less ovoid to somewhat conical or sometimes subglobose, the surface covered with rounded to irregular or somewhat elongated pits, irregularly arranged or often more or less in rows, gray-brown to yellowish brown, the edges of the pits colored like the interior or paler, and finally becoming thin and somewhat torn. STIPE 1–4 in. long, ½–1 in. thick, white to cream color or yellowish, at first cylindric, becoming more or less compressed and furrowed, sometimes much thickened at the base, glabrous to slightly floccose, mealy, hollow. ASCI cylindric, eight-spored 225–325 × (15) 18–22 (27) μ. ASCOSPORES slightly yellowish in deposits, ellipsoid, smooth, one-celled, (12) 16–22 (26) × (7.5) 11–13 (14) μ.

Singly or gregarious in open woods, orchards, or grassy places. May or early June.

This is the common morel and is highly regarded as one of the best of the edible fungi. The pitted, sponge-like pilei are very characteristic and unlikely to be mistaken for anything else. However, care must be taken to distinguish *Gyromitra esculenta*, or false morel, which occurs at the same time of year and has a wrinkled and convoluted, rather than pitted, pileus.

MORCHELLA ANGUSTICEPS Peck Edible

Figure 377, page 281; Figure 428, page 304

Narrow-capped Morel

PILEUS ¾–2½ in. long, ½–1¼ in. broad at the base, more or less elongated to narrowly conic, the surface covered with somewhat elongated pits arranged more or less in vertical rows, yellowish or yellowish brown within, the edges smoky brown to black. STIPE ¾–2½ in. long, ½–1 in. thick, white to yellowish, cylindric or enlarged at the base, often furrowed toward the base, floccose-mealy, hollow. ASCI cylindric, eight-spored, 200–300 × 16–22 (26) μ. ASCOSPORES yellowish in a deposit, ellipsoid, smooth, one-celled, 18–25 (29) × 11–15 μ.

Singly or gregarious in open woods or at the edges of woods. May or early June.

This morel is distinguished from the common morel by its more conical pileus, scurfy stipe, and elongated pits with dark edge. It is fully as good to eat as the common morel. *M. conica* Fr. is probably the same species.

VERPA BOHEMICA (Krombh.) Schröt. Edible

Figure 378, page 281

PILEUS about ½–1½ in. long and ⅜–1¼ in. in diameter, somewhat bell-shaped, attached to the apex of the stalk and hanging down around it with the margin free, yellowish brown to reddish brown, the surface usually prominently ridged and reticulated. STIPE 1–5 in. long, ⅜–1 in. thick, whitish to yellowish, glabrous to somewhat floccose, especially toward the base, stuffed becoming hollow, cylindric or somewhat compressed. ASCI cylindric, two-spored, 200–325 × 18–24 (27) μ. ASCOSPORES yellowish in deposits, ellipsoid, one-celled, smooth, (45) 50–75 (84) × 15–22 μ.

Singly or gregarious on the ground in open woods. May.

This fungus might be mistaken for a morel and it has been called *Morchella bispora* Sor. but the attachment of the pileus to the upper end of the stipe and the free margin distinguish it. *Morchella semilibera* (DC.) Fr. is attached part way up the stipe and has the margin free but is a smaller plant. The two-spored asci with very large spores are characteristic of this species.

Another species of *Verpa*, *V. conica* (Müll.) Swartz, is a smaller plant with a smooth, olive-tinged pileus, and eight-spored asci with much smaller spores. It is found at the same time of the year in about the same type of habitat.

GYROMITRA ESCULENTA Fr. — Can be poisonous

Figure 379, page 281; Figure 429, page 305

False Morel

PILEUS 1–3 in. broad, variable and irregular in shape, usually more or less lobed and the surface irregularly wrinkled, folded, or convoluted, but not pitted, reddish brown to dark brown. STIPE ¾–2 in. long, ½–1 in. thick, whitish, fragile, usually somewhat compressed and grooved, hollow, glabrous to slightly floccose. ASCI cylindric, eight-spored, 225–325 × 15–18 μ. ASCOSPORES ellipsoid, one-celled, smooth, (17) 20–28 × 11–16 (17) μ.

On the ground in woods, associated with conifers. May–June.

This fungus has been the subject of much controversy, both concerning its identity and its edible qualities. Seaver (1928, 1942) claimed that *G. esculenta* and *G. infula* are both forms of the same species but this has been disputed by Kanouse (1948) on what appears to be convincing evidence that is also borne out by my personal observation. They are, therefore, regarded as distinct species here. *G. esculenta* occurs on the ground in the spring, associated with conifers, is larger and more irregular in shape and has larger spores. *G. infula* occurs in the fall on rotten wood, probably always hardwood, is smaller than *G. esculenta*, has a more regularly saddle-shaped pileus, is less wrinkled and convoluted and has smaller spores.

Undoubtedly many people eat this species with no ill effects. I have seen it on sale in grocery stores in Finland and have eaten it myself when it was served by friends there and the flavor is excellent. However, reports of poisoning occur every once in a while and it has been known to cause death. Whether or not this is the result of personal idiosyncrasy, the occurrence of certain poisonous races of the fungus, or the development of the poisonous principle by the fungus under certain conditions is not yet certain. One significant case was reported by Dearness (1911) in which a family ate part of a collection of *Gyromitra* at one meal with no ill effects, but when they ate the rest of the collection the next day severe poisoning developed and one member of the family died. This suggests that the poison may have been produced on ageing or partial decay of the fungus. Nevertheless any fungus known to have such deadly potentialities cannot be recommended as food.

G. gigas (Krombh.) Cke. is another species that occurs in the spring. It may occur on the ground or on rotten wood and has been collected on a birch stump. It is brighter yellow than *G. esculenta* and tends to grow in clusters with the stipes irregularly fused and grown together. It can easily be distinguished microscopically by the spores, which have a small apiculus on each end.

GYROMITRA INFULA (Schaeff. ex Fr.) Quél. — Dangerous

Figure 380, page 281; Figure 430, page 306

PILEUS 1–3 in. broad, usually more or less saddle-shaped but sometimes irregular, surface usually smooth to slightly wrinkled and convoluted, usually some shade of tan to brownish cinnamon, the margin partly free. STIPE ¾–2¼ in. long, ¼–⅝ in. thick, whitish or tinged the color of the pileus, finely floccose, cylindric to compressed or with irregular furrows, hollow. ASCI cylindric, eight-spored, 225–300 × 10–14 μ. ASCOSPORES ellipsoid, one-celled, smooth, 16–18 (21) × 7–9 μ.

Singly or gregarious on or in close association with rotten wood. Sept.–Oct.

For a discussion of this species see *Gyromitra esculenta*. Its edible qualities appear to be uncertain but it is regarded as dangerous and is not recommended.

HELVELLA CRISPA (Scop.) Fr. — Edible

Figure 409, page 294

PILEUS about ½–2 in. broad, saddle-shaped to irregularly lobed, reflexed, margin free from the stipe, whitish or cream colored to buff or yellowish, smooth to slightly convoluted. STIPE ¾–2½ in. long, ¼–1 in. thick, white or colored like the pileus, very uneven and deeply fluted with longitudinal furrows. ASCI cylindric, eight-spored, 225–300 × 14–18 μ. ASCOSPORES one-celled, ellipsoid, smooth, (16) 18–20 (22.5) × 10–13 μ.

Usually gregarious on the ground in damp woods. Aug.–Oct.

The pale colors and deeply fluted stipe are the chief distinguishing characters of this species. *H. lacunosa* Afz. ex Fr. is similar in shape and stature but the pileus is smoky gray to nearly black. The stipe is also deeply fluted and usually paler than the pileus but becoming smoky gray.

HELVELLA ELASTICA Bull. ex Fr.

Figure 410, page 294

PILEUS ¾–1¼ in. in diameter, usually more or less saddled-shaped or irregularly two- to three-lobed, margin free from the stipe, smoky gray to yellowish brown or nearly black, smooth or slightly convoluted. STIPE 1¼–4 in. long, ⅛–⅜ in. thick, rather slender, even, not fluted, whitish to yellowish, cylindrical or slightly compressed, usually tapering upward, hollow. ASCI cylindric, eight-spored, 200–270 × 15–18 μ. ASCOSPORES ellipsoid, smooth, one-celled, 18–20 × 10–12 μ.

Singly or gregarious on the ground in woods. June–Oct.

This species is recognized by its dark color and smooth stipe.

PEZIZA REPANDA Pers. ex Fr.

Figure 381, page 281

APOTHECIA about 2–4 in. in diameter, at first cup-shaped, expanding and becoming nearly flat, or sometimes convex, pale brown, externally whitish, smooth, margin even or somewhat wavy, fleshy in consistency, rather brittle, sessile or short-stipitate. ASCI cylindric, eight-spored, 175–250 × 12–15 μ. ASCOSPORES ellipsoid, one-celled, smooth, 14–16 (18) × 8.5–10.5 μ.

On rotten logs or occasionally on the ground in woods. May–Oct.

There are a number of species of cup fungi but this is one of the largest and most common. Most of this group require microscopic study for their identification.

SARCOSCYPHA COCCINEA (Jacq.) Pers. — Not edible

Figure 382, page 281

APOTHECIA ¾–1½ in. in diameter, deep cup-shaped, bright scarlet within, externally whitish and densely covered with fine, long hairs, margin usually incurved and more or less fringed or torn, tough and rather leathery in consistency, more or less stipitate, the stipe ⅛ in. or slightly more in thickness and variable in length. ASCI cylindric, eight-spored, 350–450 × 14–18 μ. ASCOSPORES ellipsoid, one-celled, smooth, 28–35 × 12–15 μ.

On buried or partly buried twigs and branches. April–June.

This is one of the earliest fungi to appear in the spring. It is too tough to be of any value as food but the brilliant scarlet color of the disk is very striking.

URNULA CRATERIUM (Schw.) Fr. — Not edible

Figure 383, page 281

APOTHECIA 1–1½ in. in diameter, at first closed, becoming somewhat goblet-shaped, entirely black or brownish black, externally covered with a dense tomentum and sometimes becoming somewhat scaly, margin notched and lacerated, irregular, tough and leathery in consistency, stipitate. ASCI cylindric, eight-spored, very long 400–600 × 16–18 μ. ASCOSPORES ellipsoid, one-celled, smooth, 25–40 × 11–14 μ.

On buried or partly buried wood, probably always hardwood. April–May.

This species is too tough to be edible but the black fruit bodies are likely to attract attention.

HYPOMYCES LACTIFLUORUM (Schw.) Tul.

Figure 350, page 235

Fungus growing on mushrooms, producing a stroma, which may entirely cover the lamellae and stipe and obliterate the lamellae, which may appear only as slight ridges, scarlet to bright orange-red or finally purple-red; perithecia thickly scattered, immersed in the stroma and appearing as small pimples; asci very long and narrow, cylindrical; ascospores, fusiform, slightly curved, with an apiculus at each end, rough-walled, 35–40 × 7–8 μ.

On species of *Lactarius*. August and September.

This one representative of the Pyrenomycetes has been included because the malformed mushrooms that have been attacked by it are fairly common and always attract the attention of collectors by reason of their brilliant colors.

The parasitized mushrooms have been reported to be edible, but since it is usually not possible to identify the species of mushroom attacked and there is the possibility of a poisonous species being parasitized by the *Hypomyces*, they are not recommended as food.

TECHNICAL KEY TO THE GENERA OF MUSHROOMS

The amateur collector who is simply interested in mushrooms as food need pay no attention to this key, but for the benefit of those who might be more interested in the classification of the mushrooms it was thought desirable to include one that is more technical. The following is based largely on an unpublished manuscript of Dr. A. H. Smith, University of Michigan, and the keys of Singer (1951). These authors treat the mushrooms as an order, the Agaricales, and group the genera into families within the order. However, since there is still lack of agreement on the bases to be used for the erection of families in the Agaricales, this category has not been recognized here.

This key is intended to show better the scientific bases for the separation of genera and more emphasis is placed on microscopic characters than in the previous key (p. 257). A great many more genera are recognized than have been used elsewhere in this book. *Cantharellus* and related forms that are not regarded as true agarics are not included. It was thought desirable to indicate the type species of each genus, and where the genus is relatively unfamiliar, the older genus where the type species has been placed or would likely be sought, is indicated in brackets. This does not mean that all of the species included in the new genus were originally all in the same old genus. For example, the type of *Leucopaxillus* is *L. tricolor* which was formerly in *Tricholoma* but some other species now considered to belong in *Leucopaxillus* were formerly in *Clitocybe*. However, to those familiar with the species under the old names, this does give some idea of the concept of the newer genus.

This is by no means a complete survey of the modern genera of Agaricales; Singer (1951), for example, recognizes 145 genera excluding the Boletaceae. It does, however, include most of the genera that are likely to be found in Canada.

1. Trama of pileus and stipe composed of both sphaerocysts and filamentous hyphae; spores amyloid, more or less ornamented 2
1. Not with above combination of characters 3

2. Latex present .... *Lactarius*
   L. deliciosus (L. ex Fr.) S. F. Gray
2. Latex absent .... *Russula*
   R. lutea (Huds. ex Fr.) Fr.

3. Parasitic on other agarics and flesh of pileus breaking down into a mass of chlamydospores .... *Asterophora*
   A. (Nyctalis) lycoperdoides (Bull.) Ditmar ex S. F. Gray
3. Not parasitic on other agarics or if occasionally so then flesh of pileus not breaking up to form chlamydospores 4

4. Lamellae waxy in consistency, usually more or less decurrent 5
4. Not with above combination of characters 7

5. Spores amyloid .... *Neohygrophorus*
   N. (Hygrophorus) angelesianus (Smith & Hesler) Singer
5. Spores not amyloid 6

6. Spores echinulate (smooth in *L. trullisata*) .... *Laccaria*
   L. laccata (Scop. ex Fr.) B. & Br.
6. Spores smooth .... *Hygrophorus*
   H. eburneus (Bull. ex Fr.) Fr.

7. Trama of lamellae divergent; lamellae free or nearly so; spores white; either a partial or universal veil or both present 8
7. Not with above combination of characters 10

8. Volva absent (glutinous universal veil may be present) .... *Limacella*
   L. (Lepiota) delicata (Fr.) Earle ex H. V. Smith
8. Volva present 9

9. Annulus absent .... *Amanitopsis*
   A. vaginata Fr.
9. Annulus present .... *Amanita*
   A. phalloides (Vaill. ex Fr.) Secr.

10. Spore deposit greenish 11
10. Spore deposit white to creamy or pale dingy vinaceous 12
10. Spore deposit more deeply colored 49

11. Surface of the pileus composed of interwoven hyphae .... *Chlorophyllum*
    C. (Lepiota) molybdites (Meyer ex Fr.) Sacc.
11. Surface of the pileus composed of sphaerocysts; spore deposit becoming purplish on drying .... *Melanophyllum*
    M. (Agaricus) echinatus (Roth ex Fr.) Sing.

12. Pileus readily separable from stipe; annulus usually present 13
12. Pileus and stipe confluent or stipe lacking 15

13. Spores with lens-shaped apical pore 14
13. Spores not as above .... *Lepiota*
    L. colubrina (Pers.) ex Gray

14. Clamp connections present; pileus fleshy, not plicate-striate on margin .... *Macrolepiota*
    M. (Lepiota) procera (Scop. ex Fr.) Sing.
14. Clamp connections absent; pileus thin, margin plicate-striate .... *Leucocoprinus*
    L. (Lepiota) flavipes Pat.

15. Lamellae splitting longitudinally .... *Schizophyllum*
    S. commune Fr.
15. Lamellae not splitting longitudinally 16

16. Spores amyloid 17
16. Spores nonamyloid or pseudoamyloid 28

17. Trama of lamellae bilateral in young sporophores; lamellae decurrent; veil distinctly double; fruiting bodies usually very large .... *Catathelasma*
    C. (Armillaria) evanescens Lovej.
17. Not as above 18

18. Stipe excentric to lateral; habit pleurotoid 19
18. Stipe central 20

19. Margin of lamellae serrulate .... *Lentinellus*
    L. (Lentinus) cochleatus (Fr.) Karst.
19. Margin of lamellae even .... *Panellus*
    P. (Panus) stipticus (Bull. ex Fr.) Karst.

20. Veil present 21
20. Veil absent 23

21. Veil composed of thick-walled elements; hymenophore venose to sublamellate .... *Delicatula*
    D. (Omphalia) integrella (Pers. ex Fr.) Pat.
21. Veil not as above 22

22. Pileus covered with sphaerocysts .... *Cystoderma*
    C. (Lepiota) amianthinum (Scop. ex Fr.) Fayod
22. Pileus with a cuticle of appressed hyphae .... *Armillariella*
    A. (Armillaria) mellea (Vahl ex Fr.) Karst

23. Spores rough with a smooth spot at the hilum; cystidia usually present on edges of the lamellae and typically with a harpoon-like incrustation at apex; clamp connections absent .... *Melanoleuca*
    M. (Tricholoma) melaleuca (Pers. ex Fr.) Murr.
23. Not as above 24

24. Spores without a smooth spot at the hilum; clamp connections present; fruiting bodies fleshy .... *Leucopaxillus*
    L. (Tricholoma) tricolor (Peck) Kühner
24. Not as above 25

25. Fruiting bodies thin pliant, marasmioid; stipe cartilaginous to tough, with bright colored mycelium surrounding the base .... *Xeromphalina*
    X. (Omphalia) campanella (Batsch ex Fr.) Kühner & Maire
25. Not as above 26

26. Lamellae decurrent and margin of pileus inrolled .... *Cantharellula*
    C. (Cantharellus) umbonata (Gmelin ex Fr.) Singer
26. Lamellae variously attached but if decurrent then margin of pileus straight at first 27

27. Fruiting bodies typically small, fragile, more or less conical; trama amyloid .... *Mycena*
    M. galericulata (Scop. ex Fr.) Kummer
27. Fruiting bodies collybioid, omphalioid or clitocyboid in habit; trama not amyloid .... *Fayodia*
    F. (Omphalia) bisphaerigera (Lange) Kühner
27. Fruiting bodies larger, more fleshy; lamellae more or less sinuate .... *Tricholoma*
    T. flavovirens (Pers. ex Fr.) Lundell

28. Stipe excentric to lateral; habit pleurotoid 29
28. Stipe central 36

29. Pileus and trama of lamellae gelatinous or with well-defined gelatinous layers present in the pileus, especially the cuticle 30
29. No gelatinous layers present 31

30. Spores white, smooth .... *Resupinatus*
    R. (Pleurotus) applicatus (Batsch ex Fr. sensu Kauffm.) S. F. Gray
30. Spores rough, creamy pink .... *Rhodotus*
    R. (Pleurotus) palmatus (Bull. ex Fr.) Maire

31. Veil at first covering the hymenium; resupinate or laterally attached, not stipitate .... *Tectella*
    T. (Panus) patellaris (Fr.) Murr.
31. Veil lacking, or if present carpophore stipitate 32

32. Hyphae of the pileus and gill trama predominantly thin-walled ..... *Pleurotus*
 P. ostreatus (Jacq. ex Fr.) Kummer
32. Hyphae of trama mostly thick-walled 33

33. Trama of lamellae intricately interwoven and subhymenium inconspicuous to absent ..... *Panus*
 P. conchatus (Bull. ex Fr.) Fr.
33. Trama not intricately interwoven or if so then subhymenium very distinct 34

34. Lamellae thick on edge; consistency dry and almost leathery ..... *Plicatura*
 (= *Trogia* as used here) *P. alni* Peck
34. Lamellae thin on edge 35

35. Fruiting body tough; edge of lamellae serrate ..... *Lentinus*
 L. lepideus Fr.
35. Fruiting body fleshy; edge of lamellae even ..... *Pleurotus*
 P. ostreatus (Jacq. ex Fr.) Kummer

36. Annulus present 37
36. Annulus lacking or veil leaving a fibrillose zone on upper part of stipe 38

37. Cuticle of pileus composed of sphaerocysts ..... *Cystoderma*
 C. amianthina (Scop. ex Fr.) Fayod

37. Cuticle of pileus filamentous (see *Lentinus* also) ..... *Armillaria*
 A. luteovirens (A. & S. ex Fr.) Gill.

38. Lamellae waxy, typically flesh colored; spores echinulate (smooth in *L. trullisata*) ..... *Laccaria*
 L. laccata (Scop. ex Fr.) B. & Br.
38. Not as above 39

39. Cuticle of pileus a turf of gelatinous, narrow, branched hyphae; stipe velvety-pubescent and fulvous to dark brown below from colored tomentum ..... *Flammulina*
 F. (Collybia) velutipes (Curt. ex Fr.) Sing.
39. Not as above 40

40. Typically lignicolous in habit; cheilocystidia present, large; clamp connections present; cystidia on pileus when present decumbent as somewhat differentiated end cells of hyphae; rhizomorphs usually present at base ..... *Tricholomopsis*
 T. (Tricholoma) rutilans (Schaeff. ex Fr.) Sing.
40. Not as above 41

41. Fruiting body, especially the lamellae, staining gray, bluish, or black, or if not staining then the lamellae gray and the cuticle of the pileus filamentous; color white to black, seldom brightly colored in any part; if lamellae white at first then basidia with carminophilous granulation ..... *Lyophyllum*
 L. (Collybia) leucopheatum Karst.
41. Not as above 42

42. Stipe slender and cartilaginous to tough, or if thick, then with a clearly distinct, cartilaginous cortex 43
42. Stipe typically fleshy in consistency, if thin then it is pliant 47

43. Reviving when moistened 44
43. Not reviving when moistened 45

44. Hairs of pileus dark rusty brown in Melzer's solution .... *Crinipellis*
    C. (Collybia) stipitaria (Fr.) Pat.
44. Without hairs as above .... *Marasmius*
    M. rotula (L. ex Fr.) Fr.

45. Lamellae decurrent and margin of pileus incurved when young .... *Omphalina*
    O. (Omphalia) umbellifera (L. ex Fr.) Quél.
45. Lamellae adnate to decurrent; margin of pileus straight or incurved but not in above combination 46

46. Pileus typically convex to obtuse; margin incurved when young; hypoderm not differentiated; lamellae mostly adnate to subdecurrent .... *Collybia*
    C. dryophila (Bull. ex Fr.) Kummer
46. Pileus conic to obtuse; margin typically straight or bent in slightly; hypodermal region often of enlarged cells; lamellae adnate to decurrent .... *Mycena*
    M. galericulata (Scop. ex Fr.) Kummer

47. Spores slightly rough, creamy to vinaceous; lamellae variously attached .... *Lepista*
    L. (Tricholoma) subaequalis (Britz.) Sing.
47. Spores smooth, white to pale cream 48

48. Lamellae typically decurrent to broadly adnate .... *Clitocybe*
    C. infundibuliformis (Schaeff. ex Fr.) Quél.
48. Lamellae typically sinuate to adnexed at maturity .... *Tricholoma*
    T. flavovirens (Pers. ex Fr.) Lundell

49. Spores pink to vinaceous cinnamon in mass 50
49. Spores not as above 57

50. Stipe lateral or lacking 51
50. Stipe central 52

51. Spores angular .... *Rhodophyllus*
    (species formerly in *Claudopus*)
51. Spores not angular .... *Phyllotopsis*
    P. (Claudopus) nidulans (Pers. ex Fr.) Sing.

52. Spores angular or longitudinally striate 53
52. Spores smooth or slightly echinulate 54

53. Spores longitudinally striate .... *Clitopilus*
    C. prunulus (Scop. ex Fr.) Kummer
53. Spores angular .... *Rhodophyllus*
    R. (Entoloma) lividus (Bull. ex Fr.) Quél.
    (includes *Entoloma, Leptonia, Nolanea, Eccilia* and some species formerly placed in *Clitopilus*)

54. Volva present .... *Volvariella*
    V. argentina Speg.
54. Volva absent 55

55. Annulus present .... *Chamaeota*
    Agaricus xanthogrammus Cesati
55. Annulus lacking 56

56. Lamellae free; spores smooth ..... *Pluteus*
P. cervinus (Schaeff. ex Secr.) Fr.
56. Lamellae variously attached; spores slightly echinulate ..... *Lepista*
L. subaequalis (Britz.) Sing.

57. Spore deposit yellow-brown to purple-brown; spores truncate at apex, dull yellow-brown in KOH; cuticle of pileus not cellular 58
57. Not with above combination of characters 61

58. Spores typically purple-brown in deposit; if dull rusty brown then annulus well developed; usually with a special type of cystidia with an amorphous internal body that stains golden yellow when mounted in ammonia (chrysocystidia); if lignicolous, chrysocystidia present ..... *Stropharia*
S. aeruginosa (Curt. ex Fr.) Quél.
58. Not as above 59

59. Chrysocystidia absent; stipe typically fleshy; spores rusty brown in mass; annulus usually present; habitat typically lignicolous ..... *Kuehneromyces*
K. (Pholiota) mutabilis (Schaeff. ex Fr.) Sing. & Smith
59. Not as above 60

60. Chrysocystidia present ..... *Naematoloma*
N. (Hypholoma) sublateritium (Fr.) Karst.
60. Chrysocystidia absent ..... *Psilocybe*
P. semilanceata (Fr.) ex Kummer
61. Spore deposit typically bright rusty brown to earth-brown, spores truncate; cuticle of pileus cellular in structure 62

61. Spore deposit typically cocoa-brown to chocolate, or if rusty brown to yellow then spores have not a truncate apex 64

62. Pileus viscid and soft, often subdeliquescent; margin of pileus plicate-striate ..... *Bolbitius*
B. fragilis (L. ex Fr.) Fr.
62. Not as above 63

63. Stipe typically fleshy and spore deposit typically dull clay color to earth-brown (see *Psathyrella* also) ..... *Agrocybe*
A. (Pholiota) praecox (Pers. ex Fr.) Fayod
63. Stipe typically cartilaginous; spore deposit bright rusty brown ..... *Conocybe*
C. (Galera) tenera (Schaeff. ex Fr.) Fayod

64. Spores smooth, lamellae readily separable from pileus ..... *Paxillus*
P. involutus (Batsch ex Fr.) Fr.
64. Not as above 65

65. Spore deposit yellow to dark rusty brown 66
65. Spores cocoa-brown to chocolate or black 82

66. Stipe excentric or lacking ..... *Crepidotus*
C. mollis (Schaeff. ex Fr.) Kummer
66. Stipe typically central 67

67. Spores thin-walled (many collapsed spores usually visible in mounts); spore deposit typically pale yellow to ochraceous and spores very pale under the microscope ..... *Tubaria*
T. furfuracea (Pers. ex Fr.) Gill.
67. Spores well pigmented and with appreciably thickened walls 68

68. Spores smooth (use oil immersion) 69
68. Spores roughened or angular 73

69. Subhymenial zone typically well developed and gelatinous; typically lignicolous; often with an annulus, and stipe fibrillose to scaly below annulus or annular zone .... *Pholiota*
P. squarrosa (Pers. ex Fr.) Kummer
69. Not as above, typically terrestrial (lignicolous in *Gymnopilus*) 70

70. Stipe fleshy 71
70. Stipe brittle, typically thin 72

71. Pileus viscid; cystidia when present neither thick-walled nor encrusted .... *Hebeloma*
H. fastibile (Fr.) Kummer
71. Pileus dry or moist; if subviscid then encrusted cystidia present on sides of lamellae .... *Inocybe*
I. trechispora (Berk.) Karst.

72. Spores somewhat almond-shaped; margin of pileus appressed when young .... *Galerina*
G. (Galera) rubiginosa (Pers. ex Fr.) Kühner
72. Spores somewhat reniform in side view or elliptic; margin of pileus inrolled or incurved at first .... *Naucoria*
N. centuncula (Fr.) Kummer

73. Spores angular to nodulose or with prominent spines or compound nodules .... *Inocybe*
I. trechispora (Berk.) Karst.
73. Not as above 74

74. Spores with a smooth area around the hilum .... *Galerina*
G. rubiginosa (Pers. ex Fr.) Kummer
74. Not as above 75

75. Membranous annulus present; volva rudimentary .... *Rozites*
R. (Pholiota) caperata (Pers. ex Fr.) Karst.
75. Not as above 76

76. Typically lignicolous and spore deposit very bright rusty fulvous to orange-fulvous .... *Gymnopilus*
G. (Flammula) liquiritiae (Pers. ex Fr.) Karst.
76. Not as above 77

77. Veil lacking and stipe long-radicating .... *Phaeocollybia*
P. *(Naucoria) festiva* (Fr.) Heim. ex Sing.
77. Not as above 78

78. Lamellae separable from pileus .... *Ripartites*
R. (Inocybe) tricholoma (A. & S. ex Fr.) Karst.
78. Not as above 79

79. Stipe thin and fragile; clamp connections absent .... *Galerina*
G. rubiginosa (Pers. ex Fr.) Kummer
79. Stipe thick, or if thin then clamp connections present 80

80. Typically associated with alder, and pileus showing some differentiation of the cuticle other than a gelatinous pellicle, or cystidia present on the pileus; stipe cartilaginous-brittle .... *Alnicola*
A. (Naucoria) submelinoides Kühner
80. Not as above 81

81. Spore deposit rusty brown; cheilocystidia if present mostly clavate; pileus dry, to moist or viscid; partial veil cortinate .... *Cortinarius*
    C. violaceus (L. ex Fr.) Fr.
81. Spore deposit clay color or dull brown; cheilocystidia typically elongate to filamentose-capitate; pileus viscid; partial veil more or less cortinate to membranous or lacking .... *Hebeloma*
    H. fastibile (Fr.) Kummer

82. Cuticle of pileus cellular or lamellae deliquescing or both 83
82. Not as above 87

83. Lamellae deliquescing at maturity .... *Coprinus*
    C. comatus (Müller ex Fr.) S. F. Gray
83. Lamellae not deliquescing 84

84. Sides of lamellae mottled by maturing spores; (see *Psathyrella* also if pileus is fibrillose) .... *Panaeolus*
    P. campanulatus (L. ex Fr.) Quél.
84. Not as above 85

85. Pileus plicate-striate and pseudoparaphyses present .... *Pseudocoprinus*
    P. (Psathyrella) disseminatus (Pers. ex Fr.) Kühner
85. Never with both a plicate-striate pileus and pseudoparaphyses 86

86. Spore deposit at first greenish, becoming purplish on drying .... *Melanophyllum*
    M. (Agaricus) echinatus (Roth ex Fr.) Sing.
86. Spore deposit never greenish .... *Psathyrella*
    P. gracilis (Fr.) Quél.

87. Lamellae free or nearly so; stipe separating readily from pileus .... *Agaricus*
    A. campestris L. ex Fr.
87. Lamellae decurrent, somewhat waxy; stipe not separating readily from pileus .... *Gomphidius*
    G. glutinosus (Schaeff. ex Fr.) Fr.

GENERAL BIBLIOGRAPHY

ATKINSON, G. F. 1911. Studies of American fungi, mushrooms, edible, poisonous, etc. 3rd ed. Holt and Co., New York.

BANDONI, R. J., and SZCZAWINSKI, A. F. 1976. Guide to common mushrooms of British Columbia. Revised ed. B.C. Prov. Mus., Handbook 24. Victoria.

CHILD, G. P. 1952. The ability of coprini to sensitize man to ethyl alcohol. Mycologia 44:200-202.

CHRISTENSEN, C. M. 1943. Common edible mushrooms. Univ. Minnesota Press, Minneapolis, Minn.

COOKE, M. C. 1881-1891. Illustrations of British fungi. Williams & Norgate, London. 8 vols.

DEARNESS, J. 1911. The personal factor in mushroom poisoning. Mycologia 3:75-78.

FARLOW, W. G., and BURT, E. A. 1929. Icones Farlowianae. Farlow Herbarium of Harvard Univ., Cambridge, Mass. 120 pp., 103 pl.

GRAHAM, V. O. 1970. Mushrooms of the Great Lakes Region. Dover Publ., New York. (First published in 1944.)

GÜSSOW, H. T., and ODELL, W. S. 1927. Mushrooms and toadstools. Can. Dep. Agric., Ottawa.

HARD, M. E. 1908. Mushrooms, edible and otherwise. Mushroom Publ. Co., Columbus, Ohio.

HEIM, R. 1969. Champignons d'Europe. Editions N. Boubée, Paris.

KAUFFMAN, C. H. 1971. The gilled mushrooms of Michigan and the Great Lakes Region. Dover Publ., New York. (First published in 1918.)
KONRAD, P., and MAUBLANC. A. 1924-1937. Icones Selectae Fungorum. Lechevalier, Paris. 6 vols.
KRIEGER, L. C. 1967. The mushroom handbook. Dover Publ., New York. (First published in 1936.)
KÜHNER, R., and ROMAGNESI, H. 1953. Flore analytique des champignons supérieurs. Masson & Cie., Paris.
LANGE, J. E. 1935-1940. Flora Agaricina Danica. Soc. Adv. Mycol. Denmark & Danish Bot. Soc., Copenhagen. 5 vols.
LANGE, M., and HORA, F. B. 1963. Mushrooms and toadstools. E.P. Dutton, New York.
MCILVAINE, C., and MACADAM, R. K. 1973. One thousand American fungi. Dover Publ., New York. (First printed in 1902.)
MICHAEL, E., and HENNIG, B. 1964-1971. Handbuch für Pilzfreunde. Quelle and Meyer. Heidelberg. 5 vols.
MILLER, O. K. 1972. Mushrooms of North America. E.P. Dutton, New York.
MOSER, H. 1978. Die Röhrlinge und Blätterpilze (Polyporales, Boletales, Agaricales, Russulales). *In* H. Gams, Kleine Kryptogamenflora, Fischer Verlag, Stuttgart.
OVERHOLTS, L. O. 1938. Mycological notes for 1934-35. Mycologia 30:269-279.
PILÁT, A., and UŠÁK, O. 1958. Mushrooms. Spring Books, London.
——— 1961. Mushrooms and other fungi. Peter Nevill, London.
POMERLEAU, R. 1951. Mushrooms of Eastern Canada and the United States. Editions Chantecler, Montreal.
RAMSBOTTOM, J. 1953. Mushrooms and toadstools. Collins, London.
REA, C. 1922. British Basidiomycetae. Cambridge Univ. Press, Cambridge.
ROLFE, R. T., and ROLFE, R. W. 1928. The romance of the fungus world. Lippincott, Philadelphia.
ROMAGNESI, H. 1956-1967. Nouvel atlas des champignons. Bordas, Paris. 4 vols.
——— 1962. Petit atlas des champignons de France. Bordas, Paris. 3 vols.
SHAFFER, R. L. 1968. Keys to the genera of higher fungi. Univ. Michigan Biol. Stn., Ann Arbor.
SINGER, R. 1951. The Agaricales (mushrooms) in modern taxonomy. Lilloa 22:1-832.
——— 1961. Mushrooms and truffles. Interscience Publishers, New York.
——— 1962. The Agaricales in modern taxonomy. 2nd ed. J. Cramer, Weinheim.
SINGER, R. 1975. The Agaricales in modern taxonomy. 3rd ed. J. Cramer, Vaduz.
SMITH, A. H. 1938. Common edible and poisonous mushrooms of Southeastern Michigan. Cranbrook Inst. Sci., Bull. 14, 71 pp.
——— 1949. Mushrooms in their natural habitats. Sawyers Inc., Portland, Ore. Vol. 1. 626 pp.; Vol. 2, 33 Viewmaster reels, stereo-Kodachromes. Vol. 1 reprinted by Hafner Press, New York. 1973.
——— 1963. The mushroom hunter's field guide. Univ. Michigan Press, Ann Arbor. Revised.
SMITH, A. H. 1975. A field guide to western mushrooms. Univ. Michigan Press, Ann Arbor.
SMITH, H. V., and SMITH, A. H. 1973. How to know the non-gilled fleshy fungi. Wm. Brown Co., Dubuque.
THOMAS, W. S. 1948. Field book of common mushrooms. Putnam and Sons, New York.
WAKEFIELD, E. M., and DENNIS, R. W. G. 1950. Common British fungi. P. R. Gawthorn. London.

REFERENCES FOR SELECTED GROUPS

AGARICUS

ESSETTE, H. 1964. Les psalliotes. Lechevalier, Paris.
HOTSON, J. W., and STUNTZ, D. E. 1938. The genus *Agaricus* in western Washington. Mycologia 30:204-234.
MÖLLER, F. H. 1950-1952. Danish *Psalliota* species: I & II. Friesia 4:1-60, 135-242.
PILÁT, A. 1951. The Bohemian species of the genus *Agaricus*. Acta Mus. Nat. Pragae 7. B. 1-142.

AMANITA

BAS, C. 1969. Morphology and subdivision of *Amanita* and a monograph of its section *Lepidella*. Persoonia 5:285-579.

COKER, W. C. 1917. The Amanitas of the Eastern United States. J. Elisha Mitchell Sci. Soc. 33:1-88.

GILBERT, E. J. 1940-1941. Amanitaceae (Supplement to *Iconographia Mycologica* of Bresadola). Comitato Onoranze Bresadoliane, Milan.

HEINEMANN, P. 1964. Les Amanitées. Naturalistes belg. 45:1-15.

HOTSON, J. W. 1936. The Amanitae of Washington. Mycologia 28:63-76.

JENKINS, D. T. 1977. A taxonomic and nomenclatural study of the genus *Amanita* Section *Amanita* for North America. Biblotheca Mycologica 57. J. Cramer, Vaduz.

JOLY, P. 1967. Clés des principales amanites de la flore française. Rev. Mycol. 32 (suppl. 2):162-175.

POMERLEAU, R. 1966. Les amanites du Québec. Nat. Can. (Que.) 93:861-887.

AMANITOPSIS

See references under *Amanita*.

ARMILLARIA

HOTSON, H. H. 1940. The genus *Armillaria* in western Washington. Mycologia 32:776-790.

KAUFFMAN, C. H. 1923. The genus *Armillaria* in the United States and its relationships. Pap. Mich. Acad. Sci. Arts & Lett. 2:53-67.

MITCHEL, D. H., and SMITH, A. H. 1976. Notes on Colorado fungi. II. Species of *Armillaria* (Fr.) Kummer (Agaricales). Mycotaxon 4:513-533.

ROMAGNESI, H. 1970-1973. Observations sur les *Armillariella*. I and II. Bull. Soc. Mycol. France 86:257-265; 89:195-206.

See also Singer (1951) under general bibliography (as *Armillariella*).

ASCOMYCETES

DENNIS, R. W. G. 1977. British Ascomycetes. 3rd ed. J. Cramer, Vaduz.

SEAVER, F. J. 1961. The North American cup fungi (Operculates and Inoperculates). Hafner, New York. 2 vols. (First printed in 1928.)

WEBER, N. S. 1972. The genus *Helvella* in Michigan. Mich. Bot. 11:147-201.

See also Smith and Smith (1973) under general bibliography.

BOLETACEAE

COKER, W. C., and BEERS, A. H. 1974. The Boletaceae of North Carolina. Dover Publ., New York. (First printed in 1943.)

GRUND, D. W., and HARRISON, K. A. 1976. Nova Scotian Boletes. Biblotheca Mycologica 47. J. Cramer, Vaduz.

SINGER, R. 1945-1947. The Boletineae of Florida with notes on extralimital species: I-IV. Farlowia 2:97-141, 223-303, 527-567; Am. Midl. Nat. 37:1-135.

SLIPP, A. W., and SNELL, W. H. 1944. Taxonomic-ecologic studies of the Boletaceae in northern Idaho and adjacent Washington. Lloydia 7:1-66.

SMITH, A. H., and THIERS, H. D. 1964. A contribution toward a monograph of North American species of *Suillus*. Publ. by the authors, Ann Arbor, Mich.

——— 1971. The Boletes of Michigan. Univ. Michigan Press, Ann Arbor.

SMITH, A. H., THIERS, H. D., and WATLING, R. 1966-1967. A preliminary account of the North American species of *Leccinum*. Mich. Bot. 5:131-179 (section *Leccinum*); 6:107-154 (sections *Luteoscabra* and *Scabra*).

SNELL, W. H., 1936. Tentative keys to the Boletaceae of the United States and Canada. Rhode Island Bot. Club Publ. 1. 25 pp.

SNELL, W. H., and DICK, E. A. 1970. The Boleti of Northeastern North America. J. Cramer, Lehre.

THIERS, H. D. 1963. The bolete flora of the gulf coastal plain: I. The Strobilomycetaceae. J. Elisha Mitchell Sci. Soc. 79:32-41.

——— 1965. The genus *Xerocomus* Quelet in Northern California. Madroño 17:237-249.

——— 1965. California boletes: I. Mycologia 57:524-534.

——— 1966. California boletes: II. Mycologia 58:815-826.

——— 1967. California boletes: III. Madroño 19:148-160.

——— 1971. California boletes: IV. The genus *Leccinum*. Mycologia 63:261-276.

THIERS, H. D. 1975. California mushrooms. A field guide to the Boletes. Hafner Press, New York.

See also Smith and Smith (1973) under general bibliography.

CANTHARELLUS

BIGELOW, H. E. 1978. The cantharelloid fungi of New England and adjacent areas. Mycologia 70:707-756.

CORNER, E. J. H. 1966. A monograph of the Cantharelloid fungi. Oxford Univ. Press. London.

SMITH, A. H. 1968. The Cantharellaceae of Michigan. Mich. Bot. 7:143-183.

SMITH, A. H., and MORSE, E. 1947. The genus *Cantharellus* in the western United States. Mycologia 39:497-534.

See also Smith and Smith (1973) under general bibliography.

CLAVARIACEAE

COKER, W. C., and COUCH, J. N. 1923. The Clavarias of the United States and Canada. Univ. North Carolina Press, Chapel Hill.

CORNER, E. J. H. 1950. A monograph of *Clavaria* and allied genera. Ann. Bot. Mem. 1. 740 pp.

——— 1970. Supplement to "A monograph of *Clavaria* and allied genera." Nova Hedwigia 33:1-299.

DOTY, M. S. 1944. *Clavaria*, the species known from Oregon and the Pacific Northwest. Ore. State College Press, Corvallis.

LEATHERS, C. R. 1955. The genus *Clavaria* Fries in Michigan. Ph.D. Diss., Univ. Michigan. Unpubl. (Available from University Microfilms, Ann Arbor, Mich.)

MARR, C. D., and STUNTZ, D. E. 1973. *Ramaria* of Western Washington. Bibliotheca Mycologia, Vol. 38.

PERREAU, J. 1969. Les clavaires. Rev. Mycol. 33 (suppl. 5):396-415.

PETERSEN, R. H. 1968. The genus *Clavulinopsis* in North America. Mycologia Mem. 2.

——— 1971. The genera *Gomphus* and *Gloeocantharellus* in North America. J. Cramer. Lehre.

WELLS, V. L., and KEMPTON, P. E. 1968. A preliminary study of *Clavariadelphus* in North America. Mich. Bot. 7:35-57.

See also Smith and Smith (1973) under general bibliography.

CLITOCYBE

BIGELOW, H. E. 1965. The genus *Clitocybe* in North America: I. Section *Clitocybe*. Lloydia 28:139-180.

——— 1968. The genus *Clitocybe* in North America: II. Section *Infundibuliformes*. Lloydia 31:43-62.

BIGELOW, H. E., MILLER, O. K., and THIERS, H. D. 1976. A new species of *Omphalotus*. Mycotaxon 3:363-372.

BIGELOW, H. E., and SMITH, A. H. 1969. The status of *Lepista* – a new section of *Clitocybe*. Brittonia 21:144-177.

HARMAJA, H. 1969. The genus *Clitocybe* (Agaricales) in Fennoscandia. Karstenia 10:5-168.

KAUFFMAN, C. H. 1927. The genus *Clitocybe* in the United States, with a critical study of all the north temperate species. Pap. Mich. Acad. Sci. Arts & Lett. 8:153-214.

MÉTROD, G. 1946, 1951. Révision des Clitocybes. Bull. Soc. Mycol. France 62:42-49; 67:387-403.

CLITOPILUS

See Hesler (1967) under *Entoloma* and Singer (1951) under the general bibliography.

COLLYBIA

BIGELOW, H. E. 1973. The genus *Clitocybula*. Mycologia 65:1101-1116.

MÉTROD, G. 1952. Les Collybies. Rev. Mycol. 17 (suppl. 1):60-93.

CONOCYBE

KITS VAN WAVEREN, E. 1970. The genus *Conocybe* subgenus *Pholiotina*: I. Persoonia 6:119-165.

KÜHNER, R. 1935. Le genre *Galera*. Lechevalier, Paris.
WATLING, R. 1971. The genus *Conocybe* subgenus *Pholiotina*: II. Persoonia 6:313-339.

COPRINUS

KITS VAN WAVEREN, E. 1968. The "*stercorarius* group" of the genus *Coprinus*. Persoonia 5:131-176.
LANGE, M. 1952. Species concept in the genus *Coprinus*. Dansk Bot. Ark. 14:1-164.
LANGE, M., and SMITH, A. H. 1953. The *Coprinus ephemerus* group. Mycologia 45:747-780.
PILÁT, A., and SVRČEK, M. 1967. Revision specierum sectionis *Herbicolae* generia *Coprinus*. Ceská Mycol. 21:136-145.
VAN DE BOGART, F. 1976. The genus *Coprinus* in western North America, Part 1: Section *Coprinus*. Mycotaxon 4:233-275.

CORTINARIUS

AMMIRATI, J. F. 1972. The section *Dermocybe* of *Cortinarius* in North America. Ph.D. Diss., Univ. Michigan. Unpubl. (Available from University Microfilms, Ann Arbor, Mich.)
BERTAUX, A. 1966. Les Cortinaires. Lechevalier, Paris.
HENRY, R. 1958. Suite à l'étude des Cortinaires. Bull. Soc. Mycol. France 74:249-422.
——— 1967-1969. Étude provisoire du genre *Hydrocybe*: Hydrocybes à pied atténué à la base. Bull. Soc. Mycol. France 83:989-1046; 84:396-421; 85:385-449.
KAUFFMAN, C. H. 1932. *Cortinarius*. North Am. Flora 10:282-348.
MOSER, M. 1960. Die Gattung *Phlegmacium*. Pilze Mitteleuropas IV. Klinkhardt, Bad Heilbrunn.
——— 1969-1970. *Cortinarius* Fr. Untergattung *Leprocybe* subgen. nov., Die Rauhkopfe. Zeitschr. Pilzk. 35:213-248; 36:37-57.
SMITH, A. H. 1942. New and unusual Cortinarii from Michigan, with a key to the North American species of subgenus *Bulbopodium*. Bull. Torrey Bot. Club 69:44-64.
——— 1944. New and interesting Cortinarii from North America. Lloydia 7:163-235.

CREPIDOTUS

HESLER, L. R., and SMITH, A. H. 1965. North American species of *Crepidotus*. Hafner, New York.
PILÁT, A. 1948. Monographie des espèces européennes du genre *Crepidotus* Fr. Atlas de Champignons de L'Europe 6. Prague.

CYSTODERMA

SMITH, A. H., and SINGER, R. 1945. A monograph on the genus *Cystoderma*. Pap. Mich. Acad. Sci., Arts & Lett. 30:125-147.
See also Hotson (1940) under Armillaria.

ENTOLOMA

HESLER, L. R. 1967. *Entoloma* (*Rhodophyllus*) in Southeastern North America. Nova Hedw. Beih. 23.
PECK, C. H. 1909. New York species of *Entoloma*. Bull. N.Y. State Mus. 13:47-58.

FLAMMULA

HESLER, L. R. 1969. North American species of *Gymnopilus*. Mycologia Mem. 3.
KAUFFMAN, C. H. 1926. The genera *Flammula* and *Paxillus* and the status of the American species. Am. J. Bot. 13:11-32.
See also Smith and Hesler (1968) under *Pholiota* for a treatment of some of the species.

GASTROMYCETES

BRODIE, H. J. 1975. The bird's nest fungi. Univ. Toronto Press, Toronto, Ont.
COKER, W. C., and COUCH, J. N. 1968. The Gastromycetes of the Eastern United States and Canada. J. Cramer, Lehre. (First printed in 1928 by Univ. North Carolina Press, Chapel Hill.)
DISSING, H., and LANGE, M. 1961-1962. The genus *Geastrum* in Denmark. Bot. Tidsskr. 57:1-27; 58:64-67.

ECKBLAD, F. E. 1955. The gastromycetes of Norway. Nytt Mag. Bot. 4:19-86.

KREISEL, H. 1967. Taxonomisch-pflanzengeographische Monographie der Gattung *Bovista*. Nova Hedw. Beih. 25.

SINGER, R. 1963. Notes on secotiaceous fungi: *Galeropsis* and *Brauniella*. Proc. Kon. Ned. Akad. Wetensch., Ser. C, 66:106-117.

SINGER, R., and SMITH, A. H. 1958-1960. Studies on secotiaceous fungi: I. *Thaxterogaster*. Brittonia 10:201-216, 1958. II. *Endoptychum*. *ibid* 10:216-221, 1958. III. *Weraroa*. Bull. Torrey Bot. Club 85:324-334, 1959. IV. *Gastroboletus*, *Truncocolumella*, *Chamonixia*. Brittonia 11:205-223, 1959. V. *Nivatogastrium*. *ibid* 11:224-228, 1959. VI. *Setchelliogaster*. Madroño 15:73-79, 1960. VII. *Secotium* and *Neosecotium*. *ibid* 15:152-158, 1960. VIII. *Brauniella*. Mycologia 50:927-938, 1958. IX. Astrogastraceous series. Mem. Torrey Bot. Club 21:1-112, 1960.

SMITH, A. H. 1951. Puffballs and their allies in Michigan. Univ. Michigan Press, Ann Arbor.

SMITH, A. H., and ZELLER, S. M. 1966. A preliminary account of the North American species of *Rhizopogon*. Mem. N.Y. Bot. Gard. 14(2):1-177.

SOEHNER, E. 1962. Monographie der Gattung *Hymenogaster*. Nova Hedw. Beih. 2.

ZELLER, S. M. 1949. Keys to the orders, families, and genera of the gastromycetes. Mycologia 41:36-58.

ZELLER, S. M., and SMITH, A. H. 1964. The genus *Calvatia* in North America. Lloydia 27:148-186.

See also Shaffer (1968) and Smith and Smith (1973) under general bibliography.

GOMPHIDIUS

MILLER, O. K. 1964. Monograph of *Chroogomphus* (Gomphidiaceae). Mycologia 56:526-549.

——— 1971. The genus *Gomphidius* with a revised description of the Gomphidiaceae and a key to the genera. Mycologia 63:1129-1163.

SINGER, R. 1949. The genus *Gomphidius* Fries in North America. Mycologia 41:462-489.

HEBELOMA

BRUCHET, G. 1970. Contribution à l'étude du genre *Hebeloma* (Fr.) Kummer; partie spéciale. Bull. Soc. Linn. Lyon 39 (suppl. 6):1-132.

MOSER, M. 1970. Beiträge zur Kenntnis der Gattung *Hebeloma*. Zeitschr. Pilzk. 36:61-75.

PECK, C. H. 1910. New York species of *Hebeloma*. N.Y. State Mus. Bull. 139:67-77.

ROMAGNESI, H. 1965. Etudes sur le genre *Hebeloma*. Bull. Soc. Mycol. France 81:321-344.

HYDNACEAE

COKER, W. C., and BEERS, A. H. 1951. The stipitate hydnums of the Eastern United States. Univ. North Carolina Press, Chapel Hill.

HALL, D., and STUNTZ, D. E. 1971. Pileate Hydnaceae of the Puget Sound area. I. White-spored genera: *Auriscalpium*, *Hericium*, *Dentinum* and *Phellodon*. Mycologia 63:1099-1128.

——— 1972. Pileate Hydnaceae of the Puget Sound area. II. Brown-spored genera: *Hydnum*. Mycologia 64:15-37.

——— 1972. Pileate Hydnaceae of the Puget Sound area. III. Brown-spored genus: *Hydnellum*. Mycologia 64:560-590.

HARRISON, K. A. 1961. The stipitate hydnums of Nova Scotia. Can. Dep. Agric. Publ. 1099.

——— 1968. Studies on the hydnums of Michigan: I. Genera *Phellodon*, *Bankera*, *Hydnellum*. Mich. Bot. 7:212-264.

——— 1973. The genus *Hericium* in North America. Mich. Bot. 12:177-194.

MAAS GEESTERANUS, R. A. 1956-1959. The stipitate hydnums of the Netherlands: I-IV. Fungus 26:44-60; 27:50-71; 28:48-61. Persoonia 1:115-147.

See also Smith and Smith (1973) under general bibliography.

HYGROPHORUS

HESLER, L. R., and SMITH, A. H. 1963. North American species of *Hygrophorus*. Univ. Tennessee Press, Knoxville.

SMITH, A. H., and HESLER, L. R. 1938, 1942. Studies in North American species of *Hygrophorus*: I & II. Lloydia 2:1-62; 5:1-94.

INOCYBE

GRUND, D. W., and STUNTZ, D. E. 1968, 1970. Nova Scotian Inocybes: I & II. Mycologia 60:406-425; 62:925-939.

GRUND, D. W., and STUNTZ, D. E. 1975. Nova Scotian Inocybes. III. Mycologia 67:19-31. IV. Mycologia 69:392-408.

HEIM, R. 1931. Le Genre *Inocybe*. Lechevalier, Paris.

KAUFFMAN, C. H. 1924. *Inocybe*. North Am. Flora 10:227-260.

KÜHNER, R., and BOURSIER, J. 1928, 1932, 1933. Les Inocybes goniosporés. Bull. Soc. Mycol. France 44:170-189; 48:118-161; 49: 81-121.

MÉTROD, G. 1956. Les Inocybes leiosporés à cystides courtes. Bull. Soc. Mycol. France 72:122-131.

PECK, C. H. 1910. New York species of *Inocybe*. N.Y. State Mus. Bull. 139:48-67.

STUNTZ, D. E. 1947. Studies in the genus *Inocybe*. I. New and noteworthy species from Washington. Mycologia 39:21-55.

LACCARIA

SINGER, R. 1967. Notes sur le genre *Laccaria*. Bull. Soc. Mycol. France 83:104-123.

LACTARIUS

COKER, W. C. 1918. The Lactarias of North Carolina. J. Elisha Mitchell Sci. Soc. 34:1-61.

BURLINGHAM, G. 1910. *Lactaria*. North Am. Flora 9:172-200.

HEINEMANN, P. 1960. Les Lactaires. Nat. Belg. 41:133-156.

HESLER, L. R., and SMITH, A. H. 1960. Studies on *Lactarius*: I. The North American species of section *Lactarius*. Brittonia 12:119-139. II. The North American species of section *Scrobiculus*, *Crocei*, *Theiogali*, and *Vellus*. *ibid* 12:306-350.

HESLER, L. R., and SMITH, A. H. (in press). The North American species of the genus *Lactarius* (Russulaceae). Univ. Michigan Press, Ann Arbor.

NEUHOFF, W. 1956. Die Milchlinge. Klinkhardt Verlag, Bad Heilbrunn.

SMITH, A. H., and HESLER, L. R. 1962. Studies on *Lactarius*: III. The North American species of section *Plinthogali*. Brittonia 14:369-440.

LENTINUS

MILLER, O. K., and STEWART, L. 1971. The genus *Lentinellus*. Mycologia 63:333-369.

PILÁT, A. 1946. Monographie des espèces européennes du genre *Lentinus* Fr. Atlas des Champignons de l'Europe 5, Prague.

See also Singer (1951) in the general bibliography.

LEPIOTA

KAUFFMAN, C. H. 1924. The genus *Lepiota* in the United States. Pap. Mich. Acad. Sci., Arts & Lett. 4:311-344.

KÜHNER, R. 1936. Recherches sur le genre *Lepiota*. Bull. Soc. Mycol. France 52:177-238.

SMITH, H. V. 1954. A revision of the Michigan species of *Lepiota*. Lloydia 17:307-328.

LEPTONIA

LARGENT, D. L. 1977. The genus *Leptonia* on the Pacific coast of the United States—Including a study of the North American types. Biblotheca Mycologia 55. J. Cramer, Vaduz.

LARGENT, D. L., and BENEDICT, R. G. 1970. Studies in the rhodophylloid fungi: II. *Alboleptonia*, a new genus. Mycologia 62:437-452.

See also Hesler (1967) under *Entoloma*.

LEUCOPAXILLUS

SINGER, R., and SMITH, A. H. 1943. A monograph of the genus *Leucopaxillus*. Pap. Mich. Acad. Sci., Arts & Lett. 28:85-132.

LIMACELLA

SMITH, H. V. 1945. The genus *Limacella* in North America. Pap. Mich. Acad. Sci., Arts & Lett. 30:125-147.

MARASMIUS

GILLIAM, M. S. 1975. *Marasmius* section *Chordales* in the Northeastern United States and adjacent Canada. Contrib. Univ. Mich. Herb. 11(2):25-40.

______ 1976. The genus *Marasmius* in the Northeastern United States and adjacent Canada. Mycotaxon 4:1-144.

MELANOLEUCA

GILLMAN, L. S., and MILLER, O. K. 1977. A study of the Boreal, Alpine, and Arctic species of *Melanoleuca*. Mycologia 69:927-951.

MÉTROD, G. 1948. Essai sur le genre *Melanoleuca* Patouillard emen. Bull. Soc. Mycol. France 64:141-165.

MYCENA

KÜHNER, R. 1938. Le genre *Mycena*. Lechevalier, Paris.

SMITH. A. H. 1947. North American species of *Mycena*. Univ. Michigan Press. Ann Arbor.

NAEMATOLOMA

SMITH. A. H. 1951. North American species of *Naematoloma*. Mycologia 43:467-521.

NAUCORIA

ROMAGNESI, H. 1962. Les *Naucoria* du group centunculus (*Ramicola* Velen.). Bull. Soc. Mycol. France 78:337-358.

See also Kühner and Romagnesi (1953) under general bibliography.

NOLANEA

MAZZER, S. J. 1976. A monographic study of the genus *Pouzarella*. Biblotheca Mycologica 46. J. Cramer, Vaduz.

See also Hesler (1967) under *Entoloma*.

PANAEOLUS

OLA'H, G. M. 1970. Le genre *Panaeolus*. Rev. Mycol. Mem. 10.

PANUS

MILLER. O. K. 1970. The genus *Panellus* in North America. Mich Bot. 9:17-30.

PAXILLUS

See Kauffman (1926) under *Flammula* and Singer (1951) under general bibliography.

PHAEOLEPIOTA

Only one species known.

PHOLIOTA

OVERHOLTS, L. O. 1927. A monograph of the genus *Pholiota* in the United States. Ann. Mo. Bot. Gard. 14:87-210.

SMITH, A. H., and HESLER, L. R. 1968. The North American species of *Pholiota*. Hafner Publ. Co., New York.

PHYLLOTOPSIS

Only one species known.

PLEUROTUS

COKER, W. C. 1944. The smaller species of *Pleurotus* in North Carolina. J. Elisha Mitchell Sci. Soc. 60:71-95.

MILLER, O. K., and MANNING, D. L. 1976. Distribution of the Lignicolous Tricholomataceae in the Southern Appalachians, Pages 307-344 *in* B. C. Parker and M. K. Roane, eds. The distributional history of the Biota of the Southern Appalachians, IV., Univ. Press of Virginia, Charlottesville.

PILÁT, A. 1935. *Pleurotus*. Atlas des champignons de l'Europe 2. Prague.

See also Singer (1951) under general bibliography.

PLUTEUS

HOMOLA, R. L. 1972. Section Celluloderma of the genus *Pluteus* in North America. Mycologia 64:1211-1247.

SINGER, R. 1956. Contributions toward a monograph of the genus *Pluteus*. Trans. Brit. Mycol. Soc. 39:145-232.

POLYPORACEAE

BONDARTSEV, A. S. 1953. The Polyporaceae of the European USSR and Caucasia. Acad. Sci. USSR, Leningrad. (English translation published by U.S. Dept. Agric. 1971.)

DOMANSKI, S. 1972. Fungi (Polyporaceae I). Nat. Center Sci., Tech. & Econ. Inf., Warsaw.

DOMANSKI, S., ORTOS, H., and SKIRGIETTO, A. 1973. Fungi (Polyporaceae II). Nat. Center Sci., Tech. & Econ. Inf., Warsaw.

FERGUS, C. L. 1960. Illustrated genera of wood decay fungi. Burgess Publ. Co., Minneapolis, Minn.

LOWE, J. L. 1957. Polyporaceae of North America. The genus *Fomes*. Tech. Publ. N.Y. State Coll. For. 80.

——— 1966. The genus *Poria*. Tech. Publ. N.Y. State Coll. For. 90.

LOWE, J. L. 1975. Polyporaceae of North America: The genus *Tyromyces*. Mycotaxon 2:1-82.

LOWE, J. L., and GILBERTSON, R. L. 1961. Synopsis of the Polyporaceae of the western United States and Canada. Mycologia 53:474-511.

OVERHOLTS, L. O. 1953. The Polyporaceae of the United States, Alaska and Canada. Univ. Michigan Press, Ann Arbor.

RYVARDEN, L. 1976-1978. The Polyporaceae of North Europe. I and II. Fungiflora. Oslo.

See also Smith and Smith (1973) under general bibliography.

PSATHYRELLA

SMITH, A. H. 1972. The North American species of *Psathyrella*. Mem. N.Y. Bot. Gard., Hafner Press, N.Y.

PSEUDOCOPRINUS

See Lange and Smith (1953) under *Coprinus*.

RUSSULA

BEARDSLEE, H. C. 1918. The Russulas of North Carolina. J. Elisha Mitchell Sci. Soc. 33:147-197.

BLUM, J. 1962. Les Russules: flore monographique des Russules de France et des pays voisins. Lechevalier, Paris.

BURLINGHAM, G. 1915. *Russula*. North Am. Flora 9:201-236.

CRAWSHAY, R. 1930. The spore ornamentation of the Russulas. Baillière, Tindall and Cox. London.

HEINEMANN, P. 1962. Les Russules. Nat. Belg. 43:1-32.

RAYNER, R. W. 1968-1970. Keys to the British species of *Russula*. Bull. Brit. Mycol. Soc. 2:76-108; 3:59; 3:89-120; 4:19-46.

ROMAGNESI, H. 1967. Les Russules d'Europe et d'Afrique du Nord. Bordas, Paris.

SCHAEFFER, J. 1952. *Russula*-Monographie. Klinkhardt Verlag, Bad Heilbrunn.

SHAFFER, R. L. 1964. The subsection Lactarioideae of *Russula*. Mycologia 56:202-231.

——— 1970. Notes on subsection Crassotunicatinae and other species of *Russula*. Lloydia 33:49-96.

——— 1972. North American Russulas of the subsection Foetentinae. Mycologia 64: 1008-1053.

SHAFFER, R. L. 1975. Some common North American species of *Russula* subsection Emeticinae. Beih. Nova Hedw. 51:207-237; pls. 49-54.

SCHIZOPHYLLUM

COOKE, W. B. 1961. The genus *Schizophyllum*. Mycologia 53:575-599.

LINDER, D. H. 1933. The genus *Schizophyllum*: I. The species of the western hemisphere. Am. J. Bot. 20:552-564.

STROPHARIA

See Kühner and Romagnesi (1953) under general bibliography.

THELEPHORACEAE

CORNER, E. J. H. 1968. A monograph of *Thelephora* (Basidiomycetes). Nova Hedw. Beih. 27.

The genus *Craterellus* is treated in the Cantharellaceae by most authors.

TREMELLALES

MARTIN, G. W. 1952. Revision of the North Central Tremellales. State Univ. Iowa Stud. Nat. Hist. 19(3).

See also Smith and Smith (1973) under general bibliography.

TRICHOLOMA

BON, M. 1967-1970. Revision des Tricholomes. Bull. Soc. Mycol. France 83:324-335; 85:475-492; 86:755-763.

MALLOCH, D. 1974. *Tricholoma fulvum*. Fungi Can. 31. (Includes a key to *Tricholoma* species with brown, viscid caps and farinaceous flesh.)
MÉTROD, G. 1942. Les Tricholomes. Rev. Mycol. 7 (suppl. 2):22-50.
SMITH, A. H. 1960. *Tricholomopsis* (Agaricales) in the western hemisphere. Brittonia 12:41-70.
See also Bigelow and Smith (1969) under *Clitocybe*.

TROGIA

No other species can be confused with *T. crispa*. It is now usually placed in the genus *Plicatura*.

TUBARIA

See Kühner and Romagnesi (1953) under general bibliography.

VOLVARIELLA

SHAFFER, R. L. 1957. *Volvariella* in North America. Mycologia 49:545-579.

XEROMPHALINA

MILLER, O. K. 1968. A revision of the genus *Xeromphalina*. Mycologia 60:156-188.

Figures 374-383

374. *Geastrum triplex.*
375. *Lycoperdon perlatum.*
376. *Morchella esculenta.*
377. *M. angusticeps.*
378. *Verpa bohemica.*
379. *Gyromitra esculenta.*
380. *G. infula.*
381. *Peziza repanda*
382. *Sarcoscypha coccinea.*
383. *Urnula craterium.*

374
375
376
377
378
379
380
381
382
383

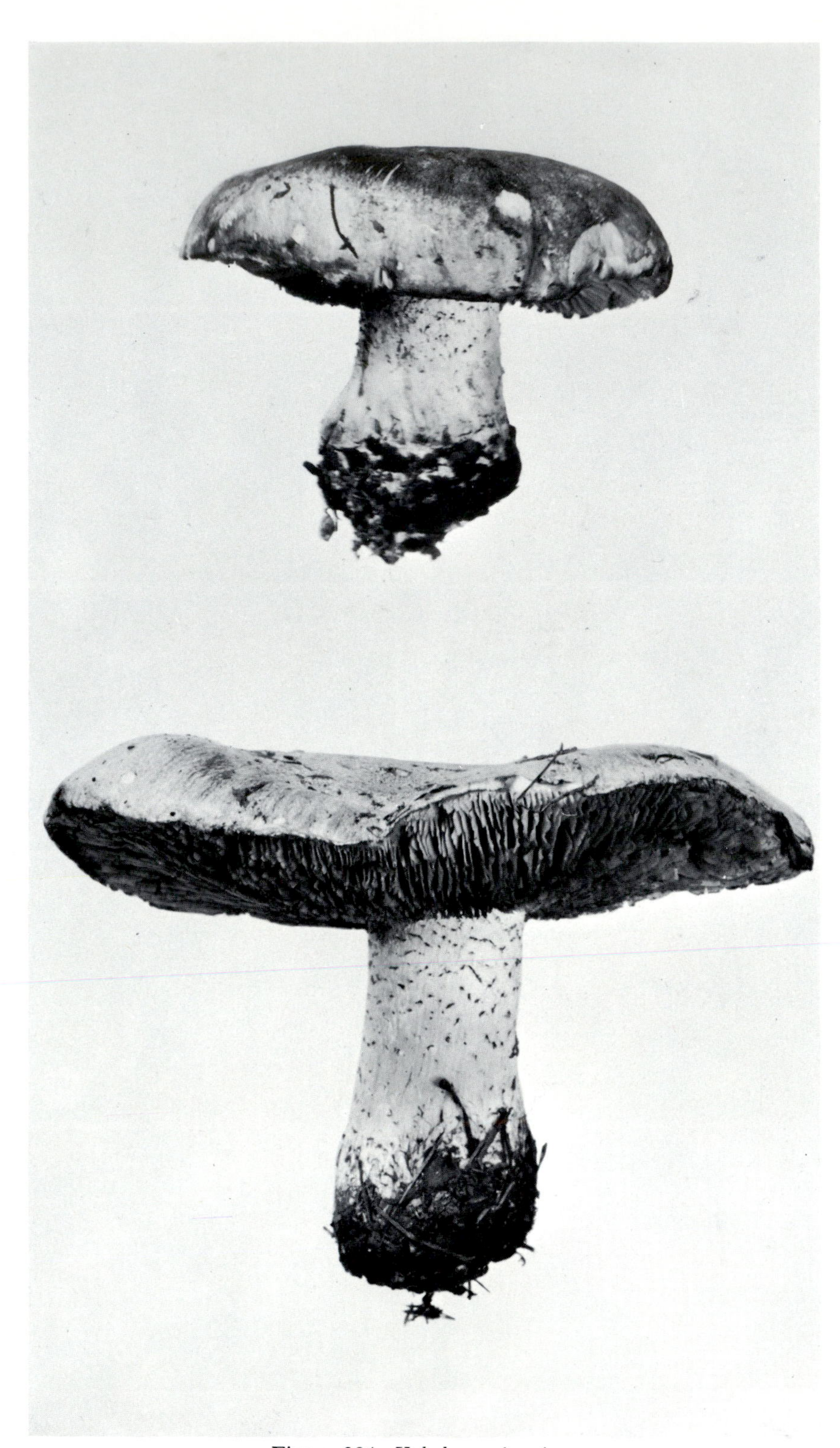

Figure 384. *Hebeloma sinapizans.*

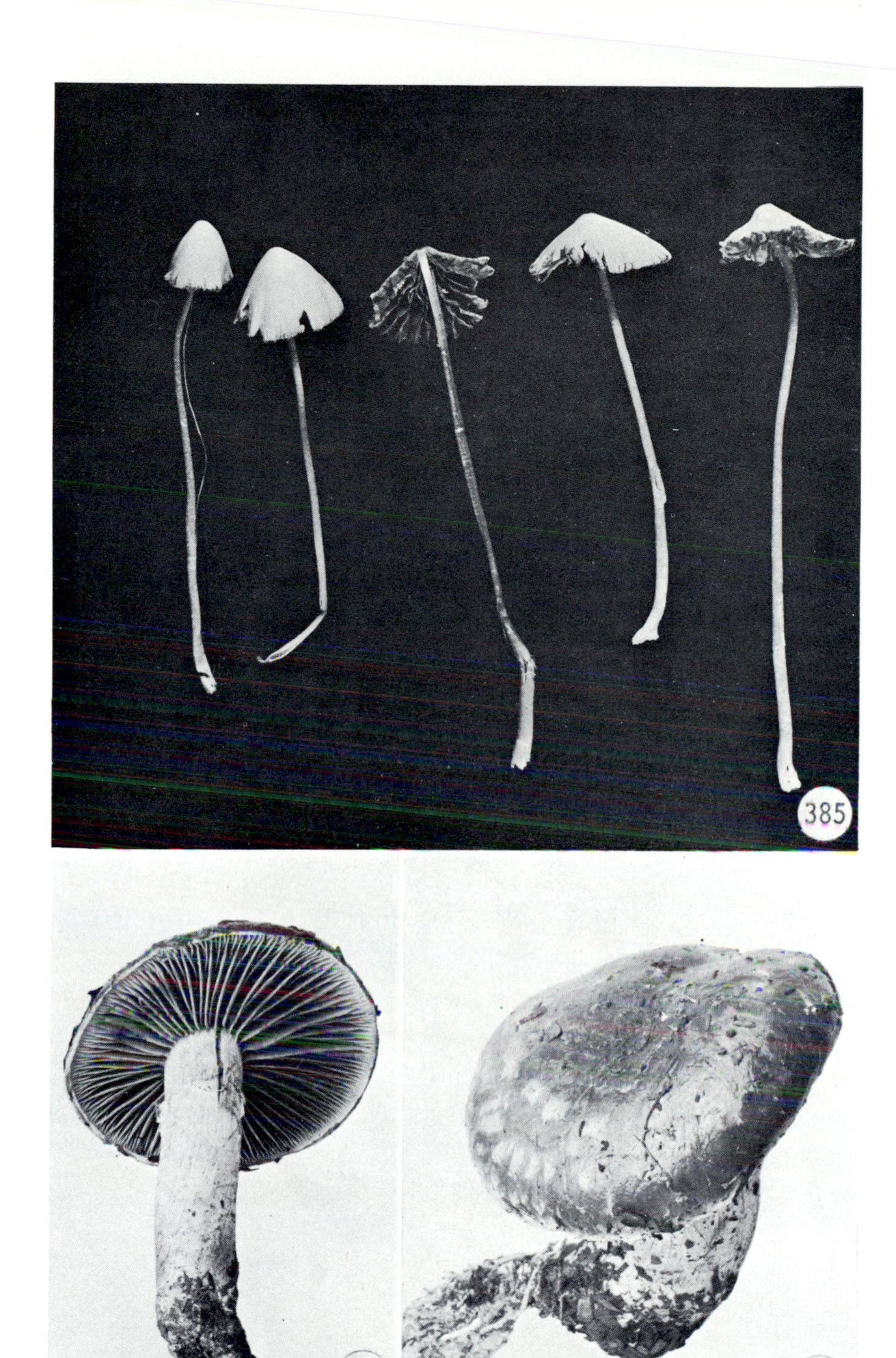

Figure 385. *Conocybe crispa.*

Figures 386-387. *Stropharia aeruginosa.*

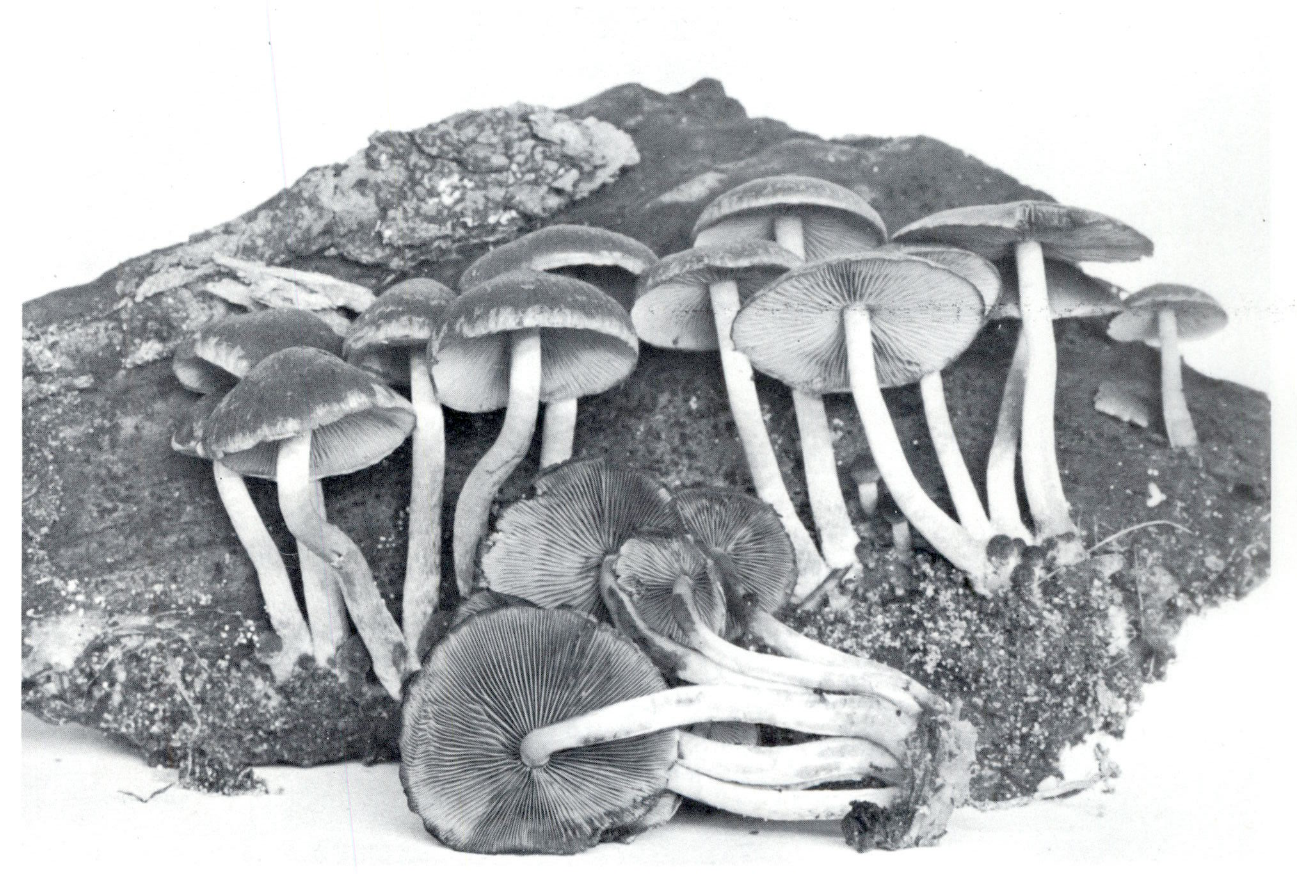

Figure 388. *Psathyrella hydrophila.*

Figure 389. *Psathyrella velutina.*

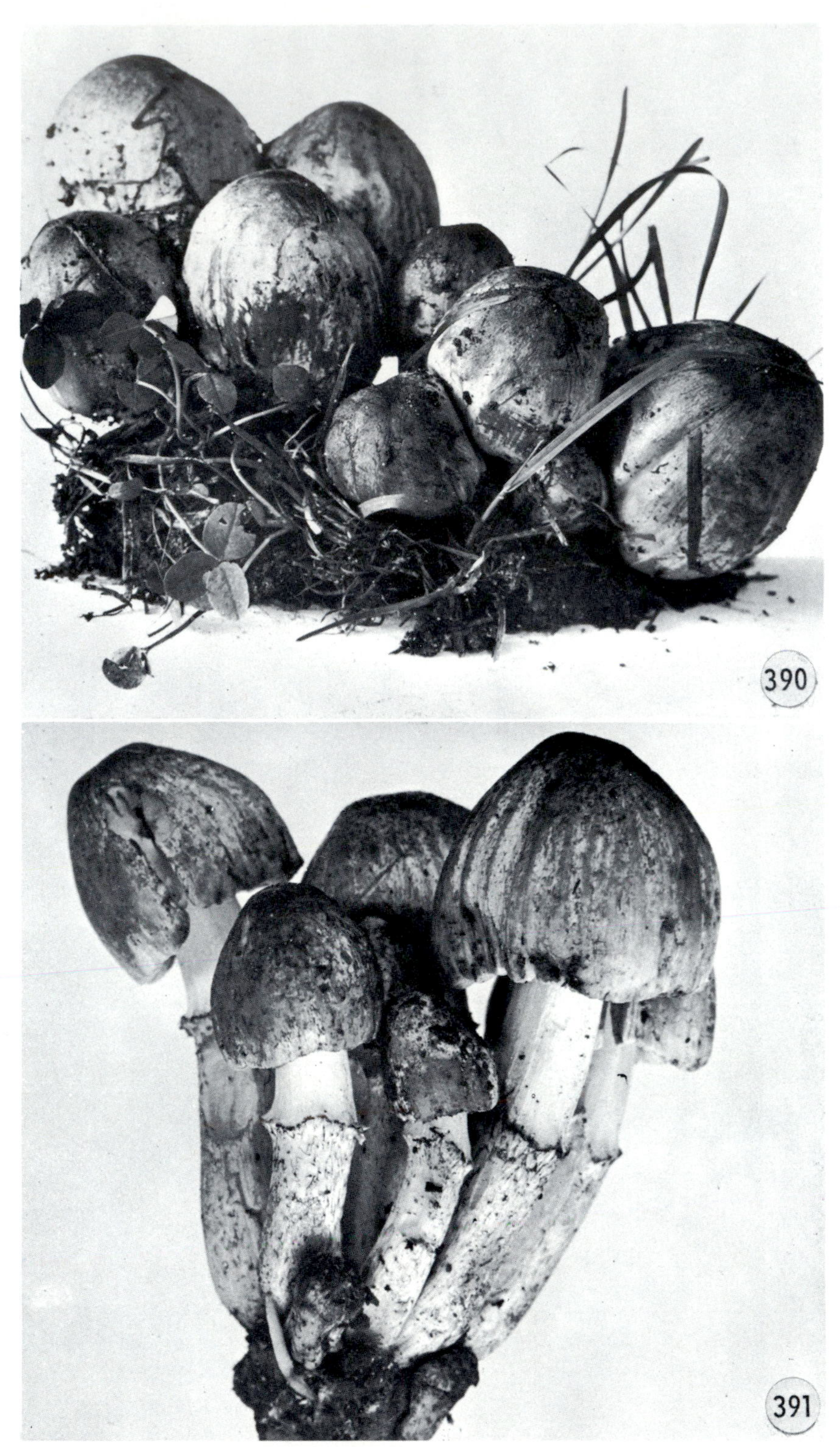

Figures 390-391. *Coprinus atramentarius.*

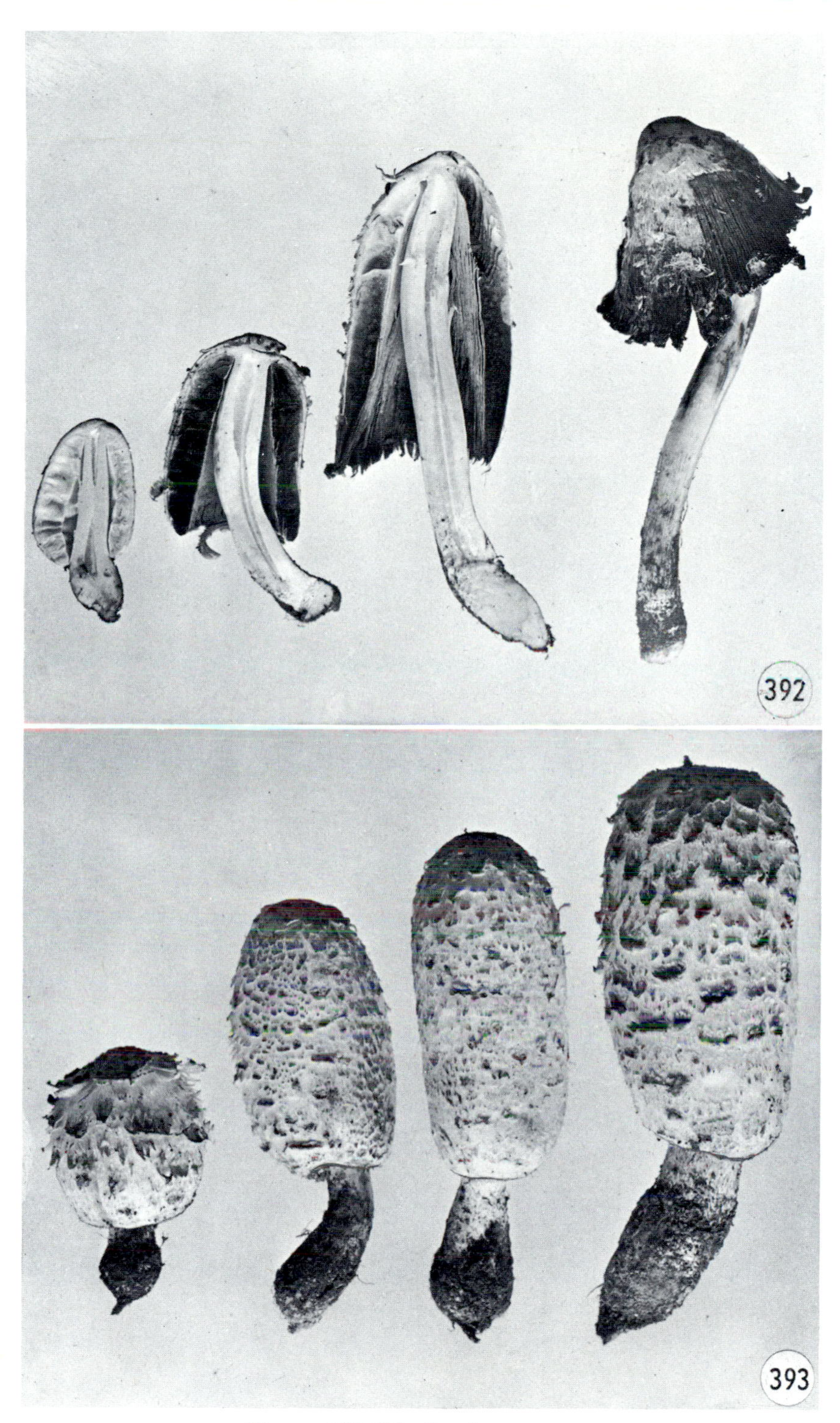

Figures 392-393. *Coprinus comatus.*

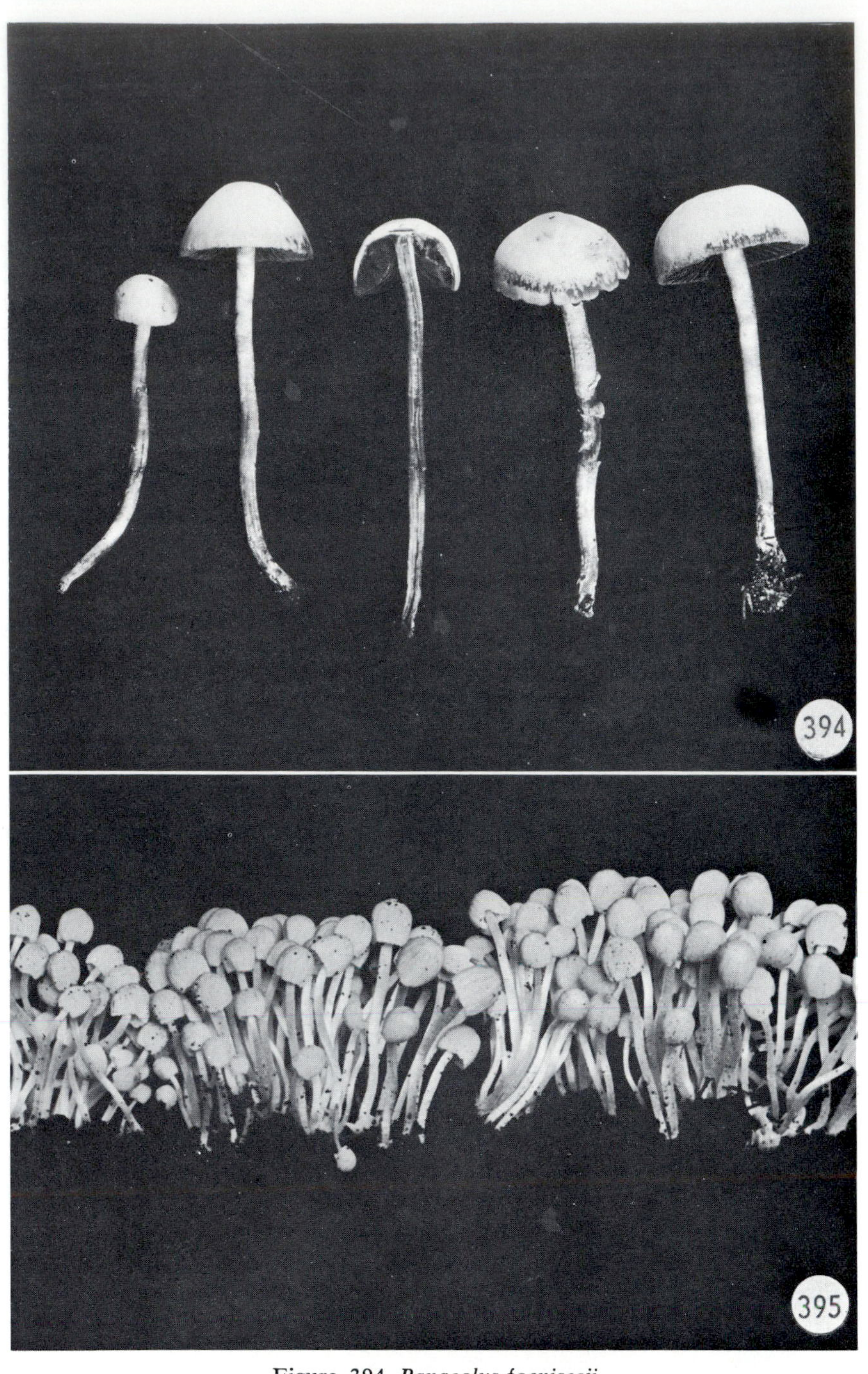

Figure 394. *Panaeolus foenisecii.*

Figure 395. *Pseudocoprinus disseminatus.*

Figure 396. *Ganoderma tsugae.*

Figure 397. *Polyporus frondosus.*

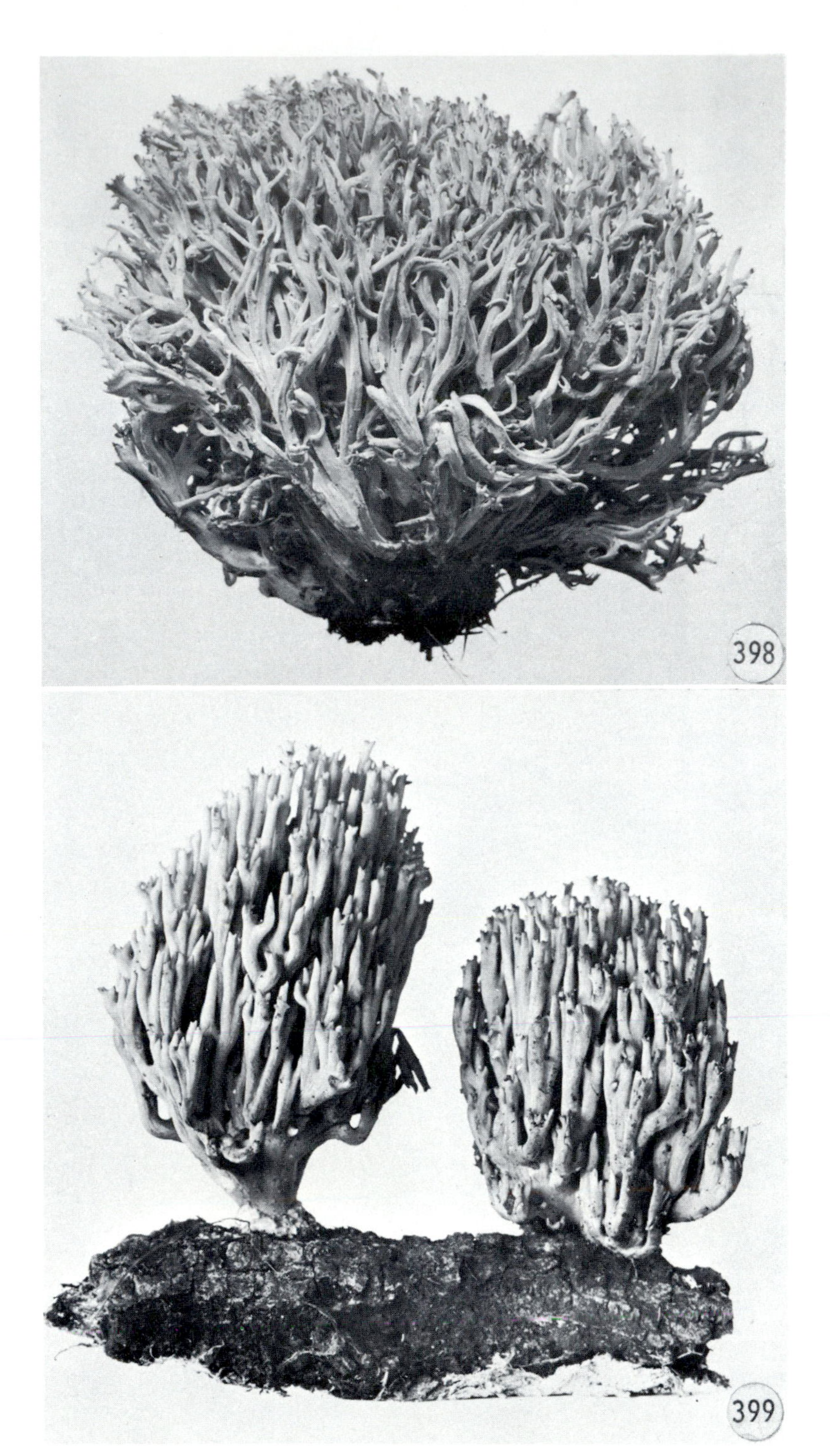

Figure 398. *Clavaria cinerea.*

Figure 399. *Clavaria flava.*

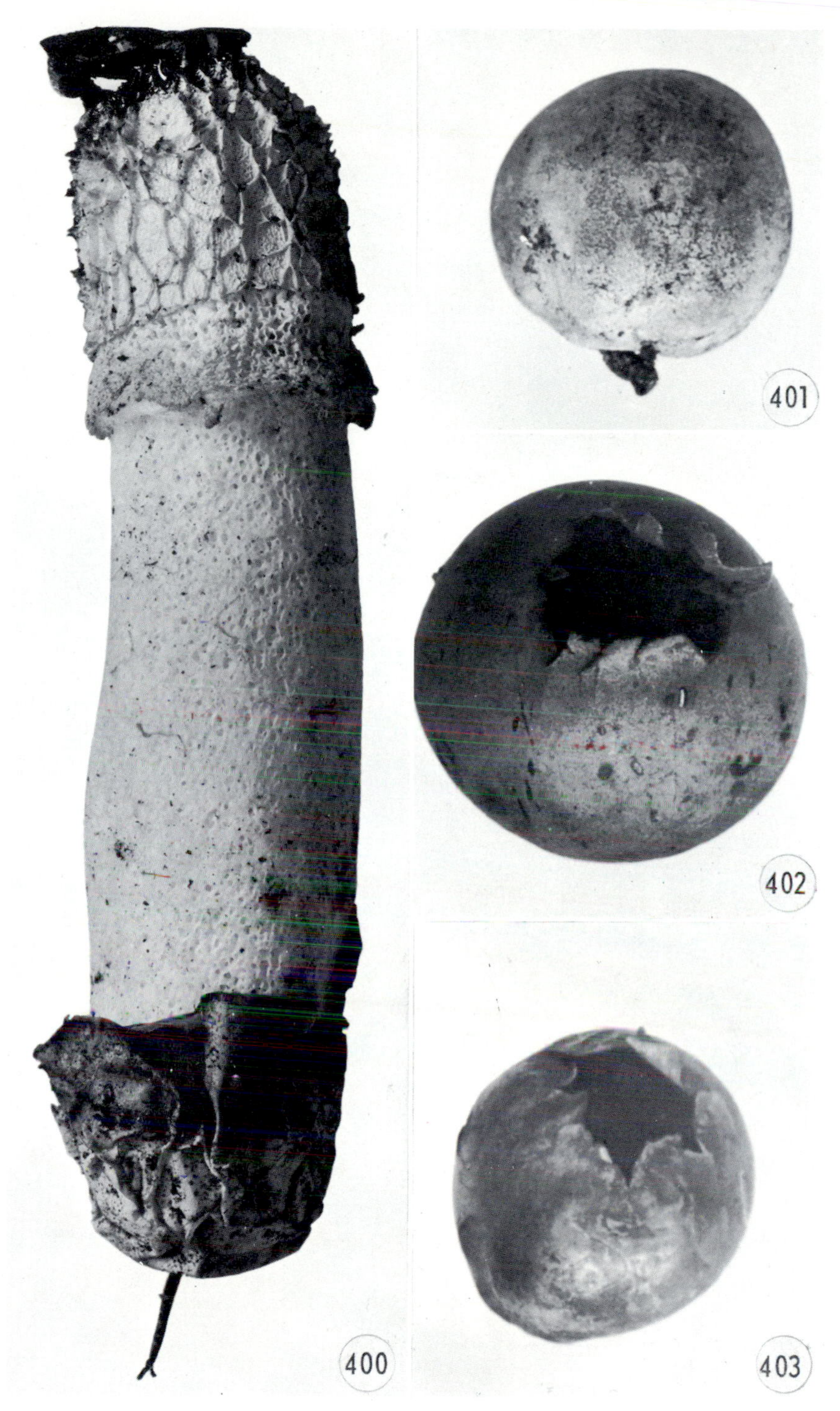

Figure 400. *Dictyophora duplicata.*

Figures 401-403. *Bovista pila.*

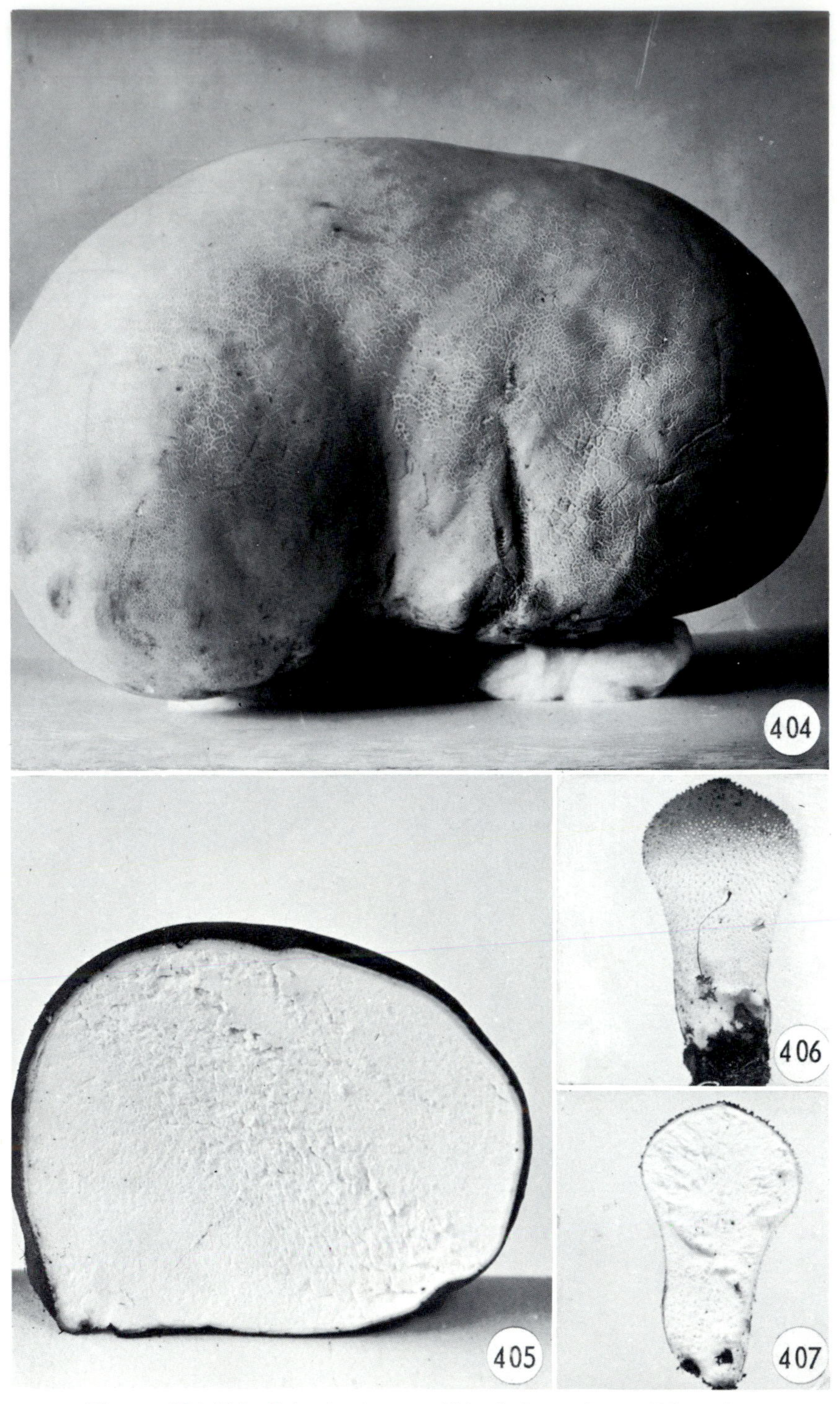

Figures 404-405. *Calvatia gigantea*. 404, whole specimen; 405, section.

Figures 406-407. *Lycoperdon perlatum*. 406, whole specimen; 407, section.

Figure 408. *Lycoperdon pyriforme.*

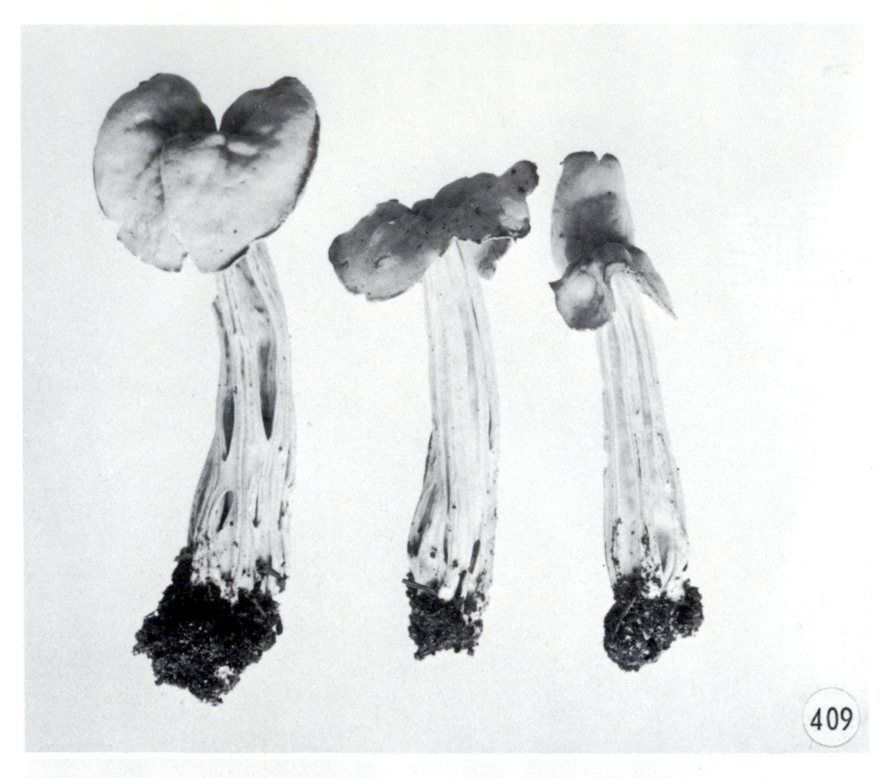

Figure 409. *Helvella crispa.*

Figure 410. *Helvella elastica.*

Figure 411. *Cantharellus cibarius.*

Figure 412. *Lactarius deliciosus.*

Figure 413. *Pleurotus serotinus.*

Figure 414. *Clitocybe clavipes.*

Figure 415. *Xeromphalina campanella*.

Figure 416. *Collybia platyphylla*.

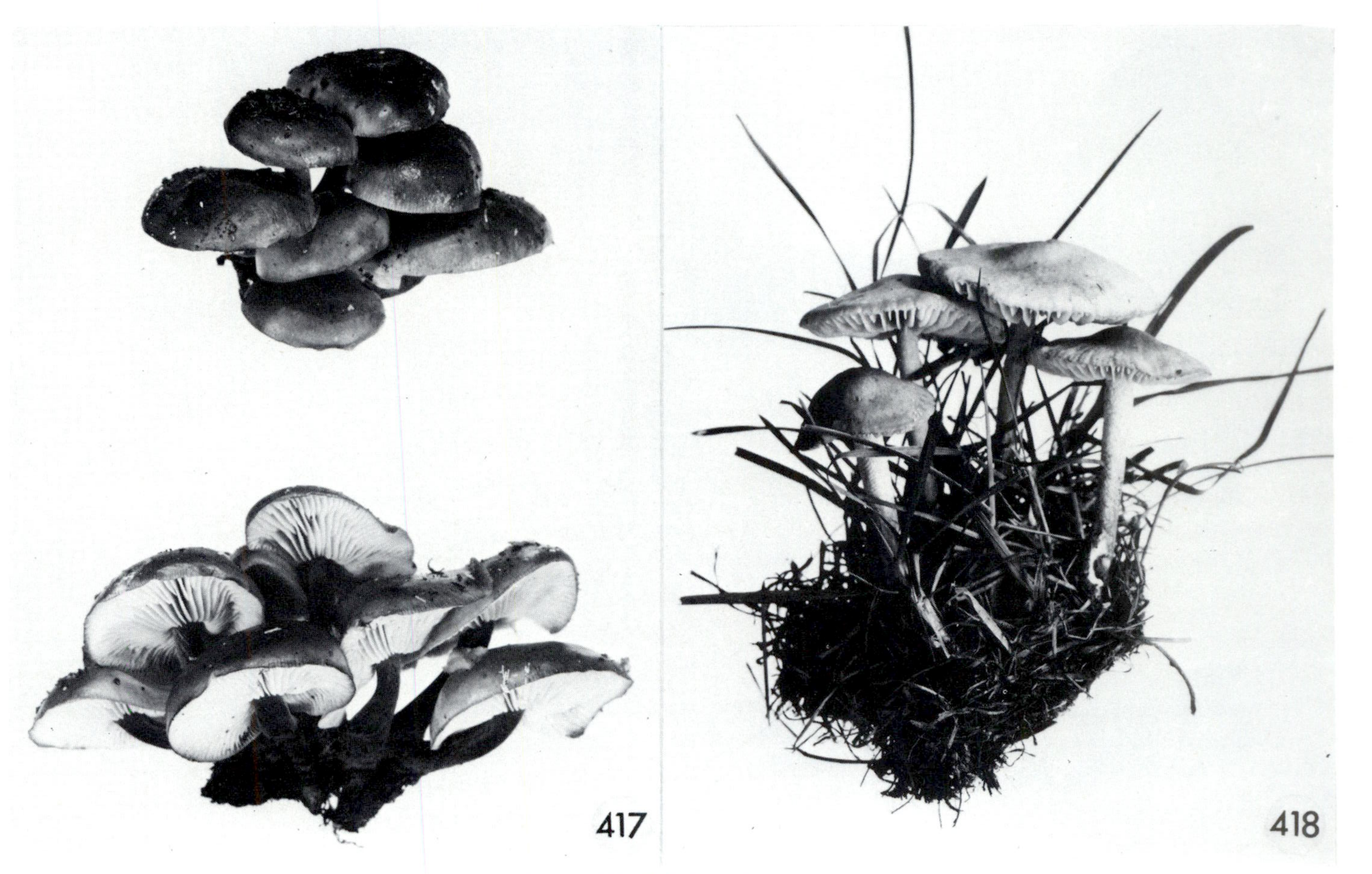

Figure 417. *Collybia velutipes.*

Figure 418. *Marasmius oreades.*

Figure 419. *Trogia crispa.*

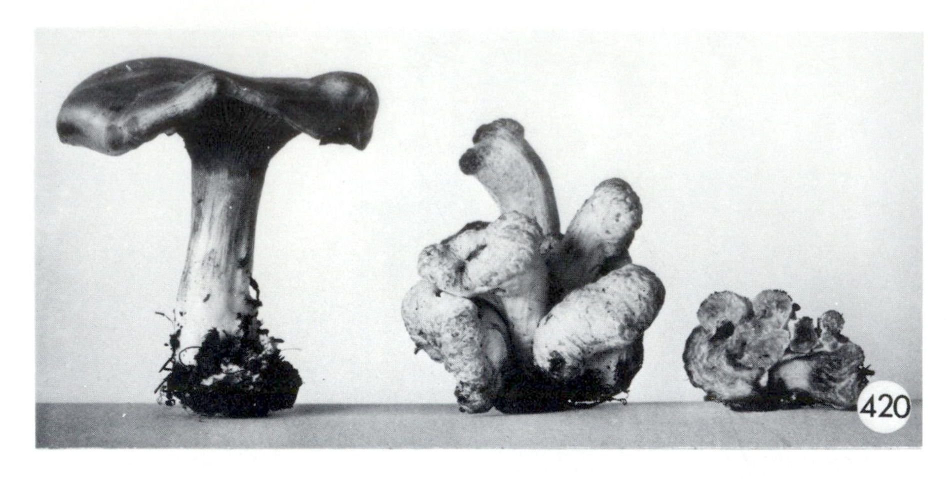

Figure 420. *Clitopilus abortivus.*

Figure 421. *Phyllotopsis nidulans.*

Figure 422. *Pholiota caperata.*

Figure 423. *Naematoloma sublateritium.*

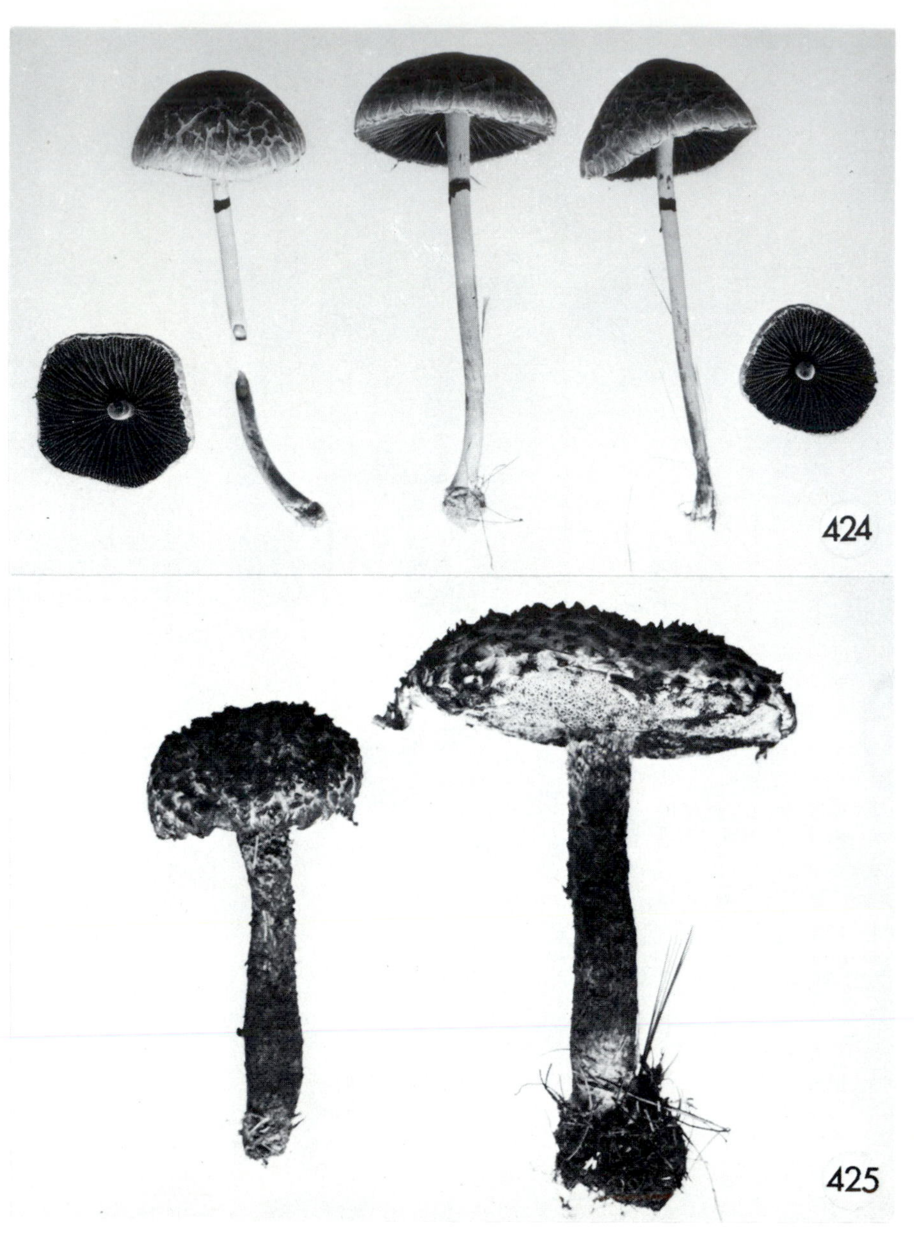

Figure 424. *Panaeolus retirugis.*

Figure 425. *Strobilomyces floccopus.*

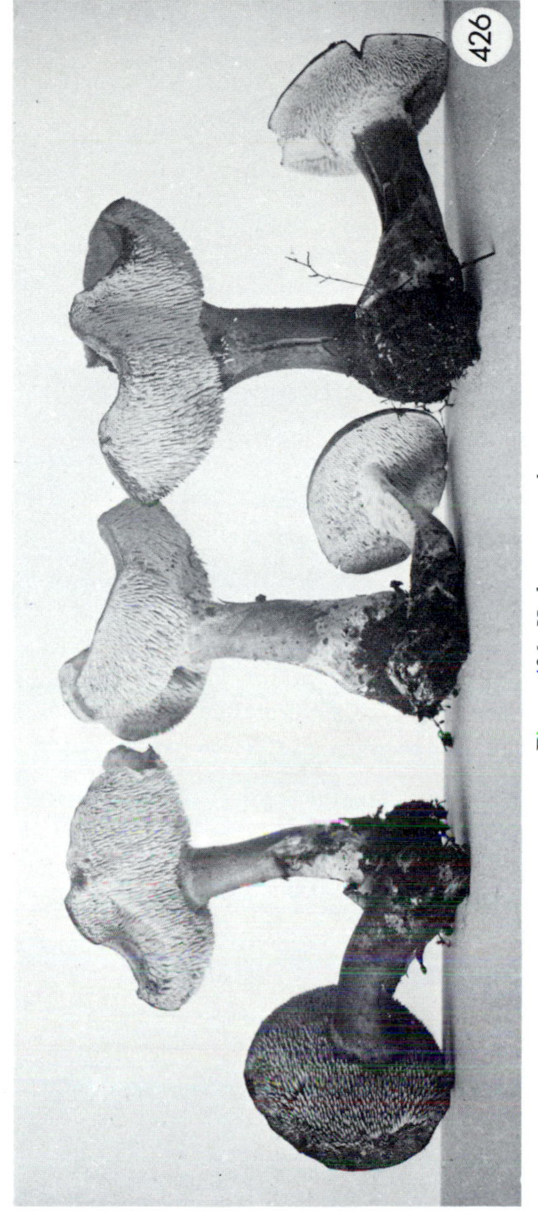

Figure 426. *Hydnum repandum.*

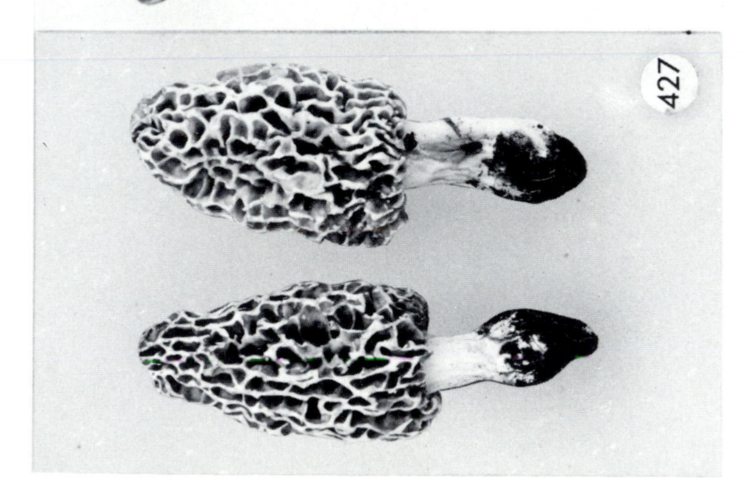

Figure 427. *Morchella esculenta.*
Figure 428. *Morchella angusticeps.*

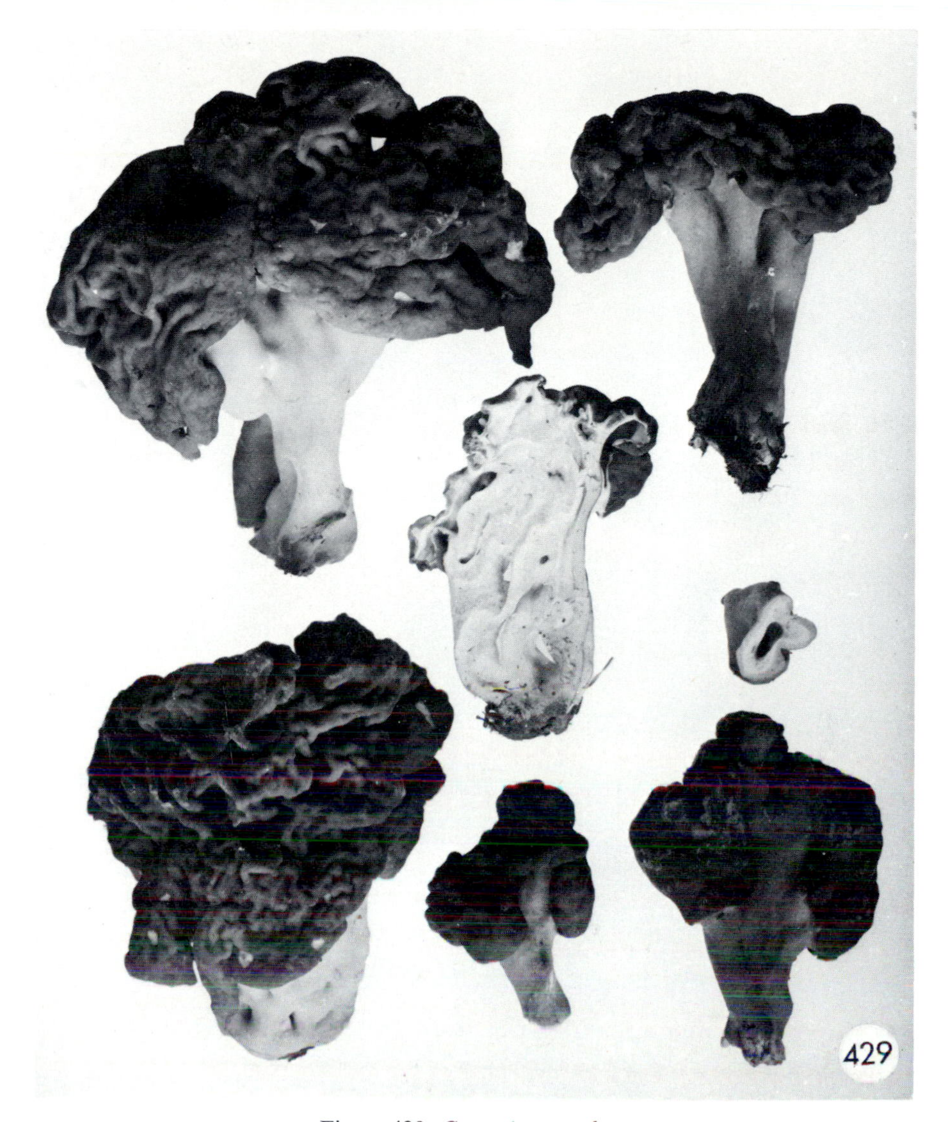

Figure 429. *Gyromitra esculenta*.

Figure 430. *Gyromitra infula.*

Figure 431. *Psilocybe semilanceata.*

ABBREVIATIONS OF NAMES OF AUTHORS

A. — J. B. von Albertini
Alb. — J. B. von Albertini
Afz. — A. Afzelius
Atk. — G. F. Atkinson
B. — M. J. Berkeley
Berk. — M. J. Berkeley
Bolt. — J. Bolton
Boud. — E. Boudier
Br. — C. E. Broome
Bref. — O. Brefeld
Bres. — G. Bresadola
Britz. — M. Britzelmayr
Bull. — J. B. F. Bulliard
Burl. — Gertrude S. Burlingham
C. — M. A. Curtis
Cke. — M. C. Cooke
Curt. — M. A. Curtis
DC. — A. P. De Candolle
Dicks. — J. Dickson
Fr. — E. M. Fries
Genev. — G. Genevier
Gill. — C. C. Gillet
Hook. — W. J. Hooker
Huds. — W. Hudson
Jacq. — N. J. von Jacquin
Jungh. — F. F. W. Junghuhn
Kalchb. — K. Kalchbrenner
Karst. — P. A. Karsten
Kauffm. — C. H. Kauffman
Kl. — J. F. Klotzch
Konr. — P. Konrad
Krombh. — J. V. von Krombholz
L. — C. Linnaeus
Lam. — J. B. A. P. M. de Lamarck
Lindbl. — M. A. Lindblad
Lovej. — Ruth H. Lovejoy
Lund. — S. Lundell
Mass. — G. Massee
Maubl. — A. Maublanc
Morg. — A. P. Morgan
Müll. — O. F. Müller
Murr. — W. A. Murrill
Nannf. — J. A. Nannfeldt
Opat. — W. Opatowski
Pat. — N. T. Patouillard
Pk. — C. H. Peck
Pers. — C. H. Persoon
Quél. — L. Quélet
Rom. — L. Romell
S. — L. D. von Schweinitz
Sacc. — P. A. Saccardo
Schaeff. — J. C. Schaeffer
Schrad. — H. A. Schrader
Schröt. — J. Schröter
Schw. — L. D. von Schweinitz
Schum. — H. D. F. Schumacher
Scop. — J. A. Scopoli
Secr. — L. Secretan
Sing. — R. Singer
Sm. — A. H. Smith
Sor. — N. V. Sorokin
Sow. — J. Sowerby
Speg. — C. L. Spegazzini
Sw. — O. P. Swartz
Tul. — E. L. R. Tulasne and C. Tulasne
Underw. — L. M. Underwood
Vaill. — S. Vaillant
Vitt. — C. Vittadini
Wahl. — G. Wahlenberg
Weinm. — J. A. Weinmann

GLOSSARY

a:- prefix signifying without or absence of.
acrid: a biting or peppery taste.
adnate: (of lamellae) broadly joined to the stipe; (of cuticle or volva) not peeling off or pulling off readily.
adnexed: (of lamellae) narrowly joined to the stipe.
allantoid: sausage-shaped.
alutaceous: light leather-colored, pale tan, pale brown.
amygdaline: (of taste) like that of peach or cherry stones.
amyloid: turning blue when treated with a solution of iodine in chloral hydrate and water.
anastomose, anastomosing: (of lamellae) joined crosswise forming angular areas or a network.
annulus: ring of tissue left on the stipe from the torn partial veil.
apical: (of stipe) the part near the attachment of the lamellae or where it joins the flesh of the pileus, the upper part.
apiculate: provided with an apiculus.
apiculus: (of spores) a short, sometimes sharp papilla or projection by which the spore was attached to the sterigma.
apothecium: the usually cup-shaped or saucer-shaped to saddle-shaped fruiting body of Discomycetes in which a layer of asci is exposed to the air.
appendiculate: (of the margin of the pileus) hung with fragments of the veil or pellicle.
appressed: closely flattened down.
arcuate: arched or curved like a bow.
areolate: (of the surface of the pileus, or stipe) marked out in little areas by cracks or crevices.
ascus (pl. **asci):** the cell in which the spores of the Ascomycetes are produced. Typically there is a fusion of nuclei in the young ascus followed by three nuclear divisions resulting in eight ascospores.
atomate: covered with minute shining particles, glistening like mica.
azonate: not zoned.
basidium (pl. **basidia):** the cell on which the spores of the Basidiomycetes are produced. Typically there is a fusion of nuclei in the young basidium followed by two nuclear divisions resulting in four spores that develop outside the cell on small stalks.
booted: (of base of stipe), closely sheathed by the volva.
campanulate: bell-shaped.
cespitose: growing in dense tufts or clusters.
chlamydospore: an asexual spore formed by the cells of the hyphae becoming rounded, thick-walled, and separated from one another.
cinereous: ashy gray.
clavate: club-shaped.
concolor, concolorous: (of lamellae or stipe), same color as the pileus.
conidium (pl. **conidia):** spore produced asexually.
cortina: the cobwebby veil found in some mushrooms.
crenate: scalloped, round-toothed.
crenulate: finely crenate.
crisped: finely curled or crinkled.
cupulate: cup-like in form.
cystidium (pl. **cystidia):** a large, sterile, more or less differentiated cell occurring among the basidia and usually projecting beyond them.
cyathiform: cup-shaped with a flaring margin.
decumbent: resting on the substratum with the end turned up.
decurrent: (of lamellae or tubes), running down the stipe.
decurved: bent down.
deliquescing: dissolving into fluid.
echinulate: covered with small pointed spines.
effused-reflexed: spread out over the substratum and turned back at the margin to form a pileus.
ellipsoid: (of spores), rounded at both ends and with sides curved.
emarginate: (of lamellae), notched near the stipe.
epiphragm: the thin membrane covering the mouth of the young peridium in the Nidulariaceae.
evanescent: soon disappearing.
excentric: (of the stipe), not attached to the center of the pileus, off-center.
farinaceous: (of odor and taste) resembling fresh meal.
ferruginous: rust colored.
fetid: stinking.
fibrillose: (of pileus or stipe), with thin thread-like filaments or fibrils, usually somewhat scattered.

filiform: very slender, thread-like.
fimbriate: with the edge or margin finely fringed or torn.
floccose: loose cottony to woolly.
floccule: small cottony tuft.
frondose: descriptive of a wood or forest of broad-leaved trees.
fulvous: reddish cinnamon-brown.
furfuraceous: covered with bran-like particles, scurfy.
fuscous: a dark smoky brown.
fusiform: spindle-shaped, tapering to both ends.
generic: of the rank of a genus or pertaining to a genus.
genus (pl. **genera**): a category used in classification; the first major grouping above the rank of species, considered to include related species.
gibbous: (of the pileus), having an unsymmetrical convexity or umbo, irregularly rounded.
glabrous: smooth, lacking scales, hairs, etc.
globose: spherical.
gloeocystidium: a special form of cystidium of gelatinous or horny consistency and with oily, resinous, granular contents.
glutinous: very sticky.
granulose: covered with granules.
hilum: (of spores), the scar marking the point of attachment.
hygrophanous: (of the pileus), with a watery appearance when moist and changing markedly in color as it dries out.
hyaline: colorless, transparent.
hymenium: the fruiting surface in fruit bodies of Ascomycetes and Basidiomycetes.
hypha (pl. **hyphae**): a single thread or filament of the vegetative structure of a fungus.
hypoderm: (of pileus), a region of differentiated hyphae just below the pellicle.
imbricate: overlapping like shingles.
infundibuliform: funnel-shaped.
intervenose: with veins between the lamellae.
involute: inrolled.
lacerate: appearing as if torn.
lamella (pl. **lamellae**): the blade-like or gill-like structure on the under side of the pileus of a mushroom.
lamellula (pl. **lamellulae**): small lamellae which do not reach the stipe.
lanceolate: lance-shaped, longer than broad and tapering.
latex: milky juice found in some mushrooms.
livid: blue-black, colored like a bruise.
marginate: (of lamellae), with the edge differently colored than the sides; (of the bulb of the stipe), having a circular ridge on the upper, exterior angle where the universal veil was attached.
membranous: thin and pliant like a membrane.
mycelioid: resembling mycelium, often applied to a mold-like growth at base of stipe.
mycelium: the vegetative part of a fungus, a collective term for the hyphae.
mycophagist: one who eats mushrooms.
obconic: inversely conic.
-oid: a suffix meaning *like* or *similar to*.
ovoid: egg shaped.
pallid: of an indefinite pale or whitish appearance.
papillate: having small nipple-shaped elevations on the surface.
papillose: same as papillate.
paraphyses: unspecialized sterile cells in the hymenium between the basidia.
pedicel: a slender stalk.
pellicle: a skin-like covering of the pileus which often peels off easily.
peridiole: the seed-like or egg-like structures in the bird's-nest fungi consisting of an inner peridium enclosing the spores.
peridium: the outer enveloping wall or coat of the puffball fruit body.
peronate: (of the stipe), sheathed by the universal veil.
pileus: the cap-like structure that bears the hymenium.
plicate: folded like a fan.
poroid: (of lamellae) becoming joined by cross veins so as to resemble pores.
pruinose: appearing as if finely powdered.
pubescent: provided with a covering of short, soft, downy hairs.
pulverulent: powdery.
punctate: marked with small point-like spots, scales, glandules, etc.
pyriform: pear-shaped.
reniform: kidney-shaped.
resupinate: (of fruit bodies), lying flat on the substratum with the hymenium facing outwards.
reticulate: marked with lines or ridges that form a network.
rhizomorph: a cord-like strand of mycelium.
rimose: cracked.
rivulose: marked with lines suggesting a river system on a map.
rugose: coarsely wrinkled.
rugulose: finely wrinkled.

scabrous: rough with short, rigid projections.

sclerotium: a resting body, usually very hard in consistency, composed of thick-walled hyphae, and sometimes with a definite rind.

scrobiculate: with shallow depressions or pits.

septum: a cross wall in a hypha or spore.

sensu: in the sense of.

serrate: (of lamellae), notched or toothed on the edge like the blade of a saw.

sessile: (of the pileus), lacking a stipe.

sinuate: (of lamellae), wavy or notched near the stipe.

species: a population of individuals with certain inherited characters in common. There are no fixed rules or standard for determining a species. The species concept is largely a matter of judgment and agreement among taxonomists. The species is designated by a Latin binomial consisting of the name of the genus (a noun) followed by the specific epithet (an adjective).

sphaerocyst: more or less globular cells found in the flesh of *Russula, Lactarius* and some other basidiomycetes.

sporangium: sac-like cell within which spores are produced.

spore: the reproductive body of a fungus or other cryptogams.

squamulose: covered with small scales.

squarrose: covered with erect, recurved scales.

sterigma (pl. **sterigmata):** the small stalk on a basidium on which the basidiospore is borne and from which it is forcibly discharged.

stipe: the stalk or stem-like part of a mushroom, bolete, etc.

stipitate: possessing a stipe.

striate: marked with minute lines or furrows.

striatulate: finely striate.

strigose: with coarse, rather long, stiff hairs.

stuffed: (of the stipe), having the central part composed of a differentiated pith that may disappear leaving the stipe hollow.

sub-: prefix meaning nearly, almost, somewhat, or under.

substrate, substratum: substance on or in which the fungus grows.

sulcate: grooved or furrowed, intermediate between striate and plicate.

taxonomy: the science of classification.

tomentose: densely matted with a covering of soft hairs.

tomentum: a covering composed of long, soft, hairy filaments or fibrils, usually more or less interlaced and matted.

trama: (of the lamellae), the tissue between the two hymenia; (of the pileus) the fleshy part.

truncate: ending abruptly as though the end were cut off.

tuberculate: covered with wart-like or knob-like projections.

turbinate: top-shaped.

umbilicate: (of the pileus), having a central navel-like depression.

umbo: a raised conical to convex swelling like the boss at the center of a shield.

umbonate: (of the pileus), having an umbo.

undulate: wavy.

ventricose: swollen or enlarged in the central part.

verrucose: warty.

vesiculose: (of cells), enlarged and swollen to globose or nearly so.

virgate: streaked, usually with fibrils of a different color.

viscid: sticky to the touch.

volva: the universal veil found in certain genera such as *Amanita.*

zonate: (of pileus), marked with concentric bands of different color than the remainder of the pileus.

INDEX

ADDENDUM

S. A. Redhead
Biosystematics Research Institute
Research Branch

Edible and Poisonous Mushrooms of Canada was first published in 1962. At that time there were many generic names that were new or had recently been restricted by using microscopic features. Their acceptance by mycologists in general was not guaranteed. For these reasons Dr. Groves chose well-established generic names and used them in broad, traditional ways. However, many of the new or restricted genera mentioned in the 1962 edition have since gained worldwide acceptance and a few additional names have been proposed. Now that some of these genera are being used in popular guides, an update of the names is appropriate. Unavoidably many genera are defined by using microscopic features, and this makes their recognition difficult for amateurs.

Only the species whose names have been changed are listed and these are in alphabetical order as they appear in the index. The updated name follows the former name as do any qualifying statements about the changes. These changes are not the last for the species treated here. Many of the genera are not universally recognized and the limits of some of them are still uncertain. However, all the names used are currently being applied in the National Mycological Herbarium, Ottawa. For the most part, Singer's (1975) *The Agaricales in Modern Taxonomy*, 3rd. ed., has been used as a standard.

The current edition of *Edible and Poisonous Mushrooms of Canada* also contains additional references, which have been published since 1975 when Dr. David Malloch enlarged the bibliography. Some of the theses mentioned in the enlarged bibliography have been replaced by their published counterparts.

Since 1974 possession of Canadian or foreign mushrooms containing the restricted drugs psilocin and psilocybin has been illegal in Canada. A number of species known from Canada in the genera *Psilocybe*, *Panaeolus*, and *Conocybe* contain these hallucinogens. Some of these species are poorly characterized and their distribution is not well known. In all cases the species are small, usually inconspicuous fungi not normally collected by persons interested in edible mushrooms. The most commonly encountered species is described below.

PSILOCYBE SEMILANCEATA (Fr.) ex Kummer — Poisonous

Figure 431, page 307

PILEUS ½–2¼ in. broad, at first obtusely conical to conico-campanulate, often becoming acutely umbonate with age, dark greenish to vinaceous brown, hygrophanous, fading to ocherous over the center and beige elsewhere, striate and viscid when moist, often slightly incurved and more conspicuously striate on the margins from heavy spore deposits between the lamellae. FLESH thin, membranous, concolorous with the pileus, with no distinctive odor. LAMELLAE ascending adnate, moderately spaced, moderately broad, narrowly ventricose, vinaceous buff when young, brown vinaceous with age, with whitish margins. STIPE 1 3/8–2 7/8 in. long, 1/16 or less in. thick, equal, slender, often wavy, dry, dull or subpolished, white to beige apically, ocherous to cinnamon basally, often partially covered with whitish scattered fibrils, darkening with age, usually developing blue to bluish green stains on the whitish silky parts when handled. SPORES 12–14.5 × 6.3–8.2 μ, smooth, purple brown, ellipsoid, slightly thick-walled, slightly truncated from a well-developed germ pore.

Singly or gregarious in pastures or other grassy areas in the eastern and western maritime regions. Sept.–Nov.

Psilocybe silvatica (Pk.) Singer & Smith and *P. pelliculosa* (Smith) Singer & Smith are similar but occur in forested areas and have smaller spores. A number of other species containing hallucinogens and having a collybioid aspect and sometimes with a prominent annulus also occur in Canada. Among the nonhallucinogenic species, *P. montana* (Pers. ex Fr.) Kummer is common on beds of *Polytrichum* (haircap mosses) and *P. atrobrunnea* (Lasch) Gillet is infrequently found on *Sphagnum* (peat moss) in bogs.

NOMENCLATURAL AND TAXONOMIC UPDATE

FORMER NAME	UPDATED NAME
abortivus, Clitopilus	*Entoloma abortivum* (B. & C.) Donk
abundans, Collybia	*Clitocybula abundans* (Pk.) Sing.
acericola, Pholiota	*Agrocybe acericola* (Pk.) Sing.
albipilata, Collybia	*Strobilurus albipilatus* (Pk.) Wells & Kempton
albogriseus, Clitopilus	*Entoloma albogrisea* (Pk.) Redhead comb. nov.[1]
albolanatus, Pleurotus	*Nothopanus porrigens* (B. & C.) Sing.
americana, Lepiota	*Leucocoprinus americana* (Pk.) Redhead comb. nov.[2]
amoenus, Hygrophorus	*Hygrocybe calyptaeformis* (Berk.) Fayod
angustatus, Panus	*Hohenbuehelia angusta* (Berk.) Sing.
angusticeps, Morchella	*Morchella elata* Fr.
applicatus, Pleurotus	*Resupinatus applicatus* (Batsch ex Fr.) S. F. Gray
aurantiaca, Clitocybe	*Hygrophoropsis aurantiaca* (Wulf. ex Fr.) Maire
aurea, Clavaria	*Ramaria aurea* (Fr.) Quél.
auricolor, Agaricus	This is most likely *Agaricus semotus* Fr.
autumnalis, Pholiota	*Galerina autumnalis* (Pk.) Smith & Sing.
benzoinus, Polyporus	*Ischnoderma benzoinum* (Wahl. ex Fr.) Karst.
betulinus, Polyporus	*Piptoporus betulinus* (Fr.) Karst.
borealis, Hygrophorus	*Camarophyllus borealis* (Pk.) Murr.
botrytis, Clavaria	*Ramaria botrytis* (Pers. ex. Fr.) Ricken
brevipes, Cantharellus	*Gomphus clavatus* (Pers. ex Fr.) S. F. Gray
brunnea, Lepiota	*Macrolepiota rachodes* (Vitt.) Sing.
candidissimus, Pleurotus	*Cheimonophyllum candidissimus* (B. & C.) Sing.
cantharellus, Hygrophorus	*Hygrocybe cantharellus* (Schw.) Murr.
caperata, Pholiota	*Rozites caperata* (Pers. ex Fr.) Karst.
capnoides, Naematoloma	*Hypholoma capnoides* (Fr. ex Fr.) Kummer
cartilaginea, Clitocybe	In the sense of Bresadola this is *Lyophyllum loricatum* (Fr.) Kühner.
ceraceus, Hygrophorus	*Hygrocybe ceracea* (Fr.) Kummer
chlorophanus, Hygrophorus	*Hygrocybe chlorophana* (Fr.) Wünsche
cinerea, Clavaria	*Clavulina cinerea* (Fr.) Schroet.
clavatus, Cantharellus	*Gomphus clavatus* (Pers. ex Fr.) S. F. Gray
coccineus, Hygrophorus	*Hygrocybe coccinea* (Fr.) Kummer
cochleatus, Lentinus	*Lentinellus cochleatus* (Fr.) Karst.
confluens, Polyporus	*Albatrellus confluens* (Fr.) Kotl. & Pouz.
conica, Morchella	*Morchella elata* Fr.
conicus, Hygrophorus	*Hygrocybe conica* (Fr.) Kummer
coralloides, Hydnum	*Hericium coralloides* (Scop. ex Fr.) S. F. Gray
cothurnata, Amanita	*Amanita pantherina* var. *multisquamosa* (Pk.) Jenkins
crispa, Conocybe	*Conocybe lactea* (Lange) Métrod
crispa, Trogia	*Plicatura crispa* ((Pers.) ex Fr.) Rea
cuspidatum, Entoloma	*Nolanea murrayi* (B. & C.) Dennis
cuspidatus, Hygrophorus	*Hygrocybe cuspidata* (Pk.) Murr.
cyathiformis, Clitocybe	*Pseudoclitocybe cyathiformis* (Bull. ex Fr.) Sing.
decora, Clitocybe	*Tricholomopsis decora* (Fr.) Sing.
delica, Russula	In Grove's sense this is *Russula brevipes* Peck.

[1]Basionym: *Agaricus albogriseus* Peck, Annu. Rep. N.Y. State Mus. Nat. Hist. 31:33 (1879).
[2]Basionym: *Agaricus americanus* Peck, Annu. Rep. N.Y. State Cabinet Nat. Hist. 23:71 (1872).

diminutivus, Agaricus	In Grove's sense this is *Agaricus semotus* Fr.
disseminatus, Pseudocoprinus	*Coprinus disseminatus* (Pers. ex Fr.) S. F. Gray
duplicata, Dictyophora	*Phallus duplicata* Bosc
ectypoides, Clitocybe	*Pseudoarmillariella ectypoides* (Pk.) Sing.
edulis, Agaricus	*Agaricus bitorquis* (Quél.) Sacc.
elongatipes, Pleurotus	*Hypsizygus elongatipes* (Pk.) Bigelow
fallax, Russula	In Grove's sense this is *Russula fragilis* (Pers. ex Fr.) Fr.
familia, Collybia	*Clitocybula familia* (Pk.) Sing.
fasciculare, Naematoloma	*Hypholoma fasciculare* (Huds. ex Fr.) Kummer
flava, Clavaria	*Ramaria flava* (Fr.) Quél.
flavescens, Hygrophorus	*Hygrocybe flavescens* (Kauff.) Sing.
flavobrunneum, Tricholoma	*Tricholoma fulvum* (Bull. ex Fr.) Sacc.
floccosus, Cantharellus	*Gomphus floccosus* (Schw.) Sing.
foenisecii, Panaeolus	*Panaeolina foenisecii* (Pers. ex Fr.) Maire
foetens, Russula	In Grove's sense this is *Russula fragrantissima* Romagnesi; the true *Russula foetens* is not definitely known from North America.
foetentula, Russula	*Russula subfoetens* W. G. Smith
fragrans, Hygrophorus	*Hygrophorus pudorinus* var. *fragrans* (Murr.) Hesler & Smith
frondosus, Polyporus	*Grifola frondosa* (Fr.) S. F. Gray
fuscogrisella, Nolanea	*Leptonia fuscogrisella* (Pk.) Largent
fusiformis, Clavaria	*Clavulinopsis fusiformis* (Fr.) Corner
gemmata, Amanita	*Amanita gemmata* (Fr.) Bertillon
gigantea, Calvatia	*Langermannia gigantea* (Batsch ex Pers.) Rostkov.
glabriceps, Amanita	*Amanita pantherina* var. *multisquamosa* (Pk.) Jenkins
griseus, Polyporus	*Boletopsis subsquamosa* (L. ex Fr.) Kotl. & Pouz.
haematopus, Lentinus	*Panus suavissimus* (Fr.) Sing.
hariolorum, Collybia	*Collybia confluens* (Pers. ex Fr.) Kummer
helvelloides, Phlogiotis	*Tremiscus helvelloides* (DC. ex Pers.) Donk
helvus, Lactarius	In Grove's sense this is *Lactarius aquifluus* Peck.
hortensis, Agaricus	*Agaricus brunnescens* Peck.
illudens, Clitocybe	*Omphalotus olearius* (DC. ex Fr.) Sing.
imperialis, Armillaria	*Catathelasma imperiale* (Fr. apud Lund) Sing.
inaurata, Amanitopsis	*Amanita strangulata* (Fr.) Roze apud Karst.
infundibuliformis, Cantharellus	*Cantharellus tubaeformis* Fr.
irinum, Tricholoma	*Lepista irina* (Fr.) Bigelow
kauffmannii, Cantharellus	*Gomphus kauffmannii* (Smith) Petersen
lachrymabundum, Hypholoma	*Psathyrella lacrymabunda* (Fr.) Moser
laetus, Hygrophorus	*Hygrocybe laeta* (Fr.) Kummer
longipes, Collybia	*Oudemansiella longipes* (Bull. ex St.-Amans) Maire
marginata, Pholiota	*Galerina marginata* (Batsch ex Fr.) Kühner
marginatus, Hygrophorus	*Humidicutis marginata* (Pk.) Sing.
marginella, Pholiota	The position is not certain; possibly it should be placed in *Kuehneromyces* or *Galerina* according to Singer.
mellea, Armillaria	*Armillariella mellea* (Fr.) Karst.
merulioides, Boletinellus	*Gyrodon merulioides* (Schw.) Sing.
micromegethus, Agaricus	This is probably *Agaricus semotus* Fr.
micropus, Clitopilus	*Entoloma micropus* (Pk.) Hesler
miniatus, Hygrophorus	*Hygrocybe miniata* (Fr.) Kummer
molybdites, Lepiota	*Chlorophyllum molybdites* (Meyer ex Fr.) Mass.
multiceps, Clitocybe	*Lyophyllum decastes* (Fr. ex Fr.) Sing.
multiplex, Cantharellus	*Polyozellus multiplex* (Underw.) Murr.
naucina, Lepiota	*Leucoagaricus naucinus* (Fr.) Sing.

nigricans, Russula — In Grove's sense this is probably *Russula dissimulans* Shaffer, however, *Russula nigricans* does occur on the west coast.
nitidus, Hygrophorus — *Hygrocybe nitida* (B. & C.) Murr.
niveus, Hygrophorus — *Hygrocybe nivea* (Scop. ex Fr.) Murr.
noveboracensis, Clitopilus — *Rhodocybe mundula* (Lasch) Sing.
nudum, Tricholoma — *Lepista nuda* (Bull. ex Fr.) Cooke
operculatus, Panus — *Tectella patellaris* (Fr.) Murr.
orcellus, Clitopilus — *Clitopilus prunulus* (Scop. ex Fr.) Kummer
ovinus, Polyporus — *Albatrellus ovinus* (Fr.) Kotl. & Pouz.
personatum, Tricholoma — *Lepista personata* (Fr. ex Fr.) Cooke
petaloides, Pleurotus — *Hohenbuehelia petaloides* (Bull ex Fr.) Schulz. apud Schulz., Kanitz & Knapp
pictus, Boletinus — *Suillus pictus* (Pk.) Smith & Thiers
piperatus, Suillus — *Chalciporus piperatus* (Bull. ex Fr.) Sing.
pistillaris, Clavaria — *Clavariadelphus pistillaris* (Fr.) Donk
platyphylla, Collybia — *Tricholomopsis platyphylla* (Pers. ex Fr.) Sing.
ponderosa, Armillaria — *Tricholoma ponderosum* (Pk.) Sing.
porrigens, Pleurotus — *Nothopanus porrigens* (Pers. ex Fr.) Sing.
praecox, Pholiota — *Agrocybe praecox* (Pers. ex Fr.) Fayod
praetensis, Hygrophorus — *Camarophyllus pratensis* (Fr.) Kummer
procera, Lepiota — *Macrolepiota procera* (Scop. ex Fr.) Sing.
pseudoclavatus, Cantharellus — *Gomphus pseudoclavatus* (Smith) Corner
psittacinus, Hygrophorus — *Hygrocybe psittacina* (Fr.) Kummer
puniceus, Hygrophorus — *Hygrocybe punicea* (Fr.) Kummer
rachodes, Lepiota — *Macrolepiota rachodes* (Vitt.) Sing.
radicata, Collybia — *Oudemansiella radicata* (Relh. ex Fr.) Sing.
repandum, Hydnum — *Dentinum repandum* (Fr.) S. F. Gray
resinosus, Polyporus — *Ischnoderma resinosum* (Fr.) Karst.
rodmani, Agaricus — *Agaricus bitorquis* (Quél.) Sacc.
rubinellus, Suillus — *Chalciporus rubinellus* (Pk.) Sing.
russuloides, Amanita — *Amanita gemmata* (Fr.) Bertillon
rutilans, Tricholoma — *Tricholomopsis rutilans* (Schaeff. ex Fr.) Sing.
salicinus, Panus — *Panellus ringens* (Fr.) Romagnesi
salmoneum, Entoloma — *Nolanea quadrata* B. & C.
semilibera, Morchella — *Mitrophora semilibera* (DC. ex Fr.) Lév.
semiorbicularis, Naucoria — *Agrocybe semiorbicularis* (Bull. ex St.-Amans) Fayod In Grove's sense this is *Agrocybe pediades* (Pers. ex Fr.) Fayod.
semiovatus, Panaeolus — *Anellaria semiovata* (Sow. ex Fr.) Pearson & Dennis
separata, Anellaria — *Anellaria semiovata* (Sow. ex Fr.) Pearson & Dennis
septentrionale, Hydnum — *Steccherinum septentrionale* (Fr.) Banker
serotinus, Pleurotus — *Panellus serotinus* (Fr.) Kühner
sordida, Russula — *Russula albonigra* (Krombh.) Fr.
spathulatus, Pleurotus — *Hohenbuehelia petaloides* (Bull. ex Fr.) Schulz. apud Schulz., Kanitz & Knapp
spectabilis, Boletinus — *Suillus spectabilis* (Pk.) O. Kuntze
spectabilis, Pholiota — *Gymnopilus spectabilis* (Fr.) Sing.
spumosa, Flammula — *Pholiota spumosa* (Fr.) Sing.
squarroso-adiposa, Pholiota — *Pholiota limonella* (Pk.) Sacc.
stipticus, Panus — *Panellus stipticus* (Bull. ex Fr.) Karst.
stricta, Clavaria — *Ramaria stricta* (Fr.) Quél.
subacutum, Tricholoma — *Tricholoma virgatum* (Fr.) Kummer

subdulcis, Lactarius	In Grove's sense this represents a complex of species such as *Lactarius carbonicola* Smith in Hesler & Smith and *Lactarius thejogalus* Fr.; the true *Lactarius subdulcis* is not definitely known from North America.
subglabripes, Leccinum	*Boletus subglabripes* Peck
sublateritium, Naematoloma	*Hypholoma sublateritium* (Fr.) Quél.
subnidulans, Phyllotopsis	*Crepidotus subnidulans* (Overh.) Hesler & Smith
subpalmatus, Pleurotus	*Rhodotus palmatus* (Bull. ex Fr.) Maire
subplanus, Clitopilus	*Entoloma subplanum* (Pk.) Hesler
sulphureus, Polyporus	*Laetiporus sulphureus* (Fr.) Murr.
tenera, Galera	*Conocybe tenera* (Schaeff. ex Fr.) Fayod
tessulatus, Pleurotus	*Hypsizygus tessulatus* (Bull. ex Fr.) Sing.
tigrinus, Lentinus	*Panus tigrinus* (Bull. ex Fr.) Sing.
tomentella, Amanita	*Amanita porphyria* (A. & S. ex Fr.) Secr.
tomentosus, Gomphidius	*Chroogomphus tomentosus* (Murr.) Miller
torulosus, Panus	*Panus conchatus* (Bull. ex Fr.) Fr.
transmutans, Tricholoma	*Tricholoma fulvum* (Bull. ex Fr.) Sacc.
ulmarius, Pleurotus	*Lyophyllum ulmarium* (Bull. ex Fr.) Kühner However, in Grove's sense it is *Hypsizygus tessulatus* (Bull. ex Fr.) Sing.
umbonatus, Cantharellus	*Cantharellula umbonata* (Fr.) Sing.
unicolor, Pholiota	*Galerina unicolor* (Fr.) Sing.
vaginata, Amanitopsis	*Amanita vaginata* (Bull. ex Fr.) Vitt.
velatipes, Amanita	*Amanita pantherina* var. *velatipes* (Atk.) Jenkins
velutipes, Collybia	*Flammulina velutipes* (Curt. ex Fr.) Sing.
ventricosa, Armillaria	*Catathelasma ventricosum* (Pk.) Sing.
vermiflua, Pholiota	*Agrocybe dura* (Bolt. ex Fr.) Sing.
vinicolor, Gomphidius	*Chroogomphus vinicolor* (Pk.) Miller
virosa, Amanita	*Amanita virosa* (Lam. ex Fr.) Gillet
vulpinus, Lentinus	*Lentinellus vulpinus* (Fr.) Kühner & Maire